T0298685

Multiplicative Differential Equations

Multiplicative Differential Equations: Volume I is the first part of a comprehensive approach to the subject. It continues a series of books written by the authors on multiplicative, geometric approaches to key mathematical topics. This volume begins with a basic introduction to multiplicative differential equations and then moves on to first- and second-order equations, as well as the question of existence and uniqueness of solutions. Each chapter ends with a section of practical problems. The book is accessible to graduate students and researchers in mathematics, physics, engineering and biology.

Svetlin G. Georgiev is a mathematician who has worked in various areas of mathematics. He currently focuses on harmonic analysis, functional analysis, partial differential equations, ordinary differential equations, Clifford and quaternion analysis, integral equations and dynamic calculus on time scales.

Khaled Zennir received his PhD in mathematics in 2013 from Sidi Bel Abbés University, Algeria (assistant professor). He obtained his highest diploma in Algeria (Habilitation, mathematics) from Constantine University, Algeria in 2015 (associate professor). He is now associate professor at Qassim University, KSA. His research interests are in nonlinear hyperbolic partial differential equations, including: global existence, blow-up and long-time behavior.

Multiplicative Differential Equations

Equations

Volume I

Svetlin G. Georgiev

Khaled Zennir

CRC Press
Taylor & Francis Group
Boca Raton London New York

CRC Press is an imprint of the
Taylor & Francis Group, an **informa** business

A CHAPMAN & HALL BOOK

First edition published 2023
by CRC Press
6000 Broken Sound Parkway NW, Suite 300, Boca Raton, FL 33487-2742

and by CRC Press
4 Park Square, Milton Park, Abingdon, Oxon, OX14 4RN

CRC Press is an imprint of Taylor & Francis Group, LLC

ISBN: 978-1-032-49137-0 (hbk)
ISBN: 978-1-032-49343-5 (pbk)
ISBN: 978-1-003-39334-4 (ebk)

DOI: 10.1201/9781003393344

Typeset in Nimbus Roman
by KnowledgeWorks Global Ltd.

Contents

Preface

Differential and integral calculus, the most applicable mathematical theories were created independently by Isaac Newton and Gottfried Wilhelm Leibnitz in the second half of the 17th century. Later, Leonard Euler redirected calculus by giving a central place to the concept of function and thus founded analysis. Two operations, differentiation and integration, are basic in calculus and analysis. In fact, they are the infinitesimal versions of the subtraction and addition operations on numbers, respectively. From 1967 to 1970, Michael Grossman and Robert Katz gave definitions of a new kind of derivative and integral, moving the roles of subtraction and addition to division and multiplication, and thus established a new calculus, called multiplicative calculus. Sometimes, it is called an alternative or non-Newtonian calculus as well. Multiplicative calculus can especially be useful as a mathematical tool for economics and finance.

This book is devoted to multiplicative differential equations. It summarizes the most recent contributions in this area. The book is intended for senior undergraduate students and beginning graduate students of engineering and science courses. The book contains five chapters. The chapters in the book are pedagogically organized. Each chapter concludes with a section with practical problems.

In Chapter 1 we introduce some basic definitions and the basic problems for multiplicative differential equations. Chapter 2 is devoted to some elementary first-order multiplicative differential equations. Multiplicative separable differential equations, multiplicative homogeneous differential equations and exact multiplicative differential equations are investigated. The multiplicative integrating factor is defined, and some of its applications are given. Chapter 3 deals with first-order multiplicative differential equations. A multiplicative integral representation of the solutions of first-order multiplicative differential equations is deducted. The multiplicative Bernoulli equations and the multiplicative Riccati equations are introduced. In Chapter 4, second-order multiplicative linear differential equations are studied. The multiplicative linear dependence and independence of functions are introduced, and the method of separation of variables for second-order multiplicative linear differential equations is developed. The existence and uniqueness of solutions of multiplicative initial value problems are investigated in Chapter 5. The Picard method of successive approximations is introduced. Some multiplicative Gronwall-type multiplicative integral inequalities are deducted. Continuous dependence of the solutions of multiplicative initial value problems and multiplicative differentiability of the solutions are studied.

This book is addressed to a wide audience of specialists, such as mathematicians, physicists, engineers and biologists. It can be used as a textbook at the graduate level and as a reference book for several disciplines.

Svetlin G. Georgiev and Khaled Zennir
Paris, November 2022

1

Introduction

In this chapter, we give a definition for multiplicative differential equations (MDEs), the order of MDEs and the solution of MDEs. We give a classification of the MDEs, and they are described the basic problems for the MDEs. Suppose that $I \subseteq \mathbb{R}_*$.

1.1 Definition for Multiplicative Differential Equations

Definition 1.1.1 *A MDE is a relation that contains one multiplicative variable $x \in \mathbb{R}_*$, the multiplicative real dependent variable y and some of its multiplicative derivatives $y^\star, y^{\star\star}, \ldots, y^{\star(n)}$.*

Example 1.1.2 *The multiplicative equation*

$$x \cdot_\star y^\star +_\star 3 \cdot_\star y = e^6 \cdot_\star x^\star$$

is a multiplicative ordinary differential equation.

Example 1.1.3 *The equation*

$$\sin_\star x \cdot_\star y^{\star(5)} +_\star y = \cos_\star x$$

is a multiplicative ordinary differential equation.

Example 1.1.4 *The equation*

$$y^{\star\star} -_\star x^{3\star} +_\star \sin_\star y^\star = y^{2\star}$$

is a multiplicative ordinary differential equation.

DOI: 10.1201/9781003393344-1

1.2 Order of Multiplicative Differential Equations

Definition 1.2.1 *The order of a multiplicative ordinary differential equation is defined as the order of the highest multiplicative derivative in the equation.*

Example 1.2.2 *The equation in Example 1.1.2 is of the first order.*

Example 1.2.3 *The equation in Example 1.1.3 is of the fifth order.*

Example 1.2.4 *The equation in Example 1.1.4 is of the second order.*

Exercise 1.2.5 *Determine the order of the following MDEs.*

1. $(y^*)^{2*} -_* e^2 \cdot_* y = 0_*, \ x \in \mathbb{R}_*.$

2. $y^{***} -_* e^4 \cdot_* y^{**} = \sin_* x, \ x \in \mathbb{R}_*.$

3. $(y^{**})^{3*} = (\sin_* x)^{4*}, \ x \in \mathbb{R}_*.$

4. $y^{*(5)} -_* e^3 \cdot_* y^{**} = \cos_* x, \ x \in \mathbb{R}_*.$

5. $y^{*(11)} -_* e^2 \cdot_* y^* -_* \cos_* xy = \tan_* x, \ x \in \mathbb{R}_*.$

Answer

1. 1.

2. 3.

3. 2.

4. 5.

5. 11.

1.3 Solution of Multiplicative Differential Equations

In general, an *n*th order MDE can be written in the form

$$F\left(x, y^*, \ldots, y^{*(n)}, \right) = 0_*, \tag{1.1}$$

where *F* is a known function.

Definition 1.3.1 *A functional relation between the multiplicative real dependent variable y and the multiplicative real dependent variable x in some interval $I \subseteq \mathbb{R}_\star$ that satisfies the given MDE is said to be a solution of the MDE. The general solution of an nth order MDE depends on n multiplicative constants $c_1, \ldots, c_n$ and on the multiplicative real independent variable x. Sometimes, the solutions of an MDE are said to be multiplicative integral curves for the corresponding MDE.*

Obviously, the general solution of an *n*th order MDE is written as follows

$$\phi(x, y, c_1, \ldots, c_n) = 0_\star. \tag{1.2}$$

Here, $c_1, \ldots, c_n$ are multiplicative constants.

Example 1.3.2 *The function*

$$y_1(x) = -_\star \cos_\star x +_\star c, \quad c \in \mathbb{R}_\star, \quad x \in \mathbb{R}_\star,$$

is a solution to the MDE

$$y^\star = \sin_\star x, \quad x \in \mathbb{R}_\star. \tag{1.3}$$

The classical analog of the equation (1.3) is the following equation

$$y' = \sin x, \quad x \in \mathbb{R}. \tag{1.4}$$

Its general solution is

$$y_2(x) = -\cos x + b,$$

where $b \in \mathbb{R}$. In Fig. 1.1, the solutions y_1 and y_2 for $c = e^2$ and $b = 2$, respectively, and for $x \in (1, 4\pi)$ are shown.

Example 1.3.3 *The function*

$$y_1(x) = \sin_\star x +_\star \tan_\star x +_\star c, \quad c \in \mathbb{R}_\star, \quad x \left(0_\star, e^{\frac{\pi}{2}} \right),$$

is a solution to the equation

$$y^\star(x) = \cos_\star x +_\star 1_\star /_\star (\cos_\star x)^{2_\star}, \quad x \in \left(1, e^{\frac{\pi}{2}} \right). \tag{1.5}$$

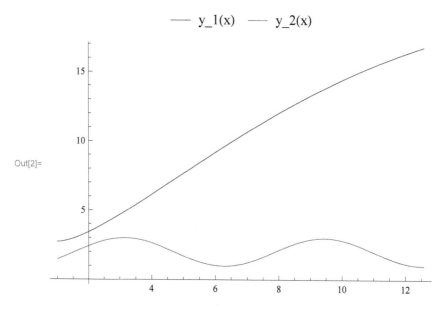

Figure 1.1
The solution y_1 of the equation (1.3) for $c = e^2$ and the solution y_2 for $b = 2$ of the equation (1.4).

The classical analog of the equation (1.5) is the following equation

$$y'(x) = \cos x + \frac{1}{(\cos x)^2}, \quad x \in \left(0, \frac{\pi}{2}\right). \tag{1.6}$$

Its solution is the function

$$y_2(x) = \sin x + \tan x + b, \quad x \in \left(0, \frac{\pi}{2}\right),$$

where $b \in \mathbb{R}$. In Fig. 1.2, the functions y_1 and y_2 for $c = e^2$ and $b = 2$, respectively, are shown.

Example 1.3.4 *Consider the equation*

$$(y^*)^{2*} +_* e^2 \cdot_* y^* = e^4 \cdot_* x^{2*} +_* e^4 \cdot_* x, \quad x \in \mathbb{R}_*. \tag{1.7}$$

Note that the function

$$y_1(x) = x^{2*} +_* c, \quad x \in \mathbb{R}_*,$$

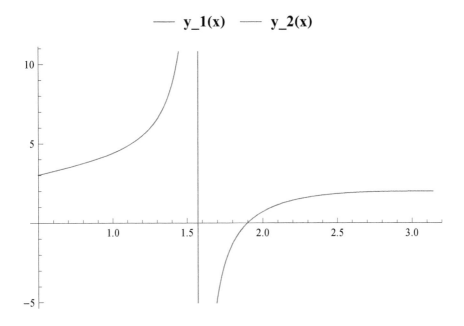

Figure 1.2
The solution y_1 of the equation (1.5) for $c = e^2$ and the solution y_2 for $b = 2$ of the equation (1.6).

where $c \in \mathbb{R}_*$, is a solution to the equation (1.7). Really, we have

$$y_1^*(x) \quad = \quad 2_* \cdot_* x$$

$$= \quad e^2 \cdot_* x$$

$$= \quad e^{2\log x}, \quad x \in \mathbb{R}_*.$$

Hence,

$$(y^*(x))^{2*} +_* e^2 \cdot_* y^*(x) \quad = \quad (2_* \cdot_* x)^{2*} +_* e^2 \cdot_* 2_* \cdot_* x$$

$$= \quad \left(e^{2\log x}\right)^{2*} +_* e^2 \cdot_* e^2 \cdot_* x$$

$$= \quad e^{4(\log x)^2} +_* e^4 \cdot_* x$$

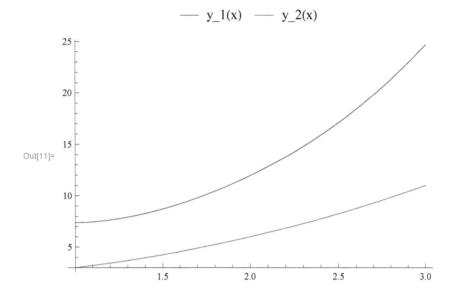

Figure 1.3
The solution y_1 of the equation (1.7) for $c = e^2$ and the solution y_2 for
$b = 2$ of the equation (1.8).

$$= \quad e^4 \cdot_* x^{2*} +_* e^4 \cdot_* x, \quad x \in \mathbb{R}_*.$$

The classical analog of the equation (1.7) is the equation

$$(y')^2 + 2y' = 4x^2 + 4x, \quad x \in \mathbb{R}. \qquad (1.8)$$

Its solution is given by the function

$$y_2(x) = x^2 + b, \quad x \in \mathbb{R},$$

where $b \in \mathbb{R}$. In Fig. 1.3, the functions y_1 and y_2 for $c = e^2$ and $b = 2$,
respectively, are shown.

Exercise 1.3.5 *Prove that*

$$y(x) = x^{2*} +_* e^x, \quad x \in \mathbb{R}_*,$$

is a solution to the MDE

$$y^* = y -_* x^{2*} +_* e^2 \cdot_* x, \quad x \in \mathbb{R}_*.$$

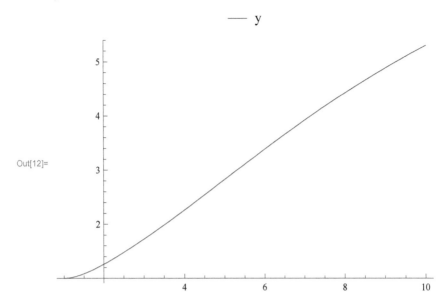

Figure 1.4
The solution $y = -_* \cos_* x +_* e$.

Definition 1.3.6 *If a solution can be obtained for some values of the constants* $c_1, \ldots, c_n$, *then it is called a particular solution. Otherwise, it is called a singular solution.*

Example 1.3.7 *The function*

$$y(x) = -_* \cos_* x +_* e, \quad x \in \mathbb{R}_*,$$

is a particular solution for the MDE given in Example 1.3.2, which is shown in Fig. 1.4.

Example 1.3.8 *Consider the equation*

$$(y^*)^{2_*} -_* x \cdot_* y^* +_* y = 0_*, \quad x \in \mathbb{R}_*.$$

Its general solution is given by

$$y_1(x) = c \cdot_* x -_* c^{2_*}, \quad c \in \mathbb{R}_*, \quad x \in \mathbb{R}_*.$$

Indeed, we have

$$y^*(x) = c, \quad x \in \mathbb{R}_*. \tag{1.9}$$

Then

$$(y_1^*(x))^{2*} -_* x \cdot_* y_1^*(x) +_* y_1(x) \quad = \quad c^{2*} -_* x \cdot_* c +_* c \cdot_* x -_* c^{2*}$$

$$= \quad 0_*, \quad x \in \mathbb{R}_*.$$

On the other hand, the function

$$y_2(x) = x^{2*}/_* e^4, \quad x \in \mathbb{R}_*, \tag{1.10}$$

is also a solution of the considered equation. Really, we have

$$y_2(x) = e^{\frac{1}{4}*} \cdot_* x^{2*}, \quad x \in \mathbb{R}_*,$$

and

$$y_2^*(x) \quad = \quad 2_* \cdot_* e^{\frac{1}{4}*} \cdot_* x$$

$$= \quad e^{\frac{1}{2}*} \cdot_* x, \quad x \in \mathbb{R}_*.$$

Hence,

$$(y_2^*(x))^{2*} -_* x \cdot_* y_2^*(x) +_* y_2(x) = \left(e^{\frac{1}{2}*} \cdot_* x \right)^{2*} -_* x \cdot_* e^{\frac{1}{2}*} \cdot_* x +_* e^{\frac{1}{4}*} \cdot_* x^{2*}$$

$$= e^{\frac{1}{4}*} \cdot_* x -_* e^{\frac{1}{2}*} \cdot_* x +_* e^{\frac{1}{4}*} \cdot_* x$$

$$= e^{\frac{1}{2}*} \cdot_* x -_* e^{\frac{1}{2}*} \cdot_* x$$

$$= 0_*, \quad x \in \mathbb{R}_*.$$

 Observe that there is any multiplicative constant c from (1.9) to obtain (1.10). Thus, (1.10) is a singular solution of the considered equation. In Fig. 1.5, the solutions y_1 and y_2 are shown.

Exercise 1.3.9 *Consider the equation*

$$x \cdot_* y \cdot_* d_* x +_* (x +_* e) \cdot_* d_* y = 0_*, \quad x \in \mathbb{R}_*.$$

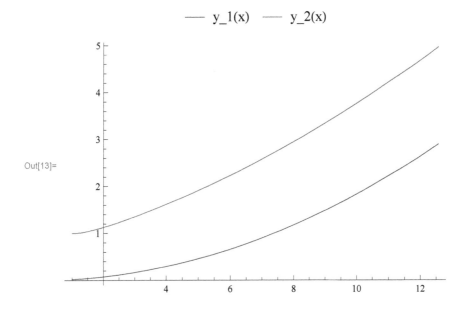

Figure 1.5
The solutions y_1 and y_2.

    1. Prove that the general solution of the considered equation is given by

$$y(x) = c \cdot_* (x +_* e) \cdot_* e^{-_* x}, \quad x \in \mathbb{R}_*.$$

Here, c is a multiplicative constant.

    2. Prove that

$$x = e^{-1}$$

is a singular solution of the considered equation.

Definition 1.3.10 *The function*

$$y = \phi(x), \quad x \in I,$$

is said to be an explicit solution of the equation (1.1) provided

$$F\left(x, \phi(x), \phi^*(x), \ldots, \phi^{*(n)}(x)\right) = 0_*, \quad x \in I. \tag{1.11}$$

Definition 1.3.11 *A relation of the form*

$$\psi(x,y) = 0_\star \qquad (1.12)$$

is said to be an implicit solution of the MDE (1.1) provided it determines one or more functions $y = \phi(x)$, $x \in I$, which satisfy (1.11).

Sometimes, it is difficult to solve (1.12). In these cases, by (1.12), we find

$$\psi_x^\star(x,y(x)) +_\star \psi_y^\star(x,y(x)) \cdot_\star y^\star(x) = 0, \quad x \in I,$$

whereupon

$$y^\star(x) = -_\star \psi_x^\star(x,y(x))/_\star \psi_y^\star(x,y(x)), \quad x \in I,$$

Hence,

$$y^{\star\star}(x) = -_\star \left(\left(\psi_{xx}^{\star\star}(x,y(x)) +_\star \psi_{xy}^{\star\star}(x,y(x)) \cdot_\star y^\star(x) \right) \cdot_\star \psi_y^\star x, y(x) \right.$$

$$\left. -_\star \psi_x^\star(x,y(x)) \cdot_\star \left(\psi_{yx}^{\star\star}(x,y(x)) +_\star \psi_{yy}^{\star\star}(x,y(x)) \cdot_\star y^\star(x) \right) \right)$$

$$/_\star \left(\psi_y^\star(x,y(x)) \right)^{2\star}, \quad x \in I,$$

and so on. Then, we check if

$$F\left(x,y(x),y^\star(x),y^{\star\star}(x),\ldots,y^{\star(n)}(x)\right) = 0_\star, \quad x \in I.$$

Definition 1.3.12 *The pair*

$$x = x(t)$$

$$y = y(t), \quad t \in I,$$

is said to be a parametric solution of the MDE (1.1), if

$$F\left(x(t),y(t),d_\star y/_\star d_\star x, d_\star^{2\star} y/_\star (d_\star x)^{2\star},\ldots,d_\star^{n\star} y/_\star (d_\star x)^{n\star}\right) = 0_\star,$$

$t \in I.$

Example 1.3.13 *We will find a MDE with solutions*

$$y(x) = e^{cx}, \quad x \in \mathbb{R}_\star,$$

where $c \in \mathbb{R}_\star$ is a constant. We have

$$y^\star(x) \quad = \quad e^{cx}$$

$$= \quad y(x), \quad x \in \mathbb{R}_\star,$$

i.e.,

$$y^\star(x) = y(x), \quad x \in \mathbb{R}_\star.$$

Example 1.3.14 *We will find a MDE with solutions given by*

$$y(x) = e^{e^c \cdot_\star x}, \quad x \in \mathbb{R}_\star,$$

where $c \in \mathbb{R}_\star$ is a given constant. We have $y(x) > 1$, $x \in \mathbb{R}_\star$, and

$$y(x) \quad = \quad e^{e^{c \log x}},$$

$$e^{c \log x} \quad = \quad \log y(x), \quad x \in \mathbb{R}_\star,$$

and

$$c = \frac{\log(\log y(x))}{\log x}, \quad x \in \mathbb{R}_\star.$$

Hence,

$$y'(x) \quad = \quad \frac{c}{x} e^{c \log x} e^{e^{c \log x}},$$

$$y^\star(x) \quad = \quad e^{x \frac{y'(x)}{y(x)}}$$

$$= \quad e^{x \frac{c}{x} e^{c \log x} \frac{e^{e^{c \log x}}}{e^{e^{c \log x}}}}$$

$$= \quad e^{c e^{c \log x}}.$$

$$= e^{\frac{\log(\log y(x))}{\log x}}$$

$$= ((\log y(x))/_\star x) \cdot_\star y(x), \quad x \in \mathbb{R}_\star,$$

and

$$x \cdot_\star (y^\star x /_\star y(x)) = \log y(x), \quad x \in \mathbb{R}_\star,$$

and

$$y(x) = e^{x \cdot_\star (y^\star(x)/_\star y(x))}, \quad x \in \mathbb{R}_\star.$$

Example 1.3.15 *We will find a MDE with solutions*

$$y(x) = (x -_\star c)^{3_\star}, \quad x \in \mathbb{R}_\star,$$

where $c \in \mathbb{R}_\star$ is a constant. We have

$$x -_\star c = (y(x))^{\frac{1}{3}_\star}, \quad x \in \mathbb{R}_\star,$$

and

$$y(x) = \left(\frac{x}{c}\right)^{3_\star}$$

$$= e^{\left(\log \frac{x}{c}\right)^3}, \quad x \in \mathbb{R}_\star.$$

Hence,

$$y'(x) = 3\left(\log\frac{x}{c}\right)^2 \frac{1}{\frac{x}{c}} \frac{1}{c} e^{\left(\log\frac{x}{c}\right)^3}$$

$$= \frac{3}{x}\left(\log\frac{x}{c}\right) e^{\left(\log\frac{x}{c}\right)^3}, \quad x \in \mathbb{R}_\star,$$

and

$$y^\star x) = e^{x\frac{y'(x)}{y(x)}}$$

$$= e^{x\frac{3}{x}\left(\log\frac{x}{c}\right)^2 \frac{e^{\left(\log\frac{x}{c}\right)^2}}{e^{\left(\log\frac{x}{c}\right)^2}}}$$

$$= e^{3\left(\log \frac{x}{c}\right)^2}$$

$$= e^3 \cdot_\star e^{\left(\log \frac{x}{c}\right)^2}$$

$$= e^3 \cdot_\star (x -_\star c)^{2_\star}$$

$$= e^3 \cdot_\star \left((y(x))^{\frac{1}{3_\star}}\right)^{2_\star}$$

$$= e^3 \cdot_\star (y(x))^{\frac{2}{3_\star}}, \quad x \in \mathbb{R}_\star,$$

i.e.,

$$y^\star(x) = e^3 \cdot_\star (y(x))^{\frac{2}{3_\star}}, \quad x \in \mathbb{R}_\star.$$

Example 1.3.16 *We will find a MDE with solutions given by*

$$y(x) = c \cdot_\star x^\star, \quad x \in \mathbb{R}_\star.$$

We have

$$x = (y(x)/_\star c)^{\overline{3}_\star}, \quad x \in \mathbb{R}_\star,$$

and

$$c = y(x)/_\star x^{3_\star}, \quad x \in \mathbb{R}_\star.$$

Hence,

$$y'(x) = \frac{3}{x}(\log c)(\log x)^2 e^{(\log c)(\log x)^3}, \quad x \in \mathbb{R}_\star,$$

and

$$y^\star(x) = e^{x \frac{y'(x)}{y(x)}}$$

$$= e^{x \frac{3}{x}(\log c)(\log x)^2 \frac{e^{(\log c)(\log x)^3}}{e^{(\log c)(\log x)^3}}}$$

$$= e^{3(\log c)(\log x)^2}$$

$$= \quad e^3 \cdot_\star c \cdot_\star e^{(\log x)^2}$$

$$= \quad e^3 \cdot_\star c \cdot_\star x^{2_\star}$$

$$= \quad e^3 \cdot_\star c \cdot_\star \left((y(x)/_\star c)^{\frac{1}{3}\star} \right)^{2_\star}$$

$$= \quad e^3 \cdot_\star c \cdot_\star \left(1_\star/_\star \left(c^{\frac{2}{3}\star} \right) \right) \cdot_\star (y(x))^{\frac{2}{3}\star}$$

$$= \quad e^3 \cdot_\star c^{\frac{1}{3}\star} \cdot_\star (y(x))^{\frac{2}{3}\star}$$

$$= \quad e^3 \cdot_\star \left((y(x))^{\frac{1}{3}\star}/_\star x \right) \cdot_\star (y(x))^{\frac{2}{3}\star}$$

$$= \quad e^3 \cdot_\star (y(x))/_\star x, \quad x \in \mathbb{R}_\star,$$

or

$$x \cdot_\star y^\star(x) = e^3 \cdot_\star y(x), \quad x \in \mathbb{R}_\star.$$

Exercise 1.3.17 *Find a MDE with solutions*

$$y(x) = c \cdot_\star \sin_\star x, \quad x \in \mathbb{R}_\star.$$

Answer

$$y^{\star\star} +_\star y = 1.$$

1.4   Classification of Multiplicative Differential Equations

In this section, we will give a classification of MDEs.

Definition 1.4.1 *A MDE is said to be a multiplicative linear MDE if it is multiplicative linear with respect to y and its multiplicative derivatives. Otherwise, the MDE is said to be a multiplicative nonlinear MDE.*

Example 1.4.2 *The MDE*

$$y^{\star\star} +_\star e^2 \cdot_\star y^\star +_\star \sin_\star x = 1, \quad x \in \mathbb{R}_\star,$$

is multiplicative linear.

Example 1.4.3 *The MDE*

$$(y^\star)^{3\star} +_\star y^{\star\star\star} +_\star \cos_\star x = 1, \quad x \in \mathbb{R}_\star,$$

is multiplicative linear.

Example 1.4.4 *The MDE*

$$y^\star \cdot_\star y^{\star\star} +_\star \sin_\star y = 1, \quad x \in \mathbb{R}_\star,$$

is multiplicative nonlinear.

Exercise 1.4.5 *Classify each of the following MDEs as multiplicative linear or multiplicative nonlinear.*

1. $y^\star +_\star \sin_\star x \cdot_\star y = \cos_\star x$, $x \in \mathbb{R}_\star$.
2. $(y^\star)^{4\star} -_\star (\cos_\star x)^{2\star} = 1$, $x \in \mathbb{R}_\star$.
3. $y^\star \cdot_\star y^{\star\star} = e^x$, $x \in \mathbb{R}_\star$.

Answer

1. *Multiplicative linear.*
2. *Multiplicative nonlinear.*
3. *Multiplicative nonlinear.*

The general form of an *n*th order multiplicative linear MDS is as follows.

$$p_0(x) \cdot_\star y^{\star(n)} +_\star p_1(x) \cdot_\star y^{\star(n-1)} +_\star \cdots +_\star p_n(x) \cdot_\star y = f(x), \quad x \in I,$$
$$(1.13)$$

where $p_j, f : I \to \mathbb{R}_\star$, $j \in \{1, \ldots, n\}$, are given functions.

Definition 1.4.6 *The equation (1.14) is said to be multiplicative homogeneous if $f(x) = 1$, $x \in I$. Otherwise, it is said to be multiplicative nonhomogeneous.*

Example 1.4.7 *The MDE*

$$y^{\star\star} +_\star y^\star = \sin_\star x, \quad x \in \mathbb{R}_\star,$$

is multiplicative nonhomogeneous.

Example 1.4.8 *The MDE*

$$y^{\star\star\star} +_\star \sin_\star x \cdot_\star y^{\star\star} +_\star y^\star = 1, \quad x \in \mathbb{R}_\star,$$

is multiplicative homogeneous.

Example 1.4.9 *The MDE*

$$y^\star +_\star \cos_\star x \cdot_\star y = 1, \quad x \in \mathbb{R}_\star,$$

is multiplicative linear.

Exercise 1.4.10 *Classify each of the following MDEs as multiplicative linear or multiplicative nonlinear.*

    1.

$$y^\star +_\star \sin_\star x \cdot_\star y = \cos_\star x, \quad x \in \mathbb{R}_\star.$$

    2.

$$(y^\star)^{4_\star} -_\star (\cos_\star x)^{2_\star} = 1, \quad x \in \mathbb{R}_\star.$$

    3.

$$y^\star \cdot_\star y^{\star\star} = e^x, \quad x \in \mathbb{R}_\star.$$

Answer

1. *Multiplicative linear.*
2. *Multiplicative nonlinear.*
3. *Multiplicative nonlinear.*

The general form of an nth order linear MDE is as follows.

$$p_0(x) \cdot_\star y^{\star(n)} +_\star p_1(x) \cdot_\star y^{\star(n-1)} +_\star \cdots +_\star p_n(x) \cdot_\star y = f(x), \quad x \in I, \tag{1.14}$$

where $p_j, f : I \to \mathbb{R}_\star$, $j \in \{1,\ldots,n\}$, are given functions.

Definition 1.4.11 *If $f(x) = 1$, $x \in I$, then the MDE (1.14) is said to be multiplicative homogeneous. Otherwise, it is said to be multiplicative nonhomogeneous.*

Example 1.4.12 *The MDE*

$$y^{\star\star\star} +_\star e^3 \cdot_\star y^{\star\star} -_\star y = \sin_\star x, \quad x \in \mathbb{R}_\star,$$

is multiplicative nonhomogeneous.

Example 1.4.13 *The MDE*

$$y^{\star\star\star\star} +_\star y^{\star\star} +_\star y = 1, \quad x \in \mathbb{R}_\star,$$

is multiplicative homogeneous.

Example 1.4.14 *The MDE*

$$y^{\star\star\star} -_\star e^x \cdot_\star y = e^x, \quad x \in \mathbb{R}_\star,$$

is multiplicative nonhomogeneous.

Exercise 1.4.15 *Classify each of the following MDEs as multiplicative homogeneous or multiplicative nonhomogeneous.*

1.
$$y^\star -_\star \sin_\star x = 1, \quad x \in \mathbb{R}_\star.$$

2.
$$y^{\star\star\star} +_\star y = 1, \quad x \in \mathbb{R}_\star.$$

3.
$$y^{\star\star} -_\star y^\star = 1, \quad x \in \mathbb{R}_\star.$$

Answer

1. *Multiplicative nonhomogeneous.*
2. *Multiplicative homogeneous.*
3. *Multiplicative nonhomogeneous.*

1.5   Basic Problems for Multiplicative Differential Equations

We will assume that the MDE (1.1) can be solved explicitly. More precisely, assume

$$y^{\star(n)} = f(x, y, y^{\star}, \ldots, y^{\star(n-1)}), \quad x \in I, \tag{1.15}$$

where f is a known function. Let $x_0 \in I$. In applications, we will study solutions of the equation (1 15) satisfying initial or boundary conditions.

Definition 1.5.1 *By initial conditions for the MDE (1.15), we mean conditions of the form*

$$
\begin{aligned}
y(x_0) &= y_0, \\
y^{\star}(x_0) &= y_1, \\
&\vdots \\
y^{\star(n-1)}(x_0) &= y_{n-1},
\end{aligned}
\tag{1.16}
$$

where $y_j \in \mathbb{R}_{\star}$, $j \in \{0, 1, \ldots, n-1\}$, are given constants.

Definition 1.5.2 *A problem consisting of the MDE (1.15) together with the initial conditions (1.16) is said to be a multiplicative initial value problem (MIVP).*

The main questions for the MIVP are as follows.

1. Existence of solutions.

2. Uniqueness of solutions.

3. Continuous dependence on initial conditions.

4. The interval under which the solutions are defined.

5. Extension of the solutions.

6. Upper and lower bounds of the unknown solutions and existence of maximal and minimal solutions.

7. Multiplicative periodicity of the solutions.

8. The behaviour of the solutions.

9. Stability of the solutions.

10. Oscillation of the solutions.

Definition 1.5.3 *Observe that the initial conditions (1.16) are prescribed at the same multiplicative point x_0. In many problems, these conditions are prescribed in two or more distinct multiplicative points. These conditions are called boundary conditions. A problem consisting of the MDE (1.15) together with boundary conditions is said to be a multiplicative boundary value problem (MBVP).*

Example 1.5.4 *The problem*

$$y^{\star\star} = \sin_\star x, \quad x \in [1, \infty),$$

$$y(1) = 2,$$

$$y^\star(1) = 3$$

is a MIVP.

Example 1.5.5 *The problem*

$$y^{\star\star} = y^{2\star} +_\star x^{2\star}, \quad x \in (2,5),$$

$$y^\star(2) = y(5) = y(2)$$

is a MBVP.

1.6 Advanced Practical Problems

Problem 1.6.1 *Determine the order of the following MDEs.*

1. $y^{\star\star\star} -_\star \sin_\star x \cdot_\star y^{\star\star} +_\star e^x \cdot_\star y = (\cos_\star x)^{2\star}, \quad x \in \mathbb{R}_\star.$

2. $y^{*(4)} -_* \cos_* x \cdot_* y^{***} = 0_*$, $x \in \mathbb{R}_*$.

3. $y^{**} -_* \tan_* x \cdot_* y = \cos_* x$, $x \in \mathbb{R}_*$.

4. $y^{*(7)} -_* \cos_* x \cdot_* y^{**} = 0_*$, $x \in \mathbb{R}_*$.

5. $(y^{**})^{7*} -_* e^x \cdot_* y^* = (\cos_* x)^{4*}$, $x \in \mathbb{R}_*$.

Answer

1. 3.

2. 4.

3. 2.

4. 7.

5. 2.

Problem 1.6.2 *Prove that*

$$y(x) = e \cdot_* e^x +_* x^{2*} +_* e^2 \cdot_* x +_* e^2, \quad x \in \mathbb{R}_*,$$

is a solution to the MDE

$$y^* = y -_* x^{2*}, \quad x \in \mathbb{R}_*.$$

Problem 1.6.3 *Prove that*

$$y(x) = x \cdot_* \tan_* (x +_* e^3), \quad x \in \left(e^{-\frac{\pi}{2}-3}, e^{\frac{\pi}{2}-3} \right),$$

is a solution to the MDE

$$x \cdot_* y^* = x^{2*} +_* y^{2*} +_* y, \quad x \in \left(e^{-\frac{\pi}{2}-3}, e^{\frac{\pi}{2}-3} \right).$$

Problem 1.6.4 *Prove that*

$$y(x) = x^{3*} +_* e /_* x^{3*}, \quad x \in (e, \infty),$$

is a solution to the equation

$$x \cdot_* y^* +_* e^3 \cdot_* y = e^6 \cdot_* x^{3*}, \quad x \in (e, \infty).$$

Problem 1.6.5 *Prove that the function y that satisfies*

$$x = -_* (y(x))^{2*} \cdot_* \log_* x, \quad x \in \mathbb{R}_*,$$

is a solution to the equation

$$e^2 \cdot_* x^{2*} \cdot_* y^* = y^{3*} +_* x \cdot_* y, \quad x \in \mathbb{R}_*.$$

Problem 1.6.6 *Prove that the function y that satisfies*

$$x^{2*} \cdot_* (y(x))^{4*} \cdot_* \log_* \left(e^2 \cdot_* x^{2*}\right) = e, \quad x \in \mathbb{R}_*,$$

is a solution to the equation

$$e^2 \cdot_* x \cdot_* d_* y +_* \left(x^{2*} \cdot_* y^{4*} +_* e\right) \cdot_* y \cdot_* d_* x = 0_*, \quad x \in \mathbb{R}_*.$$

Problem 1.6.7 *Consider the equation*

$$\left(y^{2*} +_* e\right)^{\frac{1}{2}*} \cdot_* d_* x = x \cdot_* y \cdot_* d_* y, \quad x \in \mathbb{R}_*.$$

1. *Prove that the general solution of the considered equation is given by*

$$\log_* |x|_* = c +_* \left(y^{2*} +_* e\right)^{\frac{1}{2}*}, \quad x \in \mathbb{R}_*.$$

Here c is a multiplicative constant.

2. *Prove that*
$$x = 1$$
is a singular solution of the considered equation.

Problem 1.6.8 *Consider the equation*

$$\left(x^{2*} -_* e\right) \cdot_* y^* +_* e^2 \cdot_* x \cdot_* y^{2*} = 0_*, \quad x \in \mathbb{R}_*.$$

1. *Prove that the general solution of the considered equation is given by*

$$y(x) \cdot_* \left(\log_* |x^{2*} -_* e|_* +_* c\right) = e, \quad x \in \mathbb{R}_*.$$

Here c is a multiplicative constant.

2. *Prove that*
$$y(x) = 1, \quad x \in \mathbb{R}_*,$$
is a singular solution of the considered equation.

Problem 1.6.9 *Consider the equation*

$$y^* = e^3 \cdot_* y^{\frac{2}{3}*}, \quad x \in \mathbb{R}_*.$$

1. *Prove that the general solution of the considered equation is given by*

$$y(x) = (x -_* c)^{3*}, \quad x \in \mathbb{R}_*.$$

Here c is a multiplicative constant.

2. *Prove that*

$$y(x) = 1, \quad x \in \mathbb{R}_*,$$

is a singular solution of the considered equation.

Problem 1.6.10 *Consider the equation*

$$x \cdot_* y +_* y = y^{2*}, \quad x \in \mathbb{R}_*.$$

1. *Prove that the general solution of the considered equation is given by*

$$y(x \cdot_* (e -_* c \cdot_* x) = e, \quad x \in \mathbb{R}_*.$$

Here c is a multiplicative constant.

2. *Prove that*

$$y(x) = 1, \quad x \in \mathbb{R}_*,$$

is a singular solution of the considered equation.

Problem 1.6.11 *Consider the equation*

$$y^* -_* x \cdot_* y^{2*} = e^2 \cdot_* x \cdot_* y, \quad x \in \mathbb{R}_*.$$

1. *Prove that the general solution of the considered equation is given by*

$$\left(c \cdot_* e^{-_* x^{2*}} -_* e \right) \cdot_* y(x) = e^2, \quad x \in \mathbb{R}_*.$$

Here c is a multiplicative constant.

2. *Prove that*

$$y(x) = 1, \quad x \in \mathbb{R}_*,$$

is a singular solution of the considered equation.

Problem 1.6.12 *Consider the equation*

$$e^2 \cdot_* x^{3*} \cdot_* y^* = y \cdot_* \left(e^2 \cdot_* x^{2*} -_* y^{2*} \right), \quad x \in \mathbb{R}_*.$$

1. Prove that the general solution of the considered equation is given by

$$x = e^{\pm 1} \cdot_* y(x) \cdot_* \left(\log_*(c \cdot_* x) \right)^{\frac{1}{2}*}, \quad x \in \mathbb{R}_*.$$

Here c is a multiplicative constant.

2. Prove that

$$y(x) = 1, \quad x \in \mathbb{R}_*,$$

is a singular solution of the considered equation.

Problem 1.6.13 *Consider the equation*

$$\left(e^{2*} -_* e^2 \cdot_* x \cdot_* y \right) \cdot_* d_* x +_* x^{2*} \cdot_* d_* y = 0_*, \quad x \in \mathbb{R}_*.$$

1. Prove that the general solution of the considered equation is given by

$$x \cdot_* (y(x) -_* x) = c \cdot_* y(x), \quad x \in \mathbb{R}_*.$$

Here c is a multiplicative constant.

2. Prove that

$$y(x) = 1, \quad x \in \mathbb{R}_*,$$

is a singular solution of the considered equation.

Problem 1.6.14 *Consider the equation*

$$\left(x^{2*} +_* y^{2*} \right) \cdot_* y^* = e^2 \cdot_* x \cdot_* y, \quad x \in \mathbb{R}_*.$$

1. Prove that the general solution of the considered equation is given by

$$(y(x))^{2*} -_* x^{2*} = c \cdot_* y(x), \quad x \in \mathbb{R}_*.$$

Here c is a multiplicative constant.

2. Prove that

$$y(x) = 1, \quad x \in \mathbb{R}_*,$$

is a singular solution of the considered equation.

Problem 1.6.15 *Consider the equation*

$$\left(e^2 \cdot_* x -_* e^4 \cdot_* y +_* e^6 \right) \cdot_* d_* x$$

$$+_* (x +_* y -_* e^3) \cdot_* d_* y = 0_*, \quad x \in \mathbb{R}_*.$$

1. Prove that the general solution of the considered equation is given by

$$\left(y(x) -_* e^2 \cdot_* x \right)^{3_*} = c \cdot_* \left(y(x) -_* x -_* e \right)^{2_*}, \quad x \in \mathbb{R}_*.$$

Here c is a multiplicative constant.

2. Prove that

$$y(x) = x + 1, \quad x \in \mathbb{R}_*,$$

is a singular solution of the considered equation.

Problem 1.6.16 *Consider the equation*

$$(y +_* e^2) \cdot_* d_* x = \left(e^2 \cdot_* x +_* y -_* e^4 \right) \cdot_* d_* y, \quad x \in \mathbb{R}_*.$$

1. Prove that the general solution of the considered equation is given by

$$\left(y(x) +_* e^2 \right)^{2_*} = c \cdot_* (x +_* y(x) -_* e), \quad x \in \mathbb{R}_*.$$

Here c is a multiplicative constant.

2. Prove that

$$y(x) = e -_* x, \quad x \in \mathbb{R}_*,$$

is a singular solution of the considered equation.

Problem 1.6.17 *Find a MDE with solutions*

1. $y = \sin_\star(x +_\star c)$, $x \in \mathbb{R}_\star$, *where* $c \in \mathbb{R}_\star$ *is a constant.*

2. $x^{2\star} +_\star c \cdot_\star y^{2\star} = e^2 \cdot_\star y$, $x \in \mathbb{R}_\star$, *where* $c \in \mathbb{R}_\star$ *is a constant.*

3. $y^{2\star} +_\star c \cdot_\star x = x^{2\star}$, $x \in \mathbb{R}_\star$, *where* $c \in \mathbb{R}_\star$ *is a constant.*

4. $y^\star = c \cdot_\star (x -_\star c)^{2\star}$, $x \in \mathbb{R}_\star$, *where* $c \in \mathbb{R}_\star$ *is a constant.*

5. $c \cdot_\star y = \sin_\star(c \cdot_\star x)$, $x \in \mathbb{R}_\star$, *where* $c \in \mathbb{R}_\star$ *is a constant.*

6. $y = a \cdot_\star x^{2\star} +_\star b \cdot_\star e^x$, $x \in \mathbb{R}_\star$, *where* $a, b \in \mathbb{R}_\star$ *are constants.*

7. $(x -_\star a)^{2\star} +_\star b \cdot_\star y^{2\star} = e$, $x \in \mathbb{R}_\star$, *where* $a, b \in \mathbb{R}_\star$ *are constants.*

8. $\log_\star y = a \cdot_\star x +_\star b \cdot_\star y$, $x \in \mathbb{R}_\star$, *where* $a, b \in \mathbb{R}_\star$ *are constants.*

9. $y = a \cdot_\star x^\star +_\star b \cdot_\star x^\star +_\star c \cdot_\star x$, $x \in \mathbb{R}_\star$, *where* $a, b, c \in \mathbb{R}_\star$ *are constants.*

10. $x = a \cdot_\star y^{2\star} +_\star b \cdot_\star y +_\star c$, $x \in \mathbb{R}_\star$, *where* $a, b, c \in \mathbb{R}_\star$ *are constants.*

Answer

1.
$$y^{2\star} +_\star (y^\star)^{2\star} = e.$$

2.
$$x^{2\star} \cdot_\star y^\star -_\star x \cdot_\star y = y \cdot_\star y^\star.$$

3.
$$e^2 \cdot_\star x \cdot_\star y \cdot_\star y^\star -_\star y^\star = e^2 \cdot_\star x^{3\star}.$$

4.
$$(y^\star)^{3\star} = e^4 \cdot_\star y \cdot_\star \left(x \cdot_\star y^\star -_{|star} e_\star y\right).$$

5.
$$y^\star = \cos_\star \left(\left(x \cdot_\star \left(e -_\star (y^\star)^{2\star}\right)^{\frac{1}{2}\star}\right) /_y\right).$$

6.
$$x \cdot_\star (x -_\star e^2) \cdot_\star y^{\star\star} -_\star (x^{2\star} -_\star e^2) \cdot_\star y^\star +_\star e^2 \cdot_\star (x -_\star e) \cdot_\star y = 1.$$

7.
$$\left(y \cdot_\star y^{\star\star} +_\star (y^\star)^{2\star}\right)^{2\star} = -_\star y^{3\star} \cdot_\star y^{\star\star}.$$

8.
$$y^{\star\star} \cdot_\star y^{2\star} \cdot_\star (\log_\star y -_\star e) = (y^\star)^{2\star} \cdot_\star (x \cdot_\star y^\star -_\star y).$$

9.
$$x^{3\star} \cdot_\star y^{\star\star\star} -_\star e^3 \cdot_\star x^{2\star} \cdot_\star y^{\star\star} +_\star e^6 \cdot_\star x \cdot_\star y^\star -_{|star} e^6 \cdot_\star y = 1.$$

10.
$$y^{\star\star\star} \cdot_\star y^\star = e^3 \cdot_\star (y^{\star\star})^{2\star}.$$

Problem 1.6.18 *Classify each of the following MDEs as multiplicative linear or multiplicative nonlinear.*

1.
$$(y^\star)^{2\star} -_\star (y^\star)^{3\star} = 1, \quad x \in \mathbb{R}_\star.$$

2.
$$(y^\star)^{3\star} \cdot_\star y^{\star\star} = \sin_\star x, \quad x \in \mathbb{R}_\star.$$

3.
$$y^{\star\star\star} -_\star e^3 \cdot_\star y^{\star\star} +_\star y = 1, \quad x \in \mathbb{R}_\star.$$

4.
$$(y^\star)^{7\star} -_\star \cos_\star x = 1, \quad x \in \mathbb{R}_\star.$$

5.
$$y^{\star\star\star\star} +_\star \cos_\star x = 1, \quad x \in \mathbb{R}_\star.$$

Answer

1. *Multiplicative nonlinear.*
2. *Multiplicative nonlinear.*
3. *Multiplicative linear.*
4. *Multiplicative nonlinear.*
5. *Multiplicative linear.*

Problem 1.6.19 *Classify each of the following MDEs as multiplicative homogeneous or multiplicative nonhomogeneous.*

1.
$$x^{2\star} \cdot_\star y^{\star\star\star} -_\star e^3 \cdot_\star y^{\star\star} +_\star y = 1, \quad x \in \mathbb{R}_\star.$$

2.
$$y^{\star\star} -_\star \sin_\star x \cdot_\star y = \cos_\star x, \quad x \in \mathbb{R}_\star.$$

3.
$$y^\star -_\star \cos_\star y = \tan_\star x, \quad x \in \mathbb{R}_\star.$$

4.
$$y^{\star\star\star} -_\star e^3 \cdot_\star y^{\star\star} +_\star e^3 \cdot_\star y^\star -_\star y = 1, \quad x \in \mathbb{R}_\star.$$

5.
$$y^{\star\star} -_\star \cos_\star x \cdot_\star y = e -_\star x +_\star x^{2\star}, \quad x \in \mathbb{R}_\star.$$

Answer

1. *Multiplicative homogeneous.*
2. *Multiplicative nonhomogeneous.*
3. *Multiplicative nonhomogeneous.*
4. *Multiplicative homogeneous.*
5. *Multiplicative nonhomogeneous.*

2

Elementary First-Order Multiplicative Differential Equations

2.1 Separable First-Order Multiplicative Differential Equations

Let $x_0, y_0, a, b \in \mathbb{R}_*$, $0_* < a, b < \infty$,

$$S = \left\{ (x,y) \in \mathbb{R}^2 : |x -_* x_0|_* < a, \qquad |y -_* y_0|_* < b \right\}.$$

We suppose that X_1 and X_2 are given positive continuous functions in $|x -_* x_0|_* < a$, Y_1 and Y_2 are given positive continuous functions in $|y -_* y_0|_* < b$.
We consider the equation

$$X_1(x) \cdot_* Y_1(y) \cdot_* d_* x +_* X_2(x) \cdot_* Y_2(y) \cdot_* d_* y = 0_*, \quad (x,y) \in S, \quad (2.1)$$

or

$$X_1(x) \cdot_* Y_1(y) +_* X_2(x) \cdot_* Y_2(y) \cdot_* y^* = 0_*, \quad (x,y) \in S, \quad (2.2)$$

or

$$X_1(x) \cdot_* Y_1(y) \cdot_* x^* +_* X_2(x) \cdot_* Y_2(y) = 0_*, \quad (x,y) \in S, \quad (2.3)$$

in which the variables are separated.

Definition 2.1.1 *The equation* (2.1) *is said to be separable.*

If $Y_1(y) \cdot_* X_2(x) \neq 0_*$, $(x,y) \in S$, then the equation (2.1) can be written in the following form.

$$(X_1(x)/_* X_2(x)) \cdot_* d_* x +_* (Y_2(y)/_* Y_1(y)) \cdot_* d_* y = 0_*, \quad (x,y) \in S.$$

DOI: 10.1201/9781003393344-2

Then the solution of the equation (2.1) is given by

$$\int_* (X_1(x)/_* X_2(x)) \cdot_* d_* x +_* \int_* (Y_2(y)/_* Y_1(y)) \cdot_* d_* y = c, \quad (x,y) \in S. \tag{2.4}$$

where c is a constant.

The equation (2.4) contains all solutions of (2.1) for which $Y_1(y) \cdot_* X_2(x) \neq 0_*$, $(x,y) \in S$. When we multiplicative divide the equation (2.1) by $Y_1(y) \cdot_* X_2(x)$, $(x,y) \in S$, we may leave some solutions of (2.1), and those of them which are not in (2.4) for some constant c must be coupled with (2.4) to obtain all solutions of (2.1). Note that the equation (2.4) can be rewritten in the form

$$e^{\int \frac{1}{x} \frac{\log X_1(x)}{\log X_2(x)} dx} +_* e^{\int \frac{1}{y} \frac{\log Y_1(y)}{\log Y_2(y)} dy} = c, \quad (x,y) \in S,$$

or

$$e^{\int \frac{1}{x} \frac{\log X_1(x)}{\log X_2(x)} dx + \int \frac{1}{y} \frac{\log Y_1(y)}{\log Y_2(y)} dy} = c, \quad (x,y) \in S,$$

or

$$\int \frac{1}{x} \frac{\log X_1(x)}{\log X_2(x)} dx + \int \frac{1}{y} \frac{\log Y_1(y)}{\log Y_2(y)} dy = c, \quad (x,y) \in S.$$

Example 2.1.2 *Let* $S = \{(x,y) \in \mathbb{R}_*^2 : |x -_* 1_*|_* < 3_*, \qquad |y -_* 2_*|_* < 3_*\}$. *We consider the following*

$$x \cdot_* (y^{2*} +_* 1_*) \cdot_* d_* x +_* y \cdot_* (x^{2*} +_* 1_*) \cdot_* d_* y = 0_* \tag{2.5}$$

in S.

First Way. *We multiplicative divide the considered equation by the function* $(x^{2*} +_* 1_*) \cdot_* (y^{2*} +_* 1_*) \neq 0_*$, $(x,y) \in S$, *and we get*

$$x/_* (x^{2*} +_* 1_*) \cdot_* d_* x +_* y/_* (y^{2*} +_* 1_*) \cdot_* d_* y = 0_*, \quad (x,y) \in S,$$

from where

$$\int_* (x/_* (x^{2*} +_* 1_*)) \cdot_* d_* x +_* \int_* (y/_* (y^{2*} +_* 1_*)) \cdot_* d_* y = c, \quad (x,y) \in S,$$

or

$$(1_*/_* 2_*) \cdot_* \int_* (d_* (x^{2*} +_* 1_*))/_* (x^{2*} +_* 1_*)$$

$$+_*(1_*/_*2_*)\cdot_*\int_*(d_*(y^{2*}+_*1_*))/_*(y^{2*}+_*1_*)=c,\quad (x,y)\in S,$$

or

$$\log_*(x^{2*}+_*1_*)+_*\log_*(y^{2*}+_*1_*)=c,\quad (x,y)\in S,$$

or

$$(x^{2*}+_*1_*)\cdot_*(y^{2*}+_*1_*)=c,\quad (x,y)\in S. \qquad (2.6)$$

Because $(x^{2*}+_*1_*)\cdot_*(y^{2*}+_*1_*)\neq 0_*,\ (x,y)\in S,$ *then all solutions of (2.5) are given by (2.4). Here c is an indicator for a constant.*

Second Way. *The given equation can be rewritten in the following manner.*

$$0_* = e^0$$

$$= x\cdot_*\left(e^{(\log y)^2}+_*e\right)\cdot_*e^{\frac{dx}{x}}+_*y\cdot_*\left(e^{(\log x)^2}+_*e\right)\cdot_*e^{\frac{dy}{y}}$$

$$= e^{\log x}\cdot_*e^{1+(\log y)^2}\cdot_*e^{\frac{dx}{x}}+_*e^{\log y}\cdot_*e^{1+(\log x)^2}\cdot_*e^{\frac{dy}{y}}$$

$$= e^{\frac{\log x}{x}(1+(\log y)^2)dx}+_*e^{\frac{\log y}{y}(1+(\log x)^2)dy}$$

$$= e^{\frac{\log x}{x}(1+(\log y)^2)dx+\frac{\log y}{y}(1+(\log x)^2)dy},\quad (x,y)\in S,$$

whereupon

$$\frac{\log x}{x}(1+(\log y)^2)dx+\frac{\log y}{y}(1+(\log x)^2)dy=0,\quad (x,y)\in S.$$

Then,

$$\frac{\log x}{x(1+(\log x)^2)}dx+\frac{\log y}{y(1+(\log y)^2)}dy=0,\quad (x,y)\in S.$$

We integrate the last equation and we find

$$c = \int\frac{\log x}{x(1+(\log x)^2)}dx+\int\frac{\log y}{y(1+(\log y)^2)}dy$$

$$= \int \frac{\log x}{1+(\log x)^2} d\log x + \int \frac{\log y}{1+(\log y)^2} d\log y$$

$$= \frac{1}{1}\int \frac{1}{1+(\log x)^2} d(1+(\log x)^2) + \frac{1}{2}\int \frac{1}{1+(\log y)^2} d(1+(\log y)^2)$$

$$= \frac{1}{2}\log(1+(\log x)^2) + \frac{1}{2}\log(1+(\log y)^2), \quad (x,y) \in S,$$

whereupon

$$c = \log(1+(\log x)^2) + \log(1+(\log y)^2)$$

$$= \log\left((1+(\log x)^2)(1+(\log y)^2)\right), \quad (x,y) \in S,$$

and then

$$c = (1+(\log x)^2)(1+(\log y)^2), \quad (x,y) \in S.$$

Hence,

$$c = e^{(1+(\log x)^2)(1+(\log y)^2)}$$

$$= e^{1+(\log x)^2} \cdot_* e^{1+(\log y)^2}$$

$$= \left(e +_* e^{(\log x)^2}\right) \cdot_* \left(e +_* e^{(\log y)^2}\right)$$

$$= (1_* +_* x^{2*}) \cdot_* (1_* +_* y^{2*}), \quad (x,y) \in S.$$

Example 2.1.3 *Let* $S = \{(x,y) \in \mathbb{R}_*^2 : |x -_* 2_*|_* < 1_*, |y -_* 1_*|_* < 3_*\}.$
We consider the equation

$$x \cdot_* y \cdot_* d_* x +_* (x +_* 1_*)d_* y = 0_*. \tag{2.7}$$

First Way. *We multiplicative divide the equation* (2.7) *by the function*
$y \cdot_* (x +_* 1_*),$ *for* $y \neq 0_*$ *and* $x \neq e^{-1},$ *and we get*

$$(x /_* (x +_* 1_*)) \cdot_* d_* x +_* (1_* /_* y) \cdot_* d_* y = 0_*,$$

whereupon

$$\int_* (x/_*(x+_* 1_*)) \cdot_* d_*x +_* \int_* (1_*/_*y) \cdot_* d_*y = c$$

or

$$\int_* d_*x -_* \int_* ((d_*x)/_*(x+_* 1_*)) +_* \int_* (1/_*y) \cdot_* d_*y = c,$$

from where

$$x -_* \log_* |x +_* 1_*|_* +_* \log_* |y|_* = c,$$

or

$$x -_* \log_* |(x +_* 1_*)/_*y|_* = c,$$

or

$$e^x \cdot_* (y/_*(x+_* 1_*)) = c,$$

or

$$y \cdot_* e^x = c \cdot_* (x +_* 1_*), \qquad (2.8)$$

where c is an indicator for a constant.

We note that $(x+_* 1_*) \cdot_* y = 0_*$ *if* $x = e^{-1}$ *and* $y = 0_*$. *We have that* $x = e^{-1}$ *and* $y = 0_*$ *are solutions to the equation* (2.7). *Also,* $y = 0_*$ *can be obtained from* (2.8) *for* $c = 0_*$ *and* $x = e^{-1}$ *cannot be obtained from* (2.8) *for any value of the constant c. Consequently, the solutions of the equation* (2.7) *are*

$$y \cdot_* e^x = c \cdot_* (x +_* 1_*), \qquad x = e^{-1}.$$

Second Way. *We can rewrite the given equation in the form*

$$
\begin{aligned}
0_* \ &= \ e^0 \\
&= \ e^{\log x \log y} \cdot_* e^{\frac{dx}{x}} +_* (x +_* e) \cdot_* e^{\frac{dy}{y}} \\
&= \ e^{\frac{\log x}{x} \log y dx} +_* (xe) \cdot_* e^{\frac{dy}{y}} \\
&= \ e^{\frac{\log x}{x} \log y dx} +_* e^{\log(xe)\frac{dy}{y}}
\end{aligned}
$$

$$= e^{\frac{\log x}{x}\log y\,dx} +_* e^{(1+\log x)\frac{dy}{y}}$$

$$= e^{\frac{\log x}{x}\log y\,dx + \frac{1+\log x}{y}dy}, \quad (x,y) \in S.$$

Hence, we get the equation

$$\frac{\log x}{x}\log y\,dx + \frac{1+\log x}{\log y}dy = 0, \quad (x,y) \in S. \tag{2.9}$$

We divide the last equation by $\log y(1+\log x)$ *for* $(x,y) \in S$ *for which* $x \neq e^{-1}$ *and* $y \neq 1$*, and we get*

$$\frac{\log x}{x(1+\log x)}dx + \frac{1}{y\log y}dy = 0,$$

whereupon

$$
\begin{aligned}
c &= \int \frac{\log x}{x(1+\log x)}dx + \int \frac{1}{y\log y}dy \\
&= \int \frac{\log x}{1+\log x}d\log x + \int \frac{1}{\log y}d\log y \\
&= \int d\log x - \int \frac{1}{1+\log x}d\log x + \log|\log y| \\
&= \log x - \log|1+\log x| + \log|\log y| \\
&= \log x + \log\left|\frac{\log y}{1+\log x}\right|
\end{aligned}
$$

and

$$x\frac{\log y}{1+\log x} = c, \tag{2.10}$$

from where

$$c = e^x \cdot_* e^{\frac{\log y}{1+\log x}}$$

$$= e^x \cdot_* \left(y/_* \left(e^{1+\log x} \right) \right)$$

$$= e^x \cdot_* \left(y/_* \left(e +_* e^{\log x} \right) \right)$$

$$= e^x \cdot_* \left(y/_* (1_* +_* x) \right)$$

and

$$y \cdot_* e^x = c \cdot_* (1_* +_* x).$$

Note that $x = e^{-1}$ and $y = 0_*$ are solutions to the equation (2.9). The solution $y = 0_*$ can be obtained by (2.10) for $c = 0_*$. Moreover, $x = e^{-1}$ cannot be obtained for any value $c \in \mathbb{R}_*$. Therefore, the solutions of the given equation are

$$y \cdot_* e^x = c \cdot_* (1_* +_* x), \quad (x, y) \in S, \quad x = e^{-1}.$$

Example 2.1.4 *Let $S = \{(x, y) \in \mathbb{R}_*^2 : |x|_* < 3_*, |y| < 2_*\}$. We consider the equation*

$$y^* -_* x \cdot_* y^{2*} = e^2 \cdot_* x \cdot_* y \tag{2.11}$$

in S.

First Way. *The equation (2.11) can be rewritten in the form*

$$y^* = x \cdot_* y \cdot_* (y +_* 2_*)$$

or

$$x \cdot_* y \cdot_* (y +_* 2_*) \cdot_* d_* x -_* d_* y = 0_*. \tag{2.12}$$

We multiplicative divide the last equation by the function $y \cdot_* (y +_* 2_*)$, $y \neq 0_*$, $y \neq e^{-2}$, and we get

$$x \cdot_* d_* x -_* (1_* /_* (y \cdot_* (y +_* 2_*))) \cdot_* d_* y = 0_*,$$

from where

$$\int_* x \cdot_* d_* x -_* \int_* (d_* y) /_* (y \cdot_* (y +_* 2_*)) = c,$$

or

$$\int_* x \cdot_* d_* x -_* e^{\frac{1}{2}} \cdot_* \int_* (d_* y) /_* y +_* e^{\frac{1}{2}} \cdot_* \int_* (d_* y) /_* (y +_* 2_*) = c,$$

or

$$x^{2*} - \log_* |y/_*(y+2)|_* = c,$$

or

$$e^{x^{2*}} \cdot_* (y+_* 2_*) = c \cdot_* y,$$

or

$$y +_* 2_* = c \cdot_* y \cdot_* e^{-_* x^{2*}}$$

or

$$2_* = y \cdot_* \left(c \cdot_* e^{-_* x^{2*}} -_* 1_* \right), \tag{2.13}$$

where c is an indicator for a constant. We note that $y = 0_*$ *and* $y = e^{-2}$ *are solutions to the equation (2.11), and they cannot be obtained from (2.13) for any values of the constant c. Consequently, the solutions of (2.11) are*

$$2_* = y \cdot_* \left(c \cdot_* e^{-_* x^{2*}} -_* 1_* \right), \qquad y = 0_*, \qquad y = e^{-2}. \tag{2.14}$$

Second Way. *The equation (2.12) can be rewritten in the following form*

$$
\begin{aligned}
0_* &= e^0 \\[6pt]
&= e^{\log x \log y} \cdot_* (y +_* e^2) \cdot_* e^{\frac{dx}{x}} -_* e^{\frac{dy}{y}} \\[6pt]
&= e^{\log x \log y} \cdot_* e^{2 + \log y} \cdot_* e^{\frac{dx}{x}} -_* e^{\frac{dy}{y}} \\[6pt]
&= e^{\frac{\log x}{x} \log y (2 + \log y)} -_* e^{\frac{dy}{y}} \\[6pt]
&= e^{\frac{\log x}{x} \log y (2 + \log y) dx - \frac{dy}{y}}, \qquad (x,y) \in S,
\end{aligned}
$$

whereupon

$$\frac{\log x}{x} \log y (2 + \log y) dx - \frac{dy}{y} = 0, \qquad (x,y) \in S.$$

Suppose that $(x,y) \in S$ is so that $y \neq 1$ and $y \neq e^{-2}$. We divide the last equation by $\log y(2 + \log y)$, and we find

$$\frac{\log x}{x} dx - \frac{1}{y \log y(2 + \log y)} dy = 0.$$

Hence,

$$
\begin{aligned}
c &= \int \frac{\log x}{x} dx - \int \frac{dy}{y \log y(2 + \log y)} \\
&= \int \log x\, d\log x - \int \frac{d\log y}{\log y(2 + \log y)} \\
&= \frac{1}{2}(\log x)^2 - \frac{1}{2}\int \frac{d\log y}{\log y} + \frac{1}{2}\int \frac{d\log y}{\log y + 2} \\
&= \frac{1}{2}(\log x)^2 - \frac{1}{2}\log|\log y| + \frac{1}{2}\log|2 + \log y| \\
&= \frac{1}{2}(\log x)^2 - \frac{1}{2}\log\left|\frac{\log y}{2 + \log y}\right|,
\end{aligned}
$$

whereupon

$$c = (\log x)^2 - \log\left|\frac{\log y}{2 + \log y}\right|$$

and

$$c \log y = e^{(\log x)^2(2 + \log y)},$$

or

$$e^{\log c \log y} = e^{e^{(\log x)^2}(2 + \log y)},$$

or

$$
\begin{aligned}
c \cdot_* y &= e^{e^{(\log x)^2}} \cdot_* e^{2 + \log y} \\
&= e^{x^{2*}} \cdot_* \left(e^2 \cdot_* e^{\log y}\right) \\
&= e^{x^{2*}} \cdot_* (2_* +_* y),
\end{aligned}
$$

whereupon we get (2.13). Note that $y = 0_*$ *and* $y = e^{-2}$ *are solutions of the considered equation, and they cannot be obtained by (2.13) for any value of* $c \in \mathbb{R}_*$. *Hence, the solutions of the given equation are (2.14).*

Exercise 2.1.5 *Let* $S = \{(x,y) \in \mathbb{R}_*^2 : |x|_* < 3_*, |y|_* < 2_*\}$. *Find the solutions of the equation*

$$(x^{2*} -_* 1_*) \cdot_* y^* +_* 2_* \cdot_* x \cdot_* y^{2*} = 0_* \qquad in \qquad S.$$

Answer. $y \cdot_* (c +_* \log_* |x^{2*} -_* 1_*|) = 1_*$, $y = 0_*$.

Exercise 2.1.6 *Let* $S = \{(x,y) \in \mathbb{R}_*^2 : |x -_* 3_*|_* < 2_*, |y -_* 1_*|_* < 2_*\}$. *Find the solutions of the equation*

$$y^* \cdot_* \cot_* x +_* y = 2_* \qquad in \qquad S.$$

Answer. $y = 2_* +_* c \cdot_* \cos_* x$.

Exercise 2.1.7 *Let* $S = \{(x,y) \in \mathbb{R}_*^2 : |x -_* 1_*| < 2_*, |y -_* 3_*| < 4_*\}$. *Find the solutions of the equation*

$$2_* \cdot_* x^{2*} \cdot_* y \cdot_* y^* +_* y^{2*} = 2_*.$$

Answer. $y^{2*} = 2_* +_* c \cdot_* e^{1_*/_*x}$.

Now we consider the MDE

$$y^* = f(\alpha_* \cdot_* x +_* \beta_* \cdot_* y +_* \gamma_*), \qquad (2.15)$$

where α_*, β_* and γ_* are multiplicative constants, $\beta_* \neq 0_*$, f is a given continuous function in $\mathbb{R}_*$. We set

$$z = \alpha_* \cdot_* x +_* \beta_* \cdot_* y +_* \gamma_*.$$

Then,

$$z^* = \alpha_* +_* \beta_* \cdot_* y^* \qquad \Longrightarrow y^* = (z^* -_* \alpha_*)/_*\beta_*.$$

In this way the equation (2.15) takes the form

$$(z^* -_* \alpha_*)/_*\beta_* = f(z)$$

or

$$z^* = \alpha_* +_* \beta_* \cdot_* f(z),$$

which is a separable MDE.

Example 2.1.8 *Let* $S = \{(x,y) \in \mathbb{R}_*^2 : |x|_* < 1_*, |y|_* < 2_*\}$. *We consider the MDE*

$$y^* = 2_* \cdot_* x +_* 3_* \cdot_* y +_* 2_*. \qquad (2.16)$$

We set

$$z = 2_* \cdot_* x +_* 3_* \cdot_* y +_* 2_*.$$

Then,

$$z^* = 2_* +_* 3_* \cdot_* y^* \qquad or \qquad y^* = (z^* -_* 2_*)/_* 3_*.$$

In this way the equation (2.16) takes the form

$$(z^* -_* 2_*)/_* 3_* = z,$$

or

$$z^* = 3_* \cdot_* z +_* 2_*,$$

or

$$d_* z = (3_* \cdot_* z +_* 2_*) \cdot_* d_* x.$$

We multiplicative divide the last equation by $3_* \cdot_* z +_* 2_*$, $z \neq e^{-\frac{2}{3}}$, *and we obtain*

$$(d_* z)/_* (3_* \cdot_* z +_* 2_*) = d_* x,$$

whereupon

$$\int_* (d_* z)/_* (3_* \cdot_* z +_* 2_*) = \int_* d_* x +_* c$$

or

$$1_*/_* 3_* \cdot_* \log_* |3_* \cdot_* z +_* 2_*|_* = x +_* c,$$

or

$$\log_* |3_* \cdot_* z +_* 2_*|_* = 3_* \cdot_* x +_* c,$$

or

$$3_* \cdot_* z +_* 2_* = c \cdot_* e^{3_* \cdot_* x}.$$

Therefore,

$$6_* \cdot_* x +_* 9_* \cdot_* y +_* 8_* = c \cdot_* e^{3_* \cdot_* x}$$

or

$$y = 1_*/_* 9_* \cdot_* \left(c \cdot_* e^{3_* \cdot_* x} -_* 6_* \cdot_* x -_* 8_* \right). \qquad (2.17)$$

We note that $y = -_*(2_*/_* 3_*) \cdot_* x -_* (8_*/_* 9_*)$ *is a solution to the equation (2.16), and it can be obtained from (2.17) for* $c = 0_*$. *Consequently, the solutions of (2.16) are given by (2.17).*

Example 2.1.9 *Let $S = \{(x,y) \in \mathbb{R}^2_* : |x -_* 1_*|_* < 2_*, |y -_* 1_*|_* < 3_*\}$.*
We consider the MDE

$$y^* = (\sin_*(x +_* y -_* 2_*))^{2*} -_* 1_*. \tag{2.18}$$

We set

$$z = x +_* y -_* 2_*,$$

whereupon

$$y = z -_* x +_* 2_*$$

and

$$y^* = z^* -_* 1_*.$$

In this way the equation (2.18) takes the form

$$z^* -_* 1_* = (\sin_* z)^{2*} -_* 1_*,$$

or

$$z^* = (\sin_* z)^{2*},$$

or

$$d_* z = (\sin_* z)^{2*} \cdot_* d_* x.$$

We multiplicative divide the last equation by $(\sin_* z)^{2*}$, $z \neq e^{k\pi}$, $k \in \mathbb{Z}$,
and we get

$$(d_* z)/_*(\sin_* z)^{2*} = d_* x,$$

whereupon

$$\int_* (d_* z)/_*(\sin_* z)^{2*} = \int_* d_* x +_* c,$$

and

$$-_* \cot_* z = x +_* c,$$

and

$$\cot_* z = c -_* x.$$

Therefore,

$$\cot_*(x +_* y -_* 2_*) = c -_* x,$$

or

$$x +_* y -_* 2_* = \operatorname{arccot}_*(c -_* x),$$

$$y = 2_* -_* x +_* \operatorname{arccot}_*(c -_* x). \tag{2.19}$$

We note that

$$y = 2_* -_* x -_* e^{k\pi}, \qquad k \in \mathbb{Z},$$

are solutions to the equation (2.18). They cannot be obtained from (2.19) for any values of the constant c. Consequently, the solutions of the considered equation are

$$y = 2_* -_* x +_* \text{arccot}_*(c -_* x), \qquad y = 2_* -_* x -_* e^{k\pi}, \qquad k \in \mathbb{Z}.$$

Example 2.1.10 *Let $S = \{(x,y) \in \mathbb{R}^2_* : |x -_* 1_*|_* < 1_*, |y|_* < 2_*\}$. We consider the MDE*

$$y^* = (x +_* y +_* 1_*)^{2*}.$$

We set

$$z = y +_* x +_* 1_*.$$

Then

$$y = z -_* x -_* 1_*$$

and

$$y^* = z^* -_* 1_*.$$

Hence, the considered equation takes the form

$$z^* -_* 1_* = z^{2*},$$

or

$$z^* = 1_* +_* z^{2*},$$

or

$$d_* z = (1_* +_* z^{2*}) \cdot_* d_* x.$$

We divide the last equation by $1_* +_* z^{2*}$ and we get

$$(d_* z)/_* 1_* +_* z^{2*} = d_* x,$$

from where

$$\int_* (d_* z)/_*(1_* +_* z^{2*}) = \int_* d_* x +_* c,$$

and

$$\text{arctan}_* z = x +_* c,$$

and

$$z = \tan_*(x +_* c).$$

Consequently, the solutions of the considered equation are given by

$$y = e^{-1} -_* x +_* \tan_*(x +_* c).$$

Exercise 2.1.11 *Let $S = \{(x,y) \in \mathbb{R}^2 : |x -_* 1|_* < 1_*, |y|_* < 4_*\}$. Find the solutions of the equation*

$$y^* = \cos_*(y -_* x).$$

Answer. $y = x +_* 2_* \cdot_* \text{arccot}_*(x +_* c)$.

Exercise 2.1.12 *Let $S = \{(x,y) \in \mathbb{R}^2_* : |x|_* < 2_*, |y|_* < 3_*\}$. Find the solutions of the equation*

$$y^* = y +_* 2_* \cdot_* x -_* 3_*.$$

Answer. $y = 1_* -_* 2_* \cdot_* x +_* c \cdot_* e^x$.

Exercise 2.1.13 *Let $S = \{(x,y) \in \mathbb{R}^2_* : |x|_* < 1_*, |y -_* 2_*|_* < 1_*, 4_* \cdot_* x +_* 2_* \cdot_* y -_* 1_* > 0_*\}$. Find the solutions of the equation*

$$y^* = (4_* \cdot_* x +_* 2_* \cdot_* y -_* 1_*)^{\frac{1}{2}*}.$$

Answer.

$$(4_* \cdot_* x +_* 2_* \cdot_* y -_* 1_*)^{\frac{1}{2}*}$$

$$-_* 2_* \cdot_* \log_* \left((4_* \cdot_* x +_* 2_* \cdot_* y -_* 1_1)^{\frac{1}{2}*} +_* 2_*\right) = x +_* c.$$

2.2   Multiplicative Homogeneous Functions

Let $D \subseteq \mathbb{R}^2_*$.

Definition 2.2.1 *Let $f : D \to \mathbb{R}_*$. We say that f is a multiplicative homogeneous function if*

$$f(\alpha \cdot_* x, \alpha \cdot_* y) = \alpha^{k_*} \cdot_* f(x,y), \quad (x,y) \in \mathbb{R}^2_*. \quad (2.20)$$

$\alpha \in \mathbb{R}_*$, for some $k \in \mathbb{R}$. The number k is called the degree of multiplicative homogeneity.

We can rewrite (2.20) in the following way.

$$f\left(e^{\log \alpha \log x}, e^{\log \alpha \log y}\right) = e^{(\log \alpha)^k \log f(x,y)},$$

$(x,y) \in D, \alpha \in \mathbb{R}_\star.$

Example 2.2.2 *Let*

$$f(x,y) = xy, \quad (x,y) \in \mathbb{R}_\star^2.$$

We have

$$
\begin{aligned}
f\left(e^{\log \alpha \log x}, e^{\log \alpha \log y}\right) &= e^{\log \alpha \log x} e^{\log \alpha \log y} \\[2mm]
&= e^{\log \alpha \log x + \log \alpha \log y} \\[2mm]
&= e^{\log \alpha (\log x + \log y)} \\[2mm]
&= e^{\log \alpha \log(xy)} \\[2mm]
&= e^{\log \alpha \log f(x,y)}, \quad (x,y) \in \mathbb{R}_\star^2, \quad \alpha \in \mathbb{R}_\star.
\end{aligned}
$$

Thus, f is a multiplicative homogeneous function with degree of multiplicative homogeneity 1. Also, it is a homogeneous function with degree of homogeneity 2, because

$$
\begin{aligned}
f(\alpha x, \alpha y) &= (\alpha x)(\alpha y) \\[2mm]
&= \alpha^2 xy \\[2mm]
&= \alpha^2 f(x,y), \quad (x,y) \in \mathbb{R}_\star^2, \quad \alpha \in \mathbb{R}_\star.
\end{aligned}
$$

Example 2.2.3 *Let*

$$f(x,y) = x + y, \quad (x,y) \in \mathbb{R}_\star^2.$$

We have

$$f(\alpha x, \alpha y) = \alpha x + \alpha y$$

$$= \alpha(x+y)$$

$$= \alpha f(x,y), \quad (x,y) \in \mathbb{R}^2_\star, \quad \alpha \in \mathbb{R}_\star,$$

i.e., f is a homogeneous function with degree of homogeneity 1. *Next,*

$$f\left(e^{\log \alpha \log x}, e^{\log \alpha \log y}\right) = e^{\log \alpha \log x} + e^{\log \alpha \log y},$$

$(x,y) \in \mathbb{R}^2_\star$, $\alpha \in \mathbb{R}_\star$. *Hence, f is not a multiplicative homogeneous function.*

Exercise 2.2.4 *Let*

$$f(x,y) = x^2 y^3, \quad (x,y) \in \mathbb{R}^2_\star.$$

Prove that f is a multiplicative homogeneous function.

Example 2.2.5 *Let* $D = \{(x,y) \in \mathbb{R}^2_\star : x \geq 1_\star, y \geq 1_\star\}$ *and* $f(x,y) = e^{y/\star x} +_\star \log_\star |y/_\star x|_\star$, $(x,y) \in D$. *Then, for every* $t \in \mathbb{R}_\star$, $t \neq 0_\star$, *and* $(x,y) \in D$ *we have*

$$f(\alpha \cdot_\star x, \alpha \cdot_\star y) = e^{(\alpha \cdot_\star y)/_\star(\alpha \cdot_\star x)} +_\star \log_\star |(\alpha \cdot_\star y)/_\star(\alpha \cdot_\star x)|_\star$$

$$= e^{y/_\star x} + \log_\star |y/_\star x|_\star$$

$$= f(x,y).$$

Therefore, $f(x,y)$ is a multiplicative homogeneous function of degree 0.

Example 2.2.6 *Let* $D = \mathbb{R}^2_\star$ *and* $f(x,y) = x +_\star y^{2\star}$, $(x,y) \in D$. *Then, for every* $(x,y) \in D$ *and for every* $\alpha \in \mathbb{R}_\star$, $\alpha \neq 0_\star$, *we have*

$$f(\alpha \cdot_\star x, \alpha \cdot_\star y) = \alpha \cdot_\star x +_\star \alpha^{2\star} \cdot_\star y^{2\star}$$

$$\neq \alpha^{k\star} \cdot_\star f(x,y)$$

for every $k \in \mathbb{R}$. Therefore, the function $f(x,y)$ is not a multiplicative homogeneous function.

Example 2.2.7 *Let* $D = \mathbb{R}^2_*$ *and* $f(x,y) = x^{3*} -_* 3_* \cdot_* x^{2*} \cdot_* y +_* y^{3*}$, $(x,y) \in D$. *Then, for every* $(x,y) \in D$ *and for every* $\alpha \in \mathbb{R}_*$, $\alpha \neq 0_*$, *we have*

$$f(\alpha \cdot_* x, \alpha \cdot_* y) = (\alpha \cdot_* x)^{3*} -_* 3_* \cdot_* (\alpha \cdot_* x)^{2*} (\alpha \cdot_* y) +_* (\alpha \cdot_* y)^{3*}$$

$$= \alpha^{3*} \cdot_* (x^{3*} -_* 3_* \cdot_* x^{2*} \cdot_* y +_* y^{3*})$$

$$= \alpha^{3*} \cdot_* f(x,y).$$

Therefore, the function $f(x,y)$ *is a multiplicative homogeneous function of degree 3.*

Exercise 2.2.8 *Let* $D = \{(x,y) \in \mathbb{R}^2_* : x \neq 0_* \quad$ or $\quad y \neq 0\}$. *Prove that the following functions are multiplicative homogeneous. Find the degree of homogeneity.*

1. $f(x,y) = x^{2*} -_* 2_* \cdot_* x \cdot_* y +_* y^{2*}$,
2. $f(x,y) = (x^{3*} +_* y^{3*})/_* x^{2*} +_* y^{2*}$,
3. $f(x,y) = (x -_* y)/_* (x^{4*} +_* 3_* \cdot_* x^{3*} \cdot_* y +_* y^{4*})$.

Answer

1. 2.
2. 1.
3. −3.

Exercise 2.2.9 *Let* $D = \mathbb{R}^2_*$. *Check if the function*

$$f(x,y) = x^{4*} -_* 5_* \cdot_* x^{5*} \cdot_* y +_* 6_* \cdot_* x^{6*}$$

is a multiplicative homogeneous function.

Answer *No.*

The multiplicative homogeneous functions in the domain D have the following properties.

1. If $f_1(x,y)$ and $f_2(x,y)$ are multiplicative homogeneous functions in D of degree k, then

$$a_1 \cdot_* f_1(x,y) +_* a_2 \cdot_* f_2(x,y)$$

   is a multiplicative homogeneous function in D of degree k for every $a_1, a_2 \in \mathbb{R}_*$.

   Proof 2.2.10 *We have*

   $$f_1(\alpha \cdot_* x, \alpha \cdot_* y) = \alpha^{k_*} \cdot_* f_1(x,y), \quad (x,y) \in D, \quad \alpha \in \mathbb{R}_*,$$

   and

   $$f_2(\alpha \cdot_* x, \alpha \cdot_* y) = \alpha^{k_*} \cdot_* f_2(x,y), \quad (x,y) \in D, \quad \alpha \in \mathbb{R}_*.$$

   Set

   $$g(x,y) = a_1 \cdot_* f_1(x,y) +_* a_2 \cdot_* f_2(x,y), \quad (x,y) \in D.$$

   Hence,

   $$\begin{aligned} g(\alpha \cdot_* x, \alpha \cdot_* y) &= a_1 \cdot_* f_1(\alpha \cdot_* x, \alpha_* y) \\[2mm] &\quad +_* a_2 \cdot_* f_2(\alpha \cdot_* x, \alpha \cdot_* y) \\[2mm] &= a_1 \cdot_* \alpha^{k_*} \cdot_* f_1(x,y) +_* a_2 \cdot_* \alpha^{k_*} \cdot_* f_2(x,y) \\[2mm] &= \alpha^{k_*} \cdot_* (a_1 \cdot_* f_1(x,y) +_* a_2 \cdot_* f_2(x,y)) \\[2mm] &= \alpha^{k_*} \cdot_* g(x,y), \quad (x,y) \in D, \quad \alpha \in \mathbb{R}_*. \end{aligned}$$

   Thus, g is a multiplicative homogeneous function of degree k on D. This completes the proof.

2. If $f_1(x,y)$ is a multiplicative homogeneous function in D of degree k_1 and $f_2(x,y)$ is a multiplicative homogeneous function in D of degree k_2, then $f_1(x,y) \cdot_* f_2(x,y)$ is a multiplicative homogeneous function in D of degree $k_1 + k_2$ and $f_1(x,y)/_* f_2(x,y)$ is a multiplicative homogeneous function in D of degree $k_1 - k_2$.

Proof 2.2.11 *We have*

$$f_1(\alpha \cdot_* x, \alpha \cdot_* y) = \alpha^{k_{1*}} \cdot_* f_1(x,y), \quad (x,y) \in D, \quad \alpha \in \mathbb{R}_*,$$

and

$$f_2(\alpha \cdot_* x, \alpha \cdot_* y) = \alpha^{k_{2*}} \cdot_* f_2(x,y), \quad (x,y) \in D, \quad \alpha \in \mathbb{R}_*.$$

Set

$$g_1(x,y) = f_1(x,y) \cdot_* f_2(x,y), \quad (x,y) \in D.$$

Hence,

$$g_1(\alpha \cdot_* x, \alpha \cdot_* y) = f_1(\alpha \cdot_* x, \alpha_* y) \cdot_* f_2(\alpha \cdot_* x, \alpha \cdot_* y)$$

$$= \alpha^{k_{1*}} \cdot_* f_1(x,y) \cdot_* \alpha^{k_{2*}} \cdot_* f_2(x,y)$$

$$= \alpha^{(k_1+k_2)*} \cdot_* (f_1(x,y) \cdot_* f_2(x,y))$$

$$= \alpha^{(k_1+k_2)*} \cdot_* g_1(x,y), \quad (x,y) \in D, \quad \alpha \in \mathbb{R}_*.$$

Thus, g_1 is a multiplicative homogeneous function of degree $k_1 + k_2$ on D.

Now, we denote

$$g_2(x,y) = f_1(x,y)/_* f_2(x,y), \quad (x,y) \in D.$$

Then

$$g_2(\alpha \cdot_* x, \alpha \cdot_* y) = f_1(\alpha \cdot_* x, \alpha_* y)/_* f_2(\alpha \cdot_* x, \alpha \cdot_* y)$$

$$= (\alpha^{k_{1*}} \cdot_* f_1(x,y))/_* (\alpha^{k_{2*}} \cdot_* f_2(x,y))$$

$$= \alpha^{(k_1-k_2)*} \cdot_* (((f_1(x,y))/_* (f_2(x,y))))$$

$$= \alpha^{(k_1-k_2)*} \cdot_* g_2(x,y), \quad (x,y) \in D, \quad \alpha \in \mathbb{R}_*.$$

Thus, g_2 is a multiplicative homogeneous function of degree $k_1 - k_2$ on D. This completes the proof.

3. A multiplicative homogeneous function in D of degree 0 is a function of a single multiplicative variable $y/_*x$.

Proof 2.2.12 *Let $f(x,y)$ be a multiplicative homogeneous function in D of degree zero. We take $\alpha = 1_*/_*x$ in (2.20) and we get*

$$f(1_*, y/_*x) = f(x,y), \quad (x,y) \in D.$$

4. The Lagrange identity. Let $f \in \mathscr{C}^1_*(D)$ be a multiplicative homogeneous function in D of degree k. Then

$$k_* \cdot_* f(x,y) = x \cdot_* f_x^*(x,y) +_* y \cdot_* f_y(x,y)$$

for all $(x,y) \in D$.

Proof 2.2.13 *We have*

$$f(\alpha \cdot_* x, \alpha \cdot_* y) = \alpha^{k_*} \cdot_* f(x,y), \quad \forall(x,y) \in D, \alpha \in \mathbb{R}_*, \alpha \neq 0_*.$$

The last equation we multiplicative differentiate with respect to α and we get

$$x \cdot_* f_x^*(\alpha \cdot_* x, \alpha \cdot_* y) +_* y \cdot_* f_y^*(\alpha \cdot_* x, \alpha \cdot_* y) = k_* \cdot_* \alpha^{(k-1)_*} \cdot_* f(x,y).$$

We take $\alpha = 1_*$ in the last equation and we obtain the multiplicative Lagrange identity.

2.3 Multiplicative Homogeneous Multiplicative Differential Equations

Definition 2.3.1 *Let $P, Q \in \mathscr{C}_*(D)$ be positive functions. Then the MDE*

$$P(x,y) +_* Q(x,y) \cdot_* y^* = 0_* \tag{2.21}$$

is said to be multiplicative homogeneous if the functions $P(x,y)$ and $Q(x,y)$ are multiplicative homogeneous functions in D of the same degree.

In fact, the equation (2.21) can be rewritten as follows.

$$e^0 = 0_*$$

$$= P(x,y) +_* Q(x,y) \cdot_* e^{x \frac{y'}{y}}$$

$$= P(x,y) +_* e^{x(\log(Q(x,y)))\frac{y'}{y}}$$

$$= e^{\log(P(x,y))+x(\log(Q(x,y)))\frac{y'}{y}}, \quad (x,y) \in D,$$

or

$$0 = \log(P(x,y)) + x(\log(Q(x,y)))\frac{y'}{y}, \quad (x,y) \in D,$$

or

$$0 = yP(x,y) + x(\log(Q(x,y)))y', \quad (x,y) \in D.$$

We suppose that (2.21) is a multiplicative homogeneous equation. Because P and Q are multiplicative homogeneous functions in D of the same degree, we have that $P/_*Q$ is a multiplicative homogeneous function in D of degree zero. Hence,

$$(P(x,y))/_*(Q(x,y)) = (P(1_*,y/_*x))/_*(Q(1_*,y/_*x)).$$

Therefore, we can rewrite the equation (2.21) in the form

$$P(1_*,y/_*x) +_* Q(1_*,y/_*x) \cdot_* y^* = 0_*, \quad (x,y) \in D.$$

We set $z = y/_*x$, $(x,y) \in D$. Then

$$y = z \cdot_* x$$

and

$$y^* = z^* \cdot_* x +_* z, \quad (x,y) \in D.$$

Hence, we get the equation

$$P(1_*,z_*) +_* Q(1_*,z) \cdot_* (z^* \cdot_* x +_* z) = 0_*$$

or

$$P(1_*,z) +_* z \cdot_* Q(1_*,z) +_* x \cdot_* Q(1_*,z) \cdot_* z^* = 0_*,$$

which is a separable MDE.

Example 2.3.2 *Let $D = \mathbb{R}_*^2$. Let us consider the MDE*

$$(y^{2*} -_* 2_* \cdot_* x \cdot_* y) \cdot_* d_* x +_* x^{2*} \cdot_* d_* y = 0_*.$$

Here,

$$P(x,y) = y^{2*} -_* 2_* \cdot_* x \cdot_* y, \qquad Q(x,y) = x^{2*}.$$

From here, for every $(x,y) \in D$ and for every $\alpha \in \mathbb{R}_*$, $\alpha \neq 0_*$, we have

$$
\begin{aligned}
P(\alpha \cdot_* x, \alpha \cdot_* y) &= (\alpha \cdot_* y)^{2*} -_* 2_* \cdot_* (\alpha \cdot_* x) \cdot_* (\alpha \cdot_* y) \\[2mm]
&= \alpha^{2*} \cdot_* (y^{2*} -_* 2_* \cdot_* x \cdot_* y) \\[2mm]
&= \alpha^{2*} \cdot_* P(x,y), \quad (x,y) \in D,
\end{aligned}
$$

and

$$
\begin{aligned}
Q(\alpha \cdot_* x, \alpha \cdot_* y) &= (\alpha \cdot_* x)^{2*} \\[2mm]
&= \alpha^{2*} \cdot_* x^{2*} \\[2mm]
&= \alpha^{2*} \cdot_* Q(x,y), \quad (x,y) \in D.
\end{aligned}
$$

Therefore, the functions P and Q are multiplicative homogeneous functions in D of degree 2. Consequently, the considered equation is a multiplicative homogeneous equation, which we can rewrite in the form for $x \neq 0_*$,

$$(y^{2*})/_*(x^{2*}) -_* 2_* \cdot_* (y/_* x) +_* y^* = 0_*.$$

We set $z = y/_* x$. Then

$$y = z \cdot_* x$$

and

$$y^* = z^* \cdot_* x +_* z.$$

From here, we get the following equation

$$z^{2*} -_* 2_* \cdot_* z +_* z^* \cdot_* x +_* z = 0_*$$

or

$$z^* \cdot_* x = z -_* z^{2*},$$

or

$$x \cdot_* d_* z = (z -_* z^{2*}) \cdot_* d_* x.$$

We multiplicative divide the last equation by $z -_* z^{2*}$, $z \neq 0_*$, $z \neq 1_*$, and we find

$$(1_*/_*(z -_* z^{2*})) \cdot_* d_* z = (1_*/_* x) \cdot_* d_* x.$$

Hence,

$$\int_* (d_* z)/_*(z -_* z^{2*}) = \int_* (d_* x)/_* x +_* c$$

or

$$\int_* (d_* z)/_* z +_* \int_* (d_* z)/_*(1_* -_* z) = \int_* (d_* x)/_* x +_* c,$$

or

$$\log_* |z| -_* \log_* |z - 1| = \log_* |x| +_* c,$$

or

$$c \cdot_* z = x \cdot_* (z -_* 1_*),$$

or

$$c \cdot_* (y/_* x) = x \cdot_* ((y/_* x) -_* 1_*),$$

whereupon

$$y = x/_*(x -_* c).$$

We note that $x = 0_*$, $y = 0_*$, $y = x$ are solutions of the considered equation, which cannot be obtained for any values of the constant c by the last expression. Therefore the solutions of the considered equation are

$$y = x/_*(x -_* c), \qquad x = 0_*, \qquad y = 0_*, \qquad y = x.$$

Example 2.3.3 *Let $D = \{(x,y) \in \mathbb{R}^2_* : x \neq 0_*, (x +_* y)/_* x > 0_*\}$. We consider the equation*

$$x \cdot_* y^* -_* y = (x +_* y) \cdot_* \log_*((x +_* y)/_* x),$$

which we can rewrite in the form

$$y +_* (x +_* y) \cdot_* \log_*((x +_* y)/_* x) -_* x \cdot_* y^* = 0_*.$$

Here

$$P(x,y) \;=\; y +_* (x +_* y) \log_*((x +_* y)/_* x),$$

$$Q(x,y) \;=\; -_* x.$$

Then, for every $(x,y) \in D$ *and for every* $\alpha \in \mathbb{R}_*$, $\alpha \neq 0_*$, *we have*

$$P(\alpha \cdot_* x, \alpha \cdot_* y) = \alpha \cdot_* y +_* (\alpha \cdot_* x +_* \alpha \cdot_* y)$$

$$\cdot_* \log_* ((\alpha \cdot_* x +_* \alpha \cdot_* y)/_* (\alpha \cdot_* x))$$

$$= \alpha \cdot_* (y +_* (x +_* y) \cdot_* \log_* ((x +_* y)/_* x))$$

$$= \alpha \cdot_* P(x,y), \quad (x,y) \in D,$$

and

$$Q(\alpha \cdot_* x, \alpha \cdot_* y) = -_* \alpha \cdot_* x$$

$$= \alpha \cdot_* Q(x,y).$$

Therefore, $P(x,y)$ *and* $Q(x,y)$ *are multiplicative homogeneous functions in D of degree* 1. *Consequently, the considered equation is a multiplicative homogeneous equation.*
We can rewrite the given equation in the form

$$y *-_* y/_* x = (1_* +_* y/_* x) \cdot_* \log_* (1_* +_* y/_* x).$$

We set $z = y/_* x$. *From here,*

$$y = z \cdot_* x$$

and

$$y^* = z^* \cdot_* x +_* z.$$

Thus, we get the equation

$$z^* \cdot_* x +_* z -_* z = (1_* +_* z) \cdot_* \log_* (1_* +_* z)$$

or

$$z^* \cdot_* x = (1_* +_* z) \cdot_* \log_* (1_* +_* z),$$

or

$$x \cdot_* d_* z = (1_* +_* z) \cdot_* \log_* (1_* +_* z) \cdot_* d_* x.$$

We multiplicative divide the last equation by $(1_* +_* z) \cdot_* \log_*(1_* +_* z)$, $z \neq 0_*$, *and we get*

$$(d_* z)/_*((1_* +_* z) \cdot_* \log_*(1_* +_* z)) = (d_* x)/_* x.$$

From here,

$$\int_* (d_* z)/_*((1_* +_* z) \cdot_* \log_*(1_* +_* z)) = \int_* (d_* x)/_* x +_* c$$

or

$$\int_* (d_* \log_*(1_* +_* z))/_*(\log_*(1_* +_* z)) = \log_* |x| +_* c,$$

or

$$\log_* |\log_*(1_* +_* z)| = \log_* |x|_* +_* c,$$

or

$$\log_*(1_* +_* z) = c \cdot_* x,$$

or

$$\log_*(y +_* x)/_* x = c \cdot_* x,$$

or

$$(y +_* x)/_* x = e^{c \cdot_* x},$$

or

$$y = x \cdot_* e^{c \cdot_* x} -_* x. \qquad (2.22)$$

We note that $y = 0_*$ *is a solution of the considered equation, which cannot be obtained for any values of the constant c in the expression (2.22). Therefore the solutions of the considered equation are*

$$y = x \cdot_* e^{c \cdot_* x} -_* x, \qquad y = 0_*.$$

Example 2.3.4 *Let* $D = \{(x,y) \in \mathbb{R}_*^2 : x > 0_*, y > 0_*\}$. *We consider the equation*

$$\left(y +_* (x \cdot_* y)^{\frac{1}{2}_*}\right) \cdot_* d_* x = x \cdot_* d_* y,$$

which we can rewrite in the form

$$\left(y +_* (x \cdot_* y)^{\frac{1}{2}_*}\right) \cdot_* d_* x -_* x \cdot_* d_* y = 0_*.$$

Here

$$P(x,y) = y +_* (x \cdot_* y)^{\frac{1}{2}}_*,$$

$$Q(x,y) = -_*x, \quad (x,y) \in D.$$

Then, for every $(x,y) \in D$ *and for every* $\alpha > 0_*$ *we have*

$$P(\alpha \cdot_* x, \alpha \cdot_* y) = \alpha \cdot_* y +_* (\alpha \cdot_* x \cdot_* \alpha \cdot_* y)^{\frac{1}{2}}_*$$

$$= \alpha \cdot_* y +_* \alpha \cdot_* (x \cdot_* y)^{\frac{1}{2}}_*$$

$$= \alpha \cdot_* P(x,y), \quad (x,y) \in D,$$

and

$$Q(\alpha \cdot_* x, \alpha \cdot_* y) = -_* \alpha \cdot_* x$$

$$= \alpha \cdot_* Q(x,y), \quad (x,y) \in D.$$

Therefore the functions $P(x,y)$ *and* $Q(x,y)$ *are multiplicative homogeneous functions in D of degree 1. Consequently, the considered equation is a multiplicative homogeneous equation.*
We can rewrite the given equation in the form

$$y^* = y/_*x +_* (y/_*x)^{\frac{1}{2}}_*.$$

We set $z = y/_*x$. *Then*

$$y = z \cdot_* x$$

and

$$y^* = z^* \cdot_* x +_* z.$$

In this way we get the equation

$$z^* \cdot_* x +_* z = z +_* (z)^{\frac{1}{2}}_*$$

or

$$z^* \cdot_* x = (z)^{\frac{1}{2}}_*,$$

or

$$x \cdot_* d_* z = (z)^{\frac{1}{2}*} \cdot_* d_* x.$$

We multiplicative divide the last equation by $x \cdot_* (z)^{\frac{1}{2}*}$ and we get

$$(d_* z)/_* (z)^{\frac{1}{2}*} = (d_* x)/_* x,$$

whereupon

$$\int_* (d_* z)/_* (z)^{\frac{1}{2}*} = \int_* (d_* x)/_* x +_* c$$

or

$$2_* \cdot_* (z)^{\frac{1}{2}*} = \log_*(x) +_* c,$$

or

$$y = x \left((\log_* x +_* c)/_* 2_* \right)^{2*} . \tag{2.23}$$

We note that $y = 0_*$ and $x = 0_*$ don't belong to D. Therefore all solutions of the considered equation are given by the expression (2.23).

Exercise 2.3.5 *Let $D = \{(x,y) \in \mathbb{R}_*^2 : x \neq 0_*\}$. Solve the equation*

$$y^{2*} +_* x^{2*} \cdot_* y^* = x \cdot_* y \cdot_* y^*.$$

Answer

$$y = c \cdot_* e^{y/_* x}.$$

Exercise 2.3.6 *Let $D = \{(x,y) \in \mathbb{R}_*^2 : x \neq 0_*, y \neq e^{k\pi}, \quad k \in \mathbb{Z}\}$. Solve the equation*

$$x \cdot_* y^* -_* y = x \cdot_* \tan_* (y/_* x).$$

Answer

$$y = x \cdot_* \arcsin_* (c \cdot_* x).$$

Exercise 2.3.7 *Let $D = \{(x,y) \in \mathbb{R}_*^2 : x \neq 0_*\}$. Solve the equation*

$$x \cdot_* y^* = y -_* x \cdot_* e^{y/_* x}.$$

Answer

$$y = -_* x \log_* |\log_* |c \cdot_* x|_*|_*.$$

Now we consider the equation

$$y^* = f\left((a_1 \cdot_* x +_* b_1 \cdot_* y +_* c_1)/_*(a_2 \cdot_* x +_* b_2 \cdot_* y +_* c_2)\right),$$
$$(2.24)$$

where $a_i, b_i, c_i \in \mathbb{R}_*$, $i = 1, 2$.

Let

$$\Delta_* = \left|\begin{matrix} a_1 & b_1 \\ a_2 & b_2 \end{matrix}\right|_*.$$

1. Case. $\Delta_* = 0_*$. Then there exists a constant $k \in \mathbb{R}_*$ such that

$$a_1 \cdot_* x +_* b_1 \cdot_* y = k \cdot_* (a_2 \cdot_* x +_* b_2 \cdot_* y).$$

We set

$$z = a_2 \cdot_* x +_* b_2 \cdot_* y.$$

Then

$$z^* = a_2 +_* b_2 \cdot_* y^*.$$

1.1. Case. $b_2 = 0_*$. Then the equation (2.24) takes the following form

$$y^* = f\left((a_1 \cdot_* x +_* c_1)/_*(a_2 \cdot_* x +_* c_2)\right),$$

which is a separable multiplicative differential equation.

1.2. Case. $b_2 \neq 0_*$. Then

$$y^* = (1_*/_* b_2) \cdot_* z^* -_* (a_2/_* b_2).$$

The equation (2.24) takes the form

$$(1_*/_* b_2) \cdot_* z^* -_* (a_2/_* b_2) = f\left((k \cdot_* z +_* c_1)/_*(z +_* c_2)\right)$$

or

$$z^* = a_2 +_* f\left((k \cdot_* z +_* c_1)/_*(z +_* c_2)\right),$$

which is a separable multiplicative equation.

2. Case. $\Delta_* \neq 0_*$. In this case we set

$$\begin{cases} x = u +_* h \\ y = v +_* l, \end{cases}$$

where h and l satisfy the system

$$\begin{cases} a_1 \cdot_* h +_* b_1 \cdot_* l +_* c_1 = 0_* \\ a_2 \cdot_* h +_* b_2 \cdot_* l +_* c_2 = 0_*. \end{cases}$$

The last system has unique solution (h, l) because $\Delta \neq 0_*$. Hence,

$$\begin{aligned} a_1 \cdot_* x +_* b_1 \cdot_* y +_* c_1 &= a_1 \cdot_* (u +_* h) +_* b_1 \cdot_* (v +_* l) +_* c_1 \\ &= a_1 \cdot_* u +_* b_1 \cdot_* v +_* a_1 \cdot_* h +_* b_1 \cdot_* l +_* c_1 \\ &= a_1 \cdot_* u +_* b_1 \cdot_* v, \end{aligned}$$

and

$$\begin{aligned} a_2 \cdot_* x +_* b_2 \cdot_* y +_* c_2 &= a_2 \cdot_* (u +_* h) +_* b_2 \cdot_* (v +_* l) +_* c_2 \\ &= a_2 \cdot_* u +_* b_2 \cdot_* v +_* a_2 \cdot_* h +_* b_2 \cdot_* l +_* c_2 \\ &= a_2 \cdot_* u +_* b_2 \cdot_* v, \end{aligned}$$

and the equation (2.24) takes the form

$$v^* = f\left((a_1 \cdot_* u +_* b_1 \cdot_* v)/_*(a_2 \cdot_* u +_* b_2 \cdot_* v)\right),$$

which is a multiplicative homogeneous equation.

Example 2.3.8 *We consider the equation*

$$x -_* y -_* 1_* +_* (y -_* x +_* 2_*) \cdot_* y^* = 0_*,$$

which we can rewrite in the form

$$y^* = (y -_* x +_* 1_*)/_*(y -_* x +_* 2_*), \qquad (y -_* x +_* 2 \neq 0_*).$$

Here

$$a_1 = a_2$$

$$= 1,$$

$$b_1 = b_2$$

$$= e^{-1},$$

$$c_1 = 1,$$

$$c_2 = 2.$$

Then

$$\Delta_* = \begin{vmatrix} a_1 & b_1 \\ a_2 & b_2 \end{vmatrix}_*$$

$$= \begin{vmatrix} 1_* & e^{-1} \\ 1_* & e^{-1} \end{vmatrix}_*$$

$$= 0_*.$$

We set

$$z = y -_* x,$$

whereupon

$$y = z +_* x$$

and

$$y^* = z^* +_* 1_*.$$

From here, the given equation takes the form

$$z^* +_* 1_* = (z +_* 1_*)/_*(z +_* 2_*)$$

or

$$z^* = (z +_* 1_*)/_*(z +_* 2_*) -_* 1_*$$

$$= -_*(1_*/_*(z +_* 2_*)),$$

or

$$(z +_* 2_*) \cdot_* d_* z = -_* d_* x.$$

Therefore,

$$\int_* t(z +_* 2_*) \cdot_* d_* z = -_* \int_* d_* x +_* c,$$

or

$$(z +_* 2_*)^{2*} = -_* 2_* \cdot_* x +_* c,$$

or

$$(y -_* x +_* 2_*)^{2*} = -_* 2_* \cdot_* x +_* c.$$

We note that $y = x -_* 2_*$ is not a solution to the given equation.

Example 2.3.9 *We consider the equation*

$$y^* = (2_* \cdot_* x +_* y +_* 1_*)/_*(4_* \cdot_* x +_* 2_* \cdot_* y +_* 3_*).$$

Here

$$a_1 = 2_*,$$

$$a_2 = 4_*,$$

$$b_1 = 1_*,$$

$$b_2 = 2_*,$$

$$c_1 = 1_*,$$

$$c_2 = 3_*.$$

We have that

$$\Delta_* = \begin{vmatrix} a_1 & b_1 \\ a_2 & b_2 \end{vmatrix}_*$$

$$= \begin{vmatrix} 2_* & 1_* \\ 4_* & 2_* \end{vmatrix}_*$$

$$= \; 4_* -_* 4_*$$

$$= \; 0_*.$$

We set

$$z = 4_* x +_* 2_* \cdot_* y.$$

Then

$$y = (1_*/_* 2_*) \cdot_* z -_* 2_* \cdot_* x,$$

whereupon

$$y^* = (1_*/_* 2_*) \cdot_* z^* -_* 2_*.$$

Hence, the given equation takes the form

$$(1_*/_* 2_*) \cdot_* z^* -_* 2_* = ((1_*/_* 2_*) \cdot_* z +_* 1_*)/_* (z -_* 3_*)$$

or

$$z^* -_* 4_* = (z +_* 2_*)/_* (z -_* 3_*).$$

or

$$z^* = (5_* \cdot_* z -_* 10_*)/_* (z -_* 3_*),$$

or

$$d_* z = ((5_* \cdot_* z -_* 10_*)/_* (z -_* 3_*)) \cdot_* d_* x,$$

whereupon, for $2_* \cdot_* x +_* y \neq 1_*,$

$$((z -_* 3_*)/_* (5_* \cdot_* (z -_* 2_*))) \cdot_* d_* z = d_* x.$$

From here,

$$\int_* ((z -_* 3_*)/_* (5_* \cdot_* ((z -_* 2_*)))) \cdot_* d_* z = x +_* c,$$

or

$$\int_* ((z -_* 2_*)/_* (5_* \cdot_* (z -_* 2_*))) \cdot_* dz -_*$$

$$\int_* (d_* z)/_* (5_* \cdot_* ((z -_* 2_*))) = x +_* c,$$

or

$$(1_*/_* 5_*) \cdot_* z -_* (1_*/_* 5_*) \cdot_* \log_* |z -_* 2_*|_* = x +_* c,$$

or

$$z -_* 5_* \cdot_* x -_* \log_* |z -_* 2_*|_* = c,$$

or

$$4_* \cdot_* x +_* 2_* \cdot_* y -_* 5_* \cdot_* x -_* \log_* |4_* \cdot_* x +_* 2_* \cdot_* y -_* 2_*|_* = c,$$

or

$$2_* \cdot_* y -_* x -_* \log_* |4_* \cdot_* x +_* 2_* \cdot_* y -_* 2_*|_* = c.$$

We note that $y = e^{-2} \cdot_* x +_* 1_*$ is not a solution to the considered equation.

Example 2.3.10 *Now we consider the equation*

$$y^* = (2_* \cdot_* x +_* 3_* \cdot_* y -_* 5_*)/_*(x +_* 4_* \cdot_* y).$$

Here

$$a_1 = 2,$$

$$a_2 = 1,$$

$$b_1 = 3,$$

$$b_2 = 4,$$

$$c_1 = e^{-5},$$

$$c_2 = 0_*.$$

Then

$$\Delta_* = \begin{vmatrix} a_1 & b_1 \\ a_2 & b_2 \end{vmatrix}_*$$

$$= \begin{vmatrix} 2_* & 3_* \\ 1_* & 4_* \end{vmatrix}_*$$

$$= \ 5_*$$

$$\neq \ 0_*.$$

Let

$$x \ = \ u +_* h,$$

$$y \ = \ v +_* l,$$

where (h,l) *is the solution to the system*

$$2_* \cdot_* h +_* 3_* \cdot_* l -_* 5_* \ = \ 0_*$$

$$h +_* 4_* \cdot_* l \ = \ 0_*$$

or

$$h \ = \ e^{-4} \cdot_* l$$

$$2_* \cdot_* \left(e^{-4} \cdot_* l \right) +_* 3_* \cdot_* l -_* 5_* \ = \ 0_*,$$

or

$$l \ = \ e^{-1}$$

$$h \ = \ 4_*.$$

Then

$$x \ = \ u +_* 4_*$$

$$y \ = \ v -_* 1_*$$

and

$$u \ = \ x -_* 4_*$$

$$v = y +_* 1_*.$$

The considered equation takes the form

$$(d_*(v -_* 1_*))/_*(d_*(u +_* 1_*))$$

$$= (2_* \cdot_* (u +_* 4_*) +_* 3_* \cdot_* (v -_* 1_*) -_* 5_*)/_*(u +_* 4_* +_* 4_* \cdot_* (v -_* 1_*))$$

or

$$v^* = (2_* \cdot_* u +_* 3_* \cdot_* v)/_*(u +_* 4_* \cdot_* v).$$

We set

$$z = v/_* u.$$

Then

$$z = (y +_* 1_*)/_*(x -_* 4_*),$$

and

$$v = z \cdot_* u$$

and

$$v^* = z \cdot_* u +_* z.$$

From here,

$$z^* \cdot_* +_* z = (2_* \cdot_* u +_* 3_* \cdot_* z \cdot_* u)/_*(u +_* 4_* \cdot_* z \cdot_* u)$$

$$= (2_* +_* 3_* \cdot_* z)/_*(1_* +_* 4_* \cdot_* z)$$

or

$$z^* \cdot_* u = ((2_* +_* 3_* \cdot_* z)/_*(1_* +_* 4_* \cdot_* z)) -_* z$$

$$= (2_* +_* 2_* \cdot_* z -_* 4_* \cdot_* z^{2*})/_*(1_* +_* 4_* \cdot_* z)$$

or

$$z^* \cdot_* u = e^{-2} \cdot_* (((z -_* 1_*) \cdot_* (2_* \cdot_* z +_* 1_*))/_*(4_* \cdot_* z +_* 1_*)).$$

Hence, for $z \neq 1_*$, $z \neq e^{-\frac{1}{2}}$, $u \neq 0_*$, we get to the equation

$$((4_* \cdot_* z +_* 1_*)/_*((z -_* 1_*) \cdot_* (2_* \cdot_* z +_* 1_*))) \cdot_* d_* z = -_*(2_*/_* u) \cdot_* d_* u,$$

from where

$$\int_* ((4_* z +_* 1_*)/_*((z -_* 1_*) \cdot_* (2_* z +_* 1_*))) \cdot_* d_* z = -_* \int_* (2_*/_* u) \cdot_* d_* u +_* c,$$

or

$$(5_*/_* 3_*) \cdot_* \int_* (d_* z)/_*(z -_* 1_*)$$

$$+_* (2_*/_* 3_*) \cdot_* \int_* (d_* z)/_*(2_* \cdot_* z +_* 1_*) = -_* \log_* u^{2*} +_* c,$$

or

$$(5_*/_* 3_*) \cdot_* \log_* |z -_* 1_*|_* +_* (1_*/_* 3_*) \cdot_* \log_* |2_* \cdot_* z +_* 1_*|_* = -_* \log_* u^{2*} +_* c,$$

or

$$\log_* |z -_* 1_*|_*^{5*} +_* \log_* |2_* \cdot_* z +_* 1_*|_* +_* \log_* u^{6*} = c,$$

or

$$\log_* \left| (z -_* 1_*)^{5*} \cdot_* (2_* \cdot_* z +_* 1_*) \cdot_* u^{6*} \right|_* = c,$$

or

$$(z -_* 1_*)^{2*} \cdot_* (2_* \cdot_* z +_* 1_*) \cdot_* u^{6*} = c,$$

or

$$((y +_* 1_*)/_*(x -_* 4_*) -_* 1_*)^{5*}$$

$$\cdot_* ((2_* \cdot_* y +_* 2_*)/_*(x -_* 4_*) +_* 1_*) \cdot_* (x -_* 4_*)^{6*} = c,$$

or

$$(y -_* x +_* 5_*)^{5*} \cdot_* (2_* \cdot_* y +_* x -_* 2_*) = c.$$

Exercise 2.3.11 *Solve the equations*

1. $(y^* +_* 1_*) \log_*((y +_* x)/_*(x +_* 3_*)) = (y +_* x)/_*(x +_* 3_*)$,

2. $y^* = (y +_* 2_*)/_*(x +_* 1_*) +_* \tan_*((y -_* 2_* \cdot_* x)/_*(x +_* 1_*))$,

$$3. \ y^* = (y +_* 2_*)/_* (2_* \cdot_* x +_* y -_* 4_*).$$

Answer

1. $\log_* ((y +_* x)/_* (x +_* 3_*)) = 1_* +_* (c/_* (x +_* y))$,
2. $\sin_* ((y -_* 2_* \cdot_* x)/_* (x +_* 1_*)) = c \cdot_* (x +_* 1_*)$,
3. $(y +_* 2_*)^{2*} = c \cdot_* (x +_* y -_* 1_*)$, $x +_* y -_* 1_* = 0_*$.

Some equations can be reduced to multiplicative homogeneous equations using the substitution $y = z^{m_*}$, where $m \in \mathbb{R}$ is chosen so that the transformed equation is multiplicative homogeneous.

Example 2.3.12 *We consider the equation*

$$2_* \cdot_* x^{4*} \cdot_* y \cdot_* y^* +_* y^{4*} = 4_* \cdot_* x^{6*}.$$

We set $y = z^{m_*}$. *Then*

$$y = m_* \cdot_* z^{(m-1)*} \cdot_* z^*$$

and the given equation takes the form

$$2_* \cdot_* m_* \cdot_* x^{4*} \cdot_* z^{m_*} \cdot_* z^{(m-1)*} \cdot_* z^* +_* z^{(4m)*} = 4_* \cdot_* x^{6*}$$

or

$$2_* \cdot_* m_* \cdot_* x^{4*} \cdot_* z^{(2m-1)*} \cdot_* z^* +_* z^{(4m)*} = 4_* \cdot_* x^{6*}.$$

The last equation is multiplicative homogeneous if

$$4_* +_* 2_* \cdot_* m_* -_* 1_* \ = \ 4_* \cdot_* m_*$$

$$= \ 6_*$$

or

$$m_* = 3_* /_* 2_*.$$

Hence,

$$y = z^{\frac{3}{2}*}$$

and

$$z = y^{\frac{2}{3}*}.$$

Therefore,

$$3_* \cdot_* x^{4*} \cdot_* z^{2*} \cdot_* z^* +_* z^{6*} = 4_* \cdot_* x^{6*}$$

or

$$3_* \cdot_* x^{4*} \cdot_* z^{2*} \cdot_* z^* = 4_* \cdot_* x^{6*} -_* z^{6*},$$

which we multiplicative divide of $3_* \cdot_* x^{4*} \cdot_* z^{2*} \neq 0_*$ *and we find*

$$z^* = (4_*/_*3_*) \cdot_* ((x^{2*})/_*(z^{2*})) -_* (1_*/_*3_*) \cdot_* ((z^{4*})/_*(x^{4*})).$$

The last equation is multiplicative homogeneous. We set $v = z/_*x$. *Then we have the following.*

$$z = v \cdot_* x,$$

$$v = \left(y^{\frac{2}{3}*}\right)/_*x,$$

$$z^* = v^* \cdot_* x +_* v.$$

In this way we obtain the equation

$$v^* \cdot_* x +_* v = (4_*/_*3_*) \cdot_* (1_*/_*(v^{2*})) -_* (1_*/_*3_*) \cdot_* v^{4*}$$

or

$$
\begin{aligned}
v^* \cdot_* x &= (4_*/_*3_*) \cdot_* (1_*/_*(v^{2*})) -_* (1_*/_*3_*) \cdot_* v^{4*} -_* v \\
&= (4_* -_* v^{6*} -_* 3_* \cdot_* v^{3*})/_*(3_* \cdot_* v^{2*}) \\
&= -_*((v^{6*} +_* 3_* \cdot_* v^{3*} -_* 4_*))/_*(3_* \cdot_* v^{2*}) \\
&= -_*(((v^{3*} -_* 1_*) \cdot_* (v^{3*} +_* 4_*)))/_*(3_* \cdot_* v^{2*}),
\end{aligned}
$$

whereupon

$$x \cdot_* d_* v = -_*(((v^{3*} -_* 1_*) \cdot_* (v^{3*} +_* 4_*))/_*(3_* \cdot_* v^{2*})) \cdot_* d_* x,$$

which we multiplicative divide of

$$x \cdot_* ((v^{3*} -_* 1_*) \cdot_* (v^{3*} +_* 4_*)) /_* (3_* \cdot_* v^{2*}) \neq 0_*$$

and we get

$$((3_* \cdot_* v^{2*}) /_* ((v^{3*} -_* 1_*) \cdot_* (v^{3*} +_* 4_*))) \cdot_* d_* v = -_* ((d_* x) /_* x)$$

and

$$\int_* ((3_* \cdot_* v^{2*}) /_* ((v^{3*} -_* 1_*) \cdot_* (v^{3*} +_* 4_*))) \cdot_* d_* v = -_* \int_* (d_* x) /_* x +_* c,$$

or

$$(3_* /_* 5_*) \cdot_* \int_* ((v^{2*}) /_* (v^{3*} -_* 1_*)) \cdot_* d_* v$$

$$-_* (3_* /_* 5_*) \cdot_* \int_* ((v^{2*}) /_* (v^{3*} +_* 4_*)) \cdot_* d_* v = -_* \log_* |x|_* +_* c,$$

or

$$(1_* /_* 5_*) \cdot_* \int_* (d_* (v^{3*} -_* 1_*)) /_* (v^{3*} -_* 1_*)$$

$$-_* (1_* /_* 5_*) \cdot_* \int_* (d_* (v^{3*} +_* 4_*)) /_* (v^{3*} +_* 4_*) = -_* \log_* |x|_*^{5*} +_* c,$$

or

$$\log_* \left| ((v^{3*} -_* 1_*) \cdot_* x^{5*}) /_* (v^{3*} +_* 4_*) \right|_* = c,$$

or

$$(v^{3*} -_* 1_*) \cdot_* x^{5*} = c \cdot_* (v^{3*} +_* 4_*),$$

or

$$((y^{2*}) /_* (x^{3*}) -_* 1_*) \cdot_* x^{5*} = c \cdot_* ((y^{2*}) /_* (x^{3*}) +_* 4_*),$$

or

$$(y^{2*} -_* x^{3*}) \cdot_* x^{5*} = c \cdot_* (y^{2*} +_* 4_* \cdot_* x^{3*}).$$

We note that if $z = 0_*$, then $y = 0_*$ which is not a solution to the given equation. If $v = 1_* /_* 3_*$, then

$$(y^{\frac{2}{3}*}) /_* x = 1_*,$$

or

$$y^{2*} = x^{3*},$$

which is a solution to the given equation. If $v = e^{-\sqrt[3]{4}}$, *then*

$$(y^{\frac{2}{3}*})/_* x = e^{-\sqrt[3]{4}},$$

or

$$y = -_*4_* \cdot_* x^{3*},$$

which is a solution of the given equation.

Example 2.3.13 *Now we consider the equation*

$$x^{3*} \cdot_* (y^* -_* x) = y^{2*}.$$

We set $y = z^{m*}$. *Then*

$$y^* = m_* \cdot_* z^{(m-1)*} \cdot_* z^*$$

and the given equation takes the form

$$x^{3*} \cdot_* (m_* \cdot_* z^{(m-1)*} \cdot_* z^* -_* x) = z^{(2m)*}$$

or

$$m_* \cdot_* x^{3*} \cdot_* z^{(m-1)*} \cdot_* z^* -_* x^{4*} = z^{(2m)*}.$$

The last equation is multiplicative homogeneous if

$$m_* -_* 1_* +_* 3_* \;\; = \;\; 4_*$$

$$= \;\; (2m)_*,$$

or

$$m = 2_*.$$

From here, $y = z^{2*}$ *and*

$$2_* \cdot_* x^{3*} \cdot_* z \cdot_* z^* = x^{4*} +_* z^{4*}$$

or

$$z^* = 1_*/_*2_* \cdot_* (x/_*z) +_* (1_*/_*2_*) \cdot_* (z/_*x)^{3*}.$$

We set $v = z/_*x$. Then

$$v^{2*} = y/_*(x^{2*}),$$

$$z = v \cdot_* x,$$

$$z^* = v^* \cdot_* x +_* v.$$

In this way we get the equation

$$v^* \cdot_* x +_* v = (1_*/_*2_*) \cdot_* v +_* (1_*/_*2_*) \cdot_* v^{3*},$$

or

$$v^* \cdot_* x = (1_*/_*2_*) \cdot_* (1_*/_*v) +_* (1_*/_*2_*) \cdot_* v^{3*} -_* v$$

$$= (v^{4*} -_* 2_* \cdot_* v^{2*} +_* 1_*)/_*(2_* \cdot_* v)$$

$$= ((v^{2*} -_* 1_*)^{2*})/_*(2_* \cdot_* v),$$

or

$$x \cdot_* d_* v = ((v^{2*} -_* 1_*)^{2*})/_*(2_* \cdot_* v) \cdot_* d_* x,$$

which we divide of

$$((v^{2*} -_* 1_*)^{2*})/_*(2_* \cdot_* v) \neq 0_*$$

and we find

$$(2_* \cdot_* v)/_*((v^{2*} -_* 1_*)^{2*}) \cdot_* d_* v = (d_* x)/_* x$$

and

$$\int_* (2_* \cdot_* v)/_*((v^{2*} -_* 1_*)^{2*}) \cdot_* d_* v = \int_* (d_* x)/_* x +_* c,$$

or

$$\int_* (d_*(v^{2*} -_* 1_*))/_*((v^{2*} -_* 1_*)^{2*}) = \log_* |x|_* +_* c,$$

or

$$-_*(1_*/_*(v^{2*} -_* 1_*)) = \log_* |x|_* +_* c,$$

or

$$-_*(1_*)/_*((y/_*(x^{2*}))-_*1_*) = \log_* |x|_* +_* c,$$

or

$$x^{2*} = (x^{2*} -_* y) \cdot_* (\log_* |x| +_* c).$$

If $v^{2*} = 1_*$, then $y = x^{2*}$ which is a solution of the given equation.

Exercise 2.3.14 *Solve the equations*

$$1.\ 2_* \cdot_* x^{2*} \cdot_* y^* = y^{3*} +_* x \cdot_* y,$$

$$2.\ 2_* \cdot_* x \cdot_* d_* y +_* (x^{2*} \cdot_* y^{4*} +_* 1_*) \cdot_* y \cdot_* d_* x = 0_*,$$

$$3.\ y \cdot_* d_* x +_* x \cdot_* (2_* \cdot_* x \cdot_* y +_* 1_*) \cdot_* d_* y = 0_*.$$

Answer

1. $x = -_* y^{2*} \cdot_* \log_* (c \cdot_* x),\ y = 0_*,$

2. $x^{2*} \cdot_* y^{4*} \cdot_* \log_* (c \cdot_* x^{2*}) = 1_*,\ y = 0_*,\ x = 0_*,$

3. $y^{2*} \cdot_* e^{-_*(1_*/_*(x \cdot_* y))} = c,\ y = 0_*,\ x = 0_*.$

2.4   Exact Multiplicative Differential Equations

Definition 2.4.1 *The equation (2.20) is said to be exact if there exists a function $u(x,y)$ such that*

$$
\begin{aligned}
u_x^*(x,y) &= P(x,y) \\
u_y^*(x,y) &= Q(x,y); \quad (x,y) \in D,
\end{aligned}
\tag{2.25}
$$

i.e.,

$$P(x,y) +_* Q(x,y) \cdot_* y^* = u_x^*(x,y) +_* u_y^*(x,y) \cdot_* y^*, \quad (x,y) \in D,$$

is exactly the multiplicative derivative $d_* u /_* d_* x$.

In fact, the system (2.25) can be rewritten in the form

$$e^{x \frac{u_x(x,y)}{u(x,y)}} = e^{\log P(x,y)}$$

$$e^{y \cdot \frac{u_y(x,y)}{u(x,y)}} = e^{\log Q(x,y)}, \quad (x,y) \in D,$$

or

$$x\frac{u_x(x,y)}{u(x,y)} = \log P(x,y)$$

$$y\frac{u_y(x,y)}{u(x,y)} = \log Q(x,y), \quad (x,y) \in D,$$

or

$$xu_x(x,y) = u(x,y)\log P(x,y)$$

$$yu_y(x,y) = u(x,y)\log Q(x,y), \quad (x,y) \in D.$$

Example 2.4.2 *Let us consider the equation*

$$2_* \cdot_* x \cdot_* y \cdot_* d_* x +_* (x^{2*} -_* y^{2*}) \cdot_* d_* y = 0_*.$$

Here

$$P(x,y) = 2_* \cdot_* x \cdot_* y,$$

$$Q(x,y) = x^{2*} -_* y^{2*}.$$

Then, if

$$u(x,y) = x^{2*} \cdot_* y -_* (1_*/_*3_*) \cdot_* y^{3*},$$

we have

$$u_x^*(x,y) = 2_* \cdot_* x \cdot_* y$$

$$= P(x,y),$$

$$u_y^*(x,y) = x^{2*} -_* y^{2*}$$

$$= Q(x,y),$$

i.e., the considered equation is exact.

Example 2.4.3 *Let us consider the differential equation*

$$(2_* -_* 9_* \cdot_* x \cdot_* y^{2*}) \cdot_* x \cdot_* d_* x +_* (4_* \cdot_* y^{2*} -_* 6_* \cdot_* x^{3*}) \cdot_* y \cdot_* d_* y = 0_*.$$

Let also,

$$u(x,y) = x^{2*} -_* 3_* \cdot_* x^{3*} \cdot_* y^{2*} +_* y^{4*}.$$

Here

$$P(x,y) \;=\; (2_* -_* 9_* \cdot_* x \cdot_* y^{2*}) \cdot_* x,$$

$$Q(x,y) \;=\; (4_* \cdot_* y^{2*} -_* 6_* \cdot_* x^{3*}) \cdot_* y.$$

Then

$$u_x(x,y) \;=\; 2_* \cdot_* x -_* 9_* \cdot_* x^{2*} \cdot_* y^{2*}$$

$$=\; P(x,y),$$

$$u_y^*(x,y) \;=\; -_* 6_* \cdot_* x^{3*} \cdot_* y +_* 4_* \cdot_* y^{3*}$$

$$=\; Q(x,y).$$

Therefore, the considered equation is exact.

Example 2.4.4 *Let us consider the equation*

$$(2_* \cdot_* x +_* 2_* \cdot_* y) \cdot_* d_* x +_* (2_* \cdot_* x +_* 3_* \cdot_* y^{2*}) \cdot_* d_* y = 0_*.$$

Let also,

$$u(x,y) = x^{2*} +_* 2_* \cdot_* x \cdot_* y +_* y^{3*}.$$

Here

$$P(x,y) \;=\; 2_* \cdot_* x +_* 2_* \cdot_* y,$$

$$Q(x,y) \;=\; 2_* \cdot_* x +_* 3_* \cdot_* y^{2*}.$$

Then

$$u_x^*(x,y) \;=\; 2_* \cdot_* x +_* 2_* \cdot_* y$$

$$= P(x,y),$$

$$u_y^*(x,y) = 2_* \cdot_* x +_* 3_* \cdot_* y^{2*}$$

$$= Q(x,y).$$

Consequently, the considered equation is exact.

Theorem 2.4.5 *Let the functions $P(x,y)$ and $Q(x,y)$ together with their partial multiplicative derivatives P_y^* and Q_x^* be continuous in the rectangle $S = \{(x,y) \in \mathbb{R}^2 : |x -_* x_0| < a_*, |y -_* y_0| < b_*\}$. Then the equation (2.20) is exact if and only if*

$$P_y^*(x,y) = Q_x^*(x,y) \qquad \text{in} \qquad S. \qquad (2.26)$$

Proof 2.4.6 *1. Let the equation (2.20) be an exact multiplicative differential equation. Then there exists a function $u(x,y)$ satisfying (2.25). Hence,*

$$u_{xy}^*(x,y) = P_y^*(x,y),$$

$$u_{xy}^*(x,y) = Q_y^*(x,y), \quad (x,y) \in S.$$

Therefore, (2.26) holds.

2. Let the functions $P(x,y)$ and $Q(x,y)$ satisfy the condition (2.26) in S. We multiplicative integrate both sides of the equation

$$u_x^*(x,y) = P(x,y), \quad (x,y) \in S,$$

with respect to x and we get

$$u(x,y) = \int_{*x_0}^x P(s,y) \cdot_* d_* s +_* f(y), \qquad (2.27)$$

where f is an arbitrary function and plays the role of the constant of integration. We will find the function f using the condition

$$u_y(x,y) = Q(x,y), \quad (x,y) \in S.$$

From (2.28), it follows that

$$u_y^*(x,y) \;=\; \int_{*x_0}^x \partial_*/_*\partial_* y P(s,y) \cdot_* d_* s +_* f^*(y)$$

$$=\; Q(x,y), \quad (x,y) \in S,$$

or

$$f^*(y) = Q(x,y) -_* \int_{*x_0}^x \partial_*/_*\partial_* y P(s,y) \cdot_* d_* s. \qquad (2.28)$$

Therefore,

$$Q(x,y) -_* \int_{*x_0}^x \partial_*/_*\partial_* y P(s,y) \cdot_* d_* s$$

must depend on y alone. For the function f we get

$$f(y) - f(y_0) = \int_{*y_0}^y Q(x,\tau) \cdot_* d_* \tau$$

$$-_* \int_{*y_0}^y \int_{*x_0}^x \partial_*/_*\partial_* y P(s,\tau) \cdot_* d_* s \cdot_* d_* \tau$$

$$= \int_{*y_0}^y Q(x,\tau) \cdot_* d_* \tau$$

$$-_* \int_{*x_0}^x \int_{*y_0}^y \partial_*/_*\partial_* y P(s,\tau) \cdot_* d_* \tau \cdot_* d_* s$$

$$= \int_{*y_0}^y Q(x,\tau) \cdot_* d_* \tau -_* \int_{*x_0}^x (P(s,y) -_* P(s,y_0)) \cdot_* d_* s,$$

i.e,

$$f(y) = \int_{*y_0}^y Q(x,\tau) \cdot_* d_* \tau -_* \int_{*x_0}^x (P(s,y) -_* P(s,y_0)) \cdot_* d_* s +_* f(y_0).$$

From here and from (2.27), we get

$$u(x,y) \;=\; \int_{*x_0}^x P(s,y) \cdot_* d s +_* \int_{*y_0}^y Q(x,\tau) \cdot_* d_* \tau$$

$$-_* \int_{*x_0}^x P(s,y) \cdot_* d_* s +_* \int_{*x_0}^x P(s,y_0) \cdot_* d_* s +_* f(y_0)$$

$$= \int_{*y_0}^y Q(x,\tau) \cdot_* d_* \tau +_* \int_{*x_0}^x P(s,y_0) \cdot_* d_* s +_* f(y_0),$$

whereupon a solution of the exact equation (2.20) is given by the expression

$$\int_{*y_0}^y Q(x,\tau) \cdot_* d_* \tau +_* \int_{*x_0}^x P(s,y_0) \cdot_* d_* s = c, \quad (x,y) \in S.$$

Now, we multiplicative integrate both sides of the equation

$$u_y^*(x,y) = Q(x,y)$$

with respect to y and we find

$$u(x,y) = \int_{*y_0}^y Q(x,\tau) \cdot_* d_* \tau +_* g(x), \qquad (2.29)$$

where g is a function which plays a role in the constant of integration. From (2.29) we become

$$u_x^*(x,y) = \int_{*y_0}^y \partial_* /_* \partial_* x Q(x,\tau) \cdot_* d_* \tau +_* g^*(x) = P(x,y), \quad (x,y) \in S,$$

or

$$g^*(x) = P(x,y) -_* \int_{*y_0}^y \partial_* /_* \partial_* x Q(x,\tau) \cdot_* d_* \tau, \quad (x,y) \in S.$$

We note that

$$P(x,y) - \int_{*y_0}^y \partial_* /_* \partial_* x Q(x,\tau) \cdot_* d_* \tau$$

must depend on x alone. Hence,

$$g(x) = \int_{*x_0}^x P(s,y) \cdot_* d_* s$$

$$-_* \int_{*x_0}^x \int_{y_0}^y \partial /_* \partial_* x Q(s,\tau) \cdot_* d_* \tau \cdot_* d_* s +_* g(x_0)$$

$$= \int_{*x_0}^x P(s,y) \cdot_* d_* s$$

$$-_* \int_{*y_0}^y \int_{*x_0}^x \partial_* /_* \partial_* x Q(s,\tau) \cdot_* d_* \tau \cdot_* d_* s +_* g(x_0)$$

$$= \int_{*x_0}^x P(s,y) \cdot_* d_* s$$

$$-_* \int_{*y_0}^y \left(Q(x,\tau) - Q(x_0,\tau) \right) \cdot_* d_* \tau +_* g(x_0),$$

$(x,y) \in S$. Hence and (2.29), we find

$$u(x,y) = \int_{*y_0}^y Q(x,\tau) \cdot_* d_* \tau +_* \int_{*x_0}^x P(s,y) \cdot_* d_* s$$

$$-_* \int_{*y_0}^y Q(x,\tau) \cdot_* d_* \tau +_* \int_{*y_0}^y Q(x_0,\tau) \cdot_* d_* \tau +_* g(x_0)$$

$$= \int_{x_0}^x P(s,y) \cdot_* d_* s +_* \int_{*y_0}^y Q(x_0,\tau) \cdot_* d_* \tau +_* g(x_0), \quad (x,y) \in S.$$

Consequently, a solution of the exact multiplicative differential equation (2.20) is given by

$$\int_{*x_0}^x P(s,y) \cdot_* d_* s +_* \int_{*y_0}^y Q(x_0,\tau) \cdot_* d_* \tau = c.$$

This completes the proof.

Example 2.4.7 *Let us consider the equation*

$$(2_* -_* 9_* \cdot_* x_* \cdot_* y^{2*}) \cdot_* x \cdot_* d_* x +_* (4_* \cdot_* y^{2*} -_* 6_* \cdot_* x^{3*}) \cdot_* y \cdot_* d_* y = 0_*.$$

Here

$$P(x,y) = (2_* -_* 9_* \cdot_* x \cdot_* y^{2*}) \cdot_* x,$$

$$Q(x,y) = (4_* \cdot_* y^{2*} -_* 6_* \cdot_* x^{3*}) \cdot_* y.$$

Then

$$(\partial_* P/_* \partial_* y)(x,y) \;=\; -_* 18_* \cdot_* x^{2*} \cdot_* y,$$

$$(\partial_* Q/_* \partial_* x)(x,y) \;=\; -_* 18_* \cdot_* x^{2*} \cdot_* y.$$

Consequently,

$$(\partial_* P/_* \partial_* y)(x,y) = (\partial_* Q/_* \partial_* x)(x,y)$$

and then the considered equation is an exact multiplicative differential equation. Therefore, there exists a function $u(x,y)$ such that

$$u_x^*(x,y) \;=\; P(x,y),$$

$$u_y^*(x,y) \;=\; Q(x,y)$$

or

$$u_x^*(x,y) \;=\; 2_* \cdot_* x -_* 9_* \cdot_* x^{2*} \cdot_* y^{2*},$$

$$u_y^*(x,y) \;=\; 4_* \cdot_* y^{3*} -_* 6_* \cdot_* x^{3*} \cdot_* y.$$

Hence,

$$u(x,y) \;=\; \int_{*0_*}^{x} (2_* \cdot_* s -_* 9_* \cdot_* s^{2*} \cdot_* y^{2*}) \cdot_* d_* s +_* f(y)$$

$$=\; s^{2*} \Big|_{s=0_*}^{s=x} -_* 3_* \cdot_* s^{3*} \cdot_* y^{2*} \Big|_{s=0_*}^{s=x} +_* f(y)$$

$$=\; x^{2*} -_* 3_* \cdot_* x^{3*} \cdot_* y^{2*} +_* f(y),$$

which we multiplicative differentiate with respect to y and we get

$$u_y^*(x,y) \;=\; -_* 6_* \cdot_* x^{3*} \cdot_* y +_* f^*(y)$$

$$=\; 4_* \cdot_* y^{3*} -_* 6_* \cdot_* x^{3*} \cdot_* y,$$

whereupon

$$f^*(y) = 4_* \cdot_* y^{3_*}$$

and

$$f(y) \;=\; \int_{*0}^{*y} 4_* \cdot_* s^{3_*} \cdot_* d_* s +_* f(0)$$

$$= \; y^{4_*} +_* f(0).$$

Therefore,

$$u(x,y) = x^{2_*} -_* 3_* \cdot_* x^{3_*} \cdot_* y^{2_*} +_* y^{4_*} +_* f(0),$$

i.e., the solutions of the considered equation are given by the expression

$$x^{2_*} -_* 3_* \cdot_* x^{3_*} \cdot_* y^{2_*} +_* y^{4_*} = c.$$

Example 2.4.8 *Let us consider the equation*

$$e^{-_* y} \cdot_* d_* x -_* (2_* \cdot_* y +_* x \cdot_* e^{-_* y}) \cdot_* d_* y = 0_*.$$

Here

$$P(x,y) \;=\; e^{-_* y},$$

$$Q(x,y) \;=\; -_* 2_* \cdot_* y -_* x \cdot_* e^{-_* y}.$$

Then

$$(\partial_* P /_* \partial y)(x,y) \;=\; -_* e^{-_* y},$$

$$(\partial_* Q /_* \partial_* x)(x,y) \;=\; -_* e^{-_* y}.$$

Therefore,

$$(\partial_* P /_* \partial_* y)(x,y) = (\partial_* Q /_* \partial_* x)(x,y)$$

and the considered equation is an exact MDE. From here, it follows that there is a function $u(x,y)$ such that

$$u_x^*(x,y) \;=\; e^{-_* y},$$

$$u_y^*(x,y) = -_*2_* \cdot_* y -_* x \cdot_* e^{-_*y}.$$

Hence,

$$u(x,y) = \int_{*0_*}^{*x} e^{-_*y} \cdot_* d_*x +_* f(y)$$

$$= e^{-_*y} \cdot_* x +_* f(y),$$

which we multiplicative differentiate with respect to y and we become

$$u_y^*(x,y) = -_* e^{-_*y} \cdot_* x +_* f^*(y)$$

$$= -_*2_* \cdot_* y -_* y \cdot_* e^{-_*x},$$

whereupon

$$f^*(y) = -_*2_* \cdot_* y$$

and

$$f(y) = -_* \int_{0_*}^{*y} 2_* \cdot_* sd_*s +_* f(0)$$

$$= -_* y^{2*} +_* f(0).$$

Consequently,

$$u(x,y) = x \cdot_* e^{-_*y} -_* y^{2*} +_* f(0)$$

and the solutions of the considered equation are given by the expression

$$x \cdot_* e^{-_*y} -_* y^{2*} = c.$$

Example 2.4.9 *We consider the equation*

$$((3_* \cdot_* x^{2*} +_* y^{2*})/_*(y^{2*})) \cdot_* d_*x -_* ((2_* \cdot_* x^{3*} +_* 5_* \cdot_* y)/_*(y^{3*})) \cdot_* d_*y = 0_*.$$

Here

$$P(x,y) = ((3_* \cdot_* x^{2*} +_* y^{2*})/_*(y^{2*})),$$

$$Q(x,y) \;=\; -_*((2_* \cdot_* x^{3*} +_* 5_* \cdot_* y)/_*(y^{3*})).$$

Then

$$((\partial_* P)/_*(\partial_* y))(x,y) \;=\; -_* 6_* \cdot_* (((x^{2*}))/_*(y^{3*})),$$

$$((\partial_* Q)/_*(\partial_* x))(x,y) \;=\; -_* 6_* \cdot_* ((x^{2*})/_*(y^{3*})).$$

Therefore,

$$((\partial_* P/_*(\partial_* y))(x,y) = ((\partial_* Q)/_*(\partial_* x))(x,y)$$

and the considered equation is an exact multiplicative differential equation. Therefore, there exists a function $u(x,y)$ such that

$$u_x^*(x,y) \;=\; ((2_* \cdot_* x^{2*} +_* y^{2*})/_*(y^{2*})),$$

$$u_y^*(x,y) \;=\; -_*((2_* \cdot_* x^{3*} +_* 5_* \cdot_* y)/_*(y^{3*})).$$

From the last equation, we find

$$u(x,y) \;=\; -_* \int_{*1_*}^{y} ((2_* \cdot_* x^{3*} +_* 5_* \cdot_* s)/_*(s^{3*})) \cdot_* d_* s +_* g(x)$$

$$= \; -_* 2_* \cdot_* x^{3*} \cdot_* \int_{*1_*}^{y} ((d_* s)/_*(s^{3*})) -_* 5_* \cdot_* \int_{*1_*}^{y} ((d_* s)/_*(s^{2*})) +_* g(x)$$

$$= \; ((x^{3*})/_*(s^{2*}))\Big|_{s=1_*}^{s=y} +_* (5_*/_* s)\Big|_{s=1_*}^{s=y} +_* g(x)$$

$$= \; ((x^{3*})/_*(y^{2*})) -_* x^{3*} +_* (5_*/_* y) -_* 5_* +_* g(x),$$

which we multiplicative differentiate with respect to x and we get

$$u_x^*(x,y) \;=\; 3_* \cdot_* ((x^{2*})/_*(y^{2*})) -_* 3_* \cdot_* x^{2*} +_* g^*(x)$$

$$= \; 3_* \cdot_* ((x^{2*})/_*(y^{2*})) +_* 1_*$$

or

$$g^*(x) = 3_* \cdot_* x^{2*} +_* 1_*,$$

whereupon

$$g(x) = \int_{*0_*}^{*x} (3_* \cdot_* s^{2*} +_* 1_*) \cdot_* d_* s$$

$$= s^{3*} \Big|_{s=0_*}^{s=x} +s \Big|_{s=0_*}^{s=x} = x^{3*} +_* x.$$

Therefore,

$$u(x,y) = ((x^{3*})/_*(y^{2*})) -_* x^{3*} +_* (5_*/_* y) -_* 5_* +x^{3*} +_* x$$

$$= ((x^{3*})/_*(y^{2*})) +_* 5_*/_* y +_* x -_* 5 *.$$

The solutions of the considered equations are given by the expression

$$((x^{3*})/_*(y^{2*})) +_* 5_*/_* y +_* x = c.$$

Exercise 2.4.10 *Check that the following equations are exact equations and find their solutions.*

1. $(y/_* x) \cdot_* d_* x +_* (y^{3*} +_* \log_* x) \cdot_* d_* y = 0_*,$

2. $2_* \cdot_* x \cdot_* (1_* +_* e^{(x^{2*}-_* y)\frac{1}{2}*}) \cdot_* d_* x -_* e^{(x^{2*}-_* y)\frac{1}{2}*} \cdot_* d_* y = 0_*,$

3. $(1_* +_* y^{2*} \cdot_* \sin_*(2_* \cdot_* x)) \cdot_* d_* x -_* 2_* \cdot_* y \cdot_* (\cos_* x)^{2*} \cdot_* d_* y = 0_*.$

Answer

1. $4_* \cdot_* y \cdot_* \log_* x +_* y^{4*} = c,$

2. $x^{2*} +_* (2_*/_* 3) \cdot_* (x^{2*} -_* y)^{\frac{3}{2}}_* = c,$

3. $x -_* y^{2*} \cdot_* (\cos_* x)^{2*} = c.$

2.5 Multiplicative Integrating Factor

Definition 2.5.1 *For the equation* (2.20) *a nonzero function* μ *is called a multiplicative integrating factor if the equation*

$$\mu \cdot_* P +_* \mu \cdot_* Q \cdot_* y^* = 0_* \tag{2.30}$$

is an exact multiplicative differential equation.

Theorem 2.5.2 *If $u(x,y) = c$ is a solution to the equation (2.20), then the equation (2.20) admits an infinite number of multiplicative integrating factors.*

Proof 2.5.3 *Since $u(x,y) = c$ is a solution of the equation (2.20), then*

$$P +_* Q \cdot_* y^* \;=\; u_x^* +_* u_y^* \cdot_* y^*$$

$$=\; d_* u /_* d_* x.$$

Therefore,

$$(u_x^*) /_* P \;=\; (u_y^*) /_* Q \tag{2.31}$$

$$=\; \mu,$$

where μ is some function of x and y. Hence,

$$\mu \cdot_* P +_* \mu \cdot_* Q \cdot_* y^* \;=\; u_x^* +_* u_y^* \cdot_* y^* \tag{2.32}$$

$$=\; (d_* u) /_* (d_* x).$$

From here, we conclude that the equation (2.30) is an exact multiplicative differential equation and a multiplicative integrating factor of (2.20) is given by the expression (2.31).
Let ϕ be any continuous function. Then, using (2.32), we have

$$\mu \cdot_* \phi(u) \cdot_* (P +_* Q \cdot_* y^*) \;=\; \phi(u) \cdot_* \mu \cdot_* (P +_* Q \cdot_* y^*)$$

$$=\; \phi(u) \cdot_* ((d_* u) /_* (d_* x))$$

$$=\; d_* /_* (d_* x) \cdot_* \int_{*0_*}^{*u} \phi(s) \cdot_* d_* s.$$

Consequently, $\mu \cdot_* \phi$ is a multiplicative integrating factor for the equation (2.20). This completes the proof.

Example 2.5.4 *Now we consider the equation*

$$(1_* +_* x \cdot_* e^y) \cdot_* d_* x +_* (-_* x +_* y \cdot_* e^y) \cdot_* d_* y = 0_*.$$

Here

$$P(x,y) \;=\; 1_* +_* x \cdot_* e^y,$$

$$Q(x,y) \;=\; -_* x +_* \cdot_* e^y.$$

Let $\mu(x,y) = e^{-_* y}$. *Then we multiplicative multiply the considered equation with* $\mu(x,y)$ *and we find*

$$\left(e^{-_* y} +_* x\right) \cdot_* d_* x +_* \left(-_* x \cdot_* e^{-_* y} +_* y\right) \cdot_* d_* y = 0_*.$$

We have that

$$\left(\partial_* /_* (\partial_* y)\right)\left(e^{-_* y} +_* x\right) \;=\; -e^{-_* y},$$

$$\left(\partial_* /_* (\partial_* x)\right)\left(-_* x \cdot_* e^{-_* y} +_* y\right) \;=\; -_* e^{-_* y},$$

i.e.,

$$\left(\partial_* /_* (\partial_* y)\right)(\mu \cdot_* P) = \left(\partial_* /_* (\partial_* x)\right)(\mu \cdot_* Q).$$

Consequently, $\mu(x,y) = e^{-_* y}$ *is a multiplicative integrating factor for the considered equation.*
Now we will check that the function

$$u(x,y) = x \cdot_* e^{-_* y} +_* (x^{2_*} +_* y^{2_*})/_* 2_* = c$$

is solution to the considered equation. Indeed,

$$
\begin{aligned}
d_* u(x,y) \;=\;& \left(\partial_* /_* (\partial_* x)\right)\left(x \cdot_* e^{-y} +_* (x^{2_*} +_* y^{2_*})/_* 2\right) \cdot_* d_* x \\[2mm]
&+_* \left(\partial_* /_* (\partial_* y)\right)\left(x \cdot_* e^{-y} +_* (x^{2_*} +_* y^{2_*})/_* 2\right) \cdot_* d_* y \\[2mm]
=\;& \left(e^{-y} +_* x\right) \cdot_* d_* x +_* \left(-_* x \cdot_* e^{-y} +_* y\right) \cdot_* d_* y \\[2mm]
=\;& 0_*.
\end{aligned}
$$

Hence,

$$e^{-y} \cdot_* \phi\left(x \cdot_* e^{-y} +_* (x^{2_*} +_* y^{2_*})/_* 2\right)$$

are integrating factors of the considered equation, where ϕ *is a continuous function.*

Theorem 2.5.5 *The function μ is a multiplicative integrating factor for the equation (2.20) if and only if*

$$(\mu \cdot_* P)^*_y = (\mu \cdot_* Q)^*_x. \tag{2.33}$$

Proof 2.5.6 *From (2.33) we get*

$$\mu \cdot_* P^*_y +_* \mu^*_y \cdot_* P = \mu^*_x \cdot_* Q +_* \mu \cdot_* Q^*_x$$

or

$$\mu^*_y \cdot_* P -_* \mu^*_x \cdot_* Q = \mu \cdot_* (Q^*_x -_* P^*_y). \tag{2.34}$$

If we suppose that $\mu = \mu(v)$, where v is a function of x and y, then

$$\mu^*_x = ((d_*\mu)/_*(d_*v)) \cdot_* v^*_x,$$

$$\mu^*_y = ((d_*\mu)/_*(d_*v)) \cdot_* v_y,$$

and the equation (2.34) takes the form

$$((d_*\mu)/_*(d_*v)) \cdot_* v^*_y \cdot_* P -_* ((d_*\mu)/_*(d_*v)) \cdot_* v^*_x \cdot_* Q = \mu \cdot_* (Q^*_x -_* P^*_y),$$

or

$$(1_*/_*\mu) \cdot_* ((d_*\mu)/_*(d_*v)) = ((Q^*_x -_* P^*_y)/_*(v^*_y \cdot_* P -_* v_x \cdot_* Q)).$$

Hence, if

$$((Q^*_x -_* P^*_y)/_*(v^*_y \cdot_* P -_* v^*_x \cdot_* Q)) = \phi(v),$$

for some function ϕ, we get

$$(1_*/_*\mu) \cdot_* d_*\mu = \phi(v) \cdot_* d_*v$$

or

$$\mu = c \cdot_* e^{\int \phi(v)dv}.$$

Now we will give some special cases in which v and the corresponding $\phi(v)$ are given in the following manner.

1. $v(x,y) = x$. Then

$$\phi(v) = ((-_*Q^*_x +_* P^*_y)/_*Q).$$

2. $v(x,y) = y$. Then

$$\phi(v) = ((Q_x^* -_* P_y^*)/_* P.$$

3. $v(x,y) = x -_* y$. Then

$$\phi(v) = ((P_y^* -_* Q_x^*)/_* (P +_* Q)).$$

4. $v(x,y) = x \cdot_* y$. Then

$$\phi(v) = ((Q_x^* -_* P_y^*)/_* (x \cdot_* P -_* y \cdot_* Q).$$

5. $v = x/_* y$. Then

$$\phi(v) = ((y^{2*} \cdot_* (P_y^* -_* Q_x^*))/_* (x \cdot_* P +_* y \cdot_* Q)).$$

6. $v = x^{2*} -_* y^{2*}$. Then

$$\phi(v) = ((Q_x^* -_* P_y^*)/_* (2_* \cdot_* (y \cdot_* P -_* x \cdot_* Q))).$$

Theorem 2.5.7 *Let the equation (2.21) be an exact equation. Then it has an integrating factor* $\mu(x,y) \neq const$.

Proof 2.5.8 *Since the equation (2.21) is an exact equation, we have*

$$P_y^* = Q_x^*. \tag{2.35}$$

Because μ is an integrating factor for the equation (2.21), we have

$$\mu_y^* \cdot_* P +_* \mu \cdot_* P_y^* = \mu_x^* \cdot_* Q +_* \mu \cdot_* Q_x^*.$$

Hence and (2.35), we find

$$\mu_y^* \cdot_* P = \mu_x^* \cdot_* Q.$$

We multiply (2.21) by μ_y^* and we get

$$0_* = P \cdot_* \mu_y^* +_* Q \cdot_* \mu_y^* \cdot_* y^*$$

$$= Q \cdot_* \mu_x^* +_* Q \cdot_* \mu_y^* \cdot_* y^*$$

$$= Q \cdot_* \mu_x^* +_* \mu_y^* \cdot_* y^*)$$

$$= Q \cdot_* (d_* \mu /_* d_* x).$$

From here, $\mu(x,y) = c$ is a solution to the equation (2.21).

Theorem 2.5.9 *Let $\mu_1(x,y)$ and $\mu_2(x,y)$ be two multiplicative integrating factors such that their multiplicative ratio is not a multiplicative constant. Then $\mu_1(x,y) = c \cdot_* \mu_2(x,y)$ is a solution of the equation (2.21).*

Proof 2.5.10 *We have that*

$$\mu_1 \cdot_* P +_* \mu_1 \cdot_* Q \cdot_* y^* = 0_* \tag{2.36}$$

and

$$\mu_1 \cdot_* P +_* \mu_2 \cdot_* Q \cdot_* y^* = 0_* \tag{2.37}$$

are exact equations. We multiplicative multiply (2.37) by $\mu_1/_*\mu_2$ and we get the equation (2.36), which is an exact equation. Therefore, $\mu_1/_*\mu_2$ is a multiplicative integrating factor for the equation (2.37). Hence and the previous theorem we get that $\mu_1 = c \cdot_* \mu_2$ is a solution to the equation (2.21).

Theorem 2.5.11 *Let $P_x^* = Q_y^*$ and $P_y^* = -_* Q_x^*$. Then the equation (2.21) has an integrating factor*

$$e/_*(P^{2*} +_* Q^{2*}).$$

Proof 2.5.12 *We have that*

$$\partial_* /_* \partial_* y \left(P_* /_* (P^{2*} +_* Q^{2*}) \right)$$

$$= (P_y^* \cdot_* (P^{2*} +_* Q^{2*})) -_* P \cdot_* (2_* \cdot_* P_* \cdot_* P_y^* +_* 2_* \cdot_* Q \cdot_* Q_y^*)$$

$$/_*(P^{2*} +_* Q^{2*})^{2*}$$

$$= (P_y^* \cdot_* (Q^{2*} -_* P^{2*}) -_* 2_* \cdot_* P \cdot_* Q \cdot_* P_x^*)$$

$$/_* (P^{2*} +_* Q^{2*})^{2*}$$

$$= (-_*Q_x^* \cdot_* (Q^{2*} -_* P^{2*}) -_* 2_* P \cdot_* Q \cdot_* P_x^*)$$

$$/_* (P^{2*} +_* Q^{2*})^{2*},$$

and

$$\partial_* /_* \partial_* x \left(Q /_* (P^{2*} +_* Q^{2*}) \right)$$

$$= (Q_x^* \cdot_* (P^{2*} +_* Q^{2*}) -_* Q \cdot_* (2_* \cdot_* P \cdot_* P_x^* +_* 2_* \cdot_* Q \cdot_* Q_x^*))$$

$$/_* (P^{2*} +_* Q^{2*})^{2*}$$

$$= (-_*Q_x^* \cdot_* (Q^{2*} -_* P^{2*}) -_* 2_* \cdot_* P \cdot_* Q \cdot_* P_x^*)$$

$$/_* (P^{2*} +_* Q^{2*})^{2*}.$$

Consequently,

$$\partial_* /_* \partial_* y \left(P /_* (P^{2*} +_* Q^{2*}) \right) = \partial_* /_* \partial_* x \left(Q /_* (P^{2*} +_* Q^{2*}) \right)$$

and then the function $e /_* (P^{2*} +_* Q^{2*})$ *is a multiplicative integrating factor for the equation (2.21). This completes the proof.*

Example 2.5.13 *Let us consider the equation*

$$(x^{2*} \cdot_* y +_* y +_* e) +_* x \cdot_* (e +_* x^{2*}) \cdot_* y^* = 0_*.$$

We will prove that the function

$$\mu(x,y) = e /_* (e +_* x^{2*})$$

is a multiplicative integrating factor for it. Here

$$P(x,y) \;=\; x^{2*} \cdot_* y +_* y +_* e,$$

$$Q(x,y) \;=\; x \cdot_* (e +_* x^{2*}).$$

Then

$$\mu(x,y) \cdot_* P(x,y) \;=\; (x^{2*} \cdot_* y +_* y +_* e)/_*(x^{2*} +_* e)$$

$$=\; y +_* e/_*(x^{2*} +_* e),$$

$$\mu(x,y) \cdot_* Q(x,y) \;=\; (x \cdot_* (e +_* x^{2*}))/_*(e +_* x^{2*})$$

$$=\; x.$$

Then

$$\partial_*/_*\partial_* y(\mu(x,y) \cdot_* P(x,y)) \;=\; e,$$

$$\partial_*/_*\partial_* x(\mu(x,y) \cdot_* Q(x,y)) \;=\; e.$$

Example 2.5.14 *We consider the equation*

$$(x \cdot_* y^{3*} +_* 2_* \cdot_* x^{2*} \cdot_* y^{2*} -_* y^{2*})$$

$$+_* (x^{2*} \cdot_* y^{2*} +_* 2_* \cdot_* x^{3*} \cdot_* y -_* 2_* \cdot_* x^{2*}) \cdot_* y^* = 0_*.$$

We will prove that the function

$$\mu(x,y) = (e^{x \cdot_* y})/_*(x^{2*} \cdot_* y^{2*})$$

is its multiplicative integrating factor. Here,

$$P(x,y) \;=\; x \cdot_* y^{3*} +_* 2_* \cdot_* x^{2*} \cdot_* y^{2*} -_* y^{2*},$$

$$Q(x,y) \;=\; x^{2*} \cdot_* y^{2*} +_* 2_* \cdot_* x^{3*} \cdot_* y -_* 2_* \cdot_* x^{2*}.$$

Then

$$\mu(x,y)\cdot_* P(x,y) \;=\; ((e^{x\cdot_* y})/_*(x^{2*}\cdot_* y^{2*}))$$

$$\cdot_*(x\cdot_* y^{3*}+_*2_*\cdot_* x^{2*}\cdot_* y^{2*}-_* y^{2*})$$

$$=\; (e^{x\cdot_* y})\cdot_*\left(y/_* x+_*2_*-_* e/_* x^{2*}\right),$$

$$\mu(x,y)\cdot_* Q(x,y) \;=\; ((e^{x\cdot_* y})/_*(x^{2*}\cdot_* y^{2*}))$$

$$\cdot_*(x^{2*}\cdot_* y^{2*}+_*2_*\cdot_* x^{3*}\cdot_* y-_*2_*\cdot_* x^{2*})$$

$$=\; e^{x\cdot_* y}\left(e+_*(2_*\cdot_* x)/_* y-_*2_*/_* y^{2*}\right),$$

$$\partial_*/_*\partial_* y(\mu(x,y)\cdot_* P(x,y)) \;=\; \partial_*/_*\partial_* y\left(e^{x\cdot_* y}\cdot_*\left(y/_* x+_*2_*-_* e/_* y^{2*}\right)\right)$$

$$=\; x\cdot_* e^{x\cdot_* y}\cdot_*\left(y/_* x+_*2_*-_* e/_* x^{2*}\right)$$

$$+_* e^{x\cdot_* y}\cdot_*(e/_* x)$$

$$=\; e^{x\cdot_* y}\cdot_*(y+_*2_*\cdot_* x),$$

$$\partial_*/_*\partial_* x(\mu(x,y)\cdot_* Q(x,y)) \;=\; \partial_*/_*\partial_* x\left(e^{x\cdot_* y}\cdot_*\left(e+_*2_* x/_* y-_*2_*/_* y^{2*}\right)\right)$$

$$=\; y\cdot_* e^{x\cdot_* y}\cdot_*\left(e+_*2_* x/_* y-_*2_*/_* y^{2*}\right)$$

$$+_*2_*/_* y\cdot_* e^{x\cdot_* y}$$

$$=\; e^{x\cdot_* y}\cdot_*(y+_*2_*\cdot_* x-_*2_*/_* y)$$

$$+_*(2_*/_* y)\cdot_* e^{x\cdot_* y}$$

$$=\; e^{x\cdot_* y}\cdot_*(y+_*2_*\cdot_* x).$$

Example 2.5.15 *Now we consider the equation*

$$(y +_* x \cdot_* f(x^{2*} +_* y^{2*})) +_* (y \cdot_* f(x^{2*} +_* y^{2*}) -_* x) \cdot_* y^* = 0_*,$$

where f is a multiplicative continuously differentiable function. We will prove that the function

$$\mu(x,y) = e/_*(x^{2*} +_* y^{2*})$$

is its multiplicative integrating factor. Here

$$P(x,y) \;=\; y +_* x \cdot_* f(x^{2*} +_* y^{2*}),$$

$$Q(x,y) \;=\; y \cdot_* f(x^{2*} +_* y^{2*}) -_* x.$$

Then

$$\mu(x,y) \cdot_* P(x,y) \;=\; (y +_* x \cdot_* f(x^{2*} +_* y^{2*}))/_*(x^{2*} +_* y^{2*}),$$

$$\mu(x,y) \cdot_* Q(x,y) \;=\; (y \cdot_* f(x^{2*} +_* y^{2*}) -_* x)/_*(x^{2*} +_* y^{2*}).$$

Hence,

$$\partial_*/_*\partial_* y(\mu(x,y) \cdot_* P(x,y))$$

$$= \;\; \partial_*/_*\partial_* y \left((y +_* x \cdot_* f(x^{2*} +_* y^{2*}))/_*(x^{2*} +_* y^{2*})\right)$$

$$= \;\; ((e +_* x \cdot_* f^*(x^{2*} +_* y^{2*}) \cdot_* 2_* \cdot_* y) \cdot_* (x^{2*} +_* y^{2*})$$

$$-_* 2_* \cdot_* y \cdot_* (y +_* x \cdot_* f(x^{2*} +_* y^{2*})))/_*(x^{2*} +_* y^{2*})^{2*}$$

$$= \;\; (x^{2*} -_* y^{2*} +_* 2_* \cdot_* x \cdot_* y \cdot_* f^*(x^{2*} +_* y^{2*})$$

$$-_* 2_* \cdot_* x \cdot_* y \cdot_* f(x^{2*} +_* y^{2*}))/_*(x^{2*} +_* y^{2*})^{2*},$$

$$\partial_*/_*\partial_* x(\mu(x,y) \cdot_* (Q(x,y)))$$

$$= \ \partial_*/_*\partial_{*x}\left((y\cdot_* f(x^{2*}+_* y^{2*})-_* x)/_*(x^{2*}+_* y^{2*})\right)$$

$$= \ ((y\cdot_* f^*(x^{2*}+_* y^{2*})\cdot_* 2_*\cdot_* x-_* e)\cdot_*(x^{2*}-_* y^{2*})$$

$$-_* 2_*\cdot_* x\cdot_*(y\cdot_* f(x^{2*}+_* y^{2*})-_* x))/_*(x^{2*}+_* y^{2*})^{2*}$$

$$= \ (x^{2*}-_* y^{2*}+_* 2_*\cdot_* x\cdot_* y\cdot_* f^*(x^{2*}+_* y^{2*})$$

$$-_* 2_*\cdot_* x\cdot_* y\cdot_* f(x^{2*}+_* y^{2*}))/_*(x^{2*}+_* y^{2*})^{2*}.$$

Exercise 2.5.16 *Prove that the function* $\mu(x)=e^{e^{2}\cdot_* x}$ *is a multiplicative integrating factor of the equation*

$$(x^{2*}+_* y^{2*}+_* x)\cdot_* d_* x+_* y\cdot_* d_* y=0_*.$$

Exercise 2.5.17 *Prove that the function*

$$\mu(y)=e/_*(e+_* y^{2*})^{\frac{1}{2}*}$$

is a multiplicative integrating factor of the equation

$$y\cdot_* d_* y=(x\cdot_* d_* y+_* y\cdot_* d_* x)\cdot_*(e+_* y^{2*})^{\frac{1}{2}*}.$$

Exercise 2.5.18 *Prove that the function* $\mu(x)=x$ *is a multiplicative integrating factor of the equation*

$$x\cdot_* y^{2*}\cdot_*(x\cdot_* y^*+_* y)=e.$$

Example 2.5.19 *Now we consider the equation*

$$(y-_* y^{2*})+_* x\cdot_* y^*=0_*.$$

We will search a multiplicative integrating factor for this equation in the form

$$\mu(x,y)=x^{m*}\cdot_* y^{n*}.$$

We can rewrite the given equation in the form

$$(y -_* y^{2*}) \cdot_* d_* x +_* x \cdot_* d_* y = 0_*.$$

Here

$$P(x,y) \;=\; y -_* y^{2*},$$

$$Q(x,y) \;=\; x.$$

Then

$$\mu(x,y) \cdot_* P(x,y) \;=\; x^{m*} \cdot_* y^{n*} \cdot_* (y -_* y^{2*})$$

$$= \; x^{m*} \cdot_* y^{(n+1)*} -_* x^{m*} \cdot_* y^{(n+2)*},$$

$$\mu(x,y) \cdot_* Q(x,y) \;=\; x^{(m+1)*} y^{n*}.$$

Hence,

$$\partial_* /_* \partial_* y(\mu(x,y) \cdot_* P(x,y)) \;=\; (n+1)_* \cdot_* x^{m*} \cdot_* y^{n*}$$

$$-_* (n+2)_* \cdot *x^{m*} \cdot_* y^{(n+1)*},$$

$$\partial_* /_* \partial_* x(\mu(x,y) \cdot_* Q(x,y)) \;=\; (m+1)_* \cdot_* x^{m*} \cdot_* y^{n*}.$$

We assume

$$\partial_* /_* \partial_* y(\mu(x,y) \cdot_* P(x,y)) = \partial_* /_* \partial_* x(\mu(x,y) \cdot_* Q(x,y)).$$

Therefore,

$$(n+1)_* \cdot_* x^{m*} \cdot_* y^{n*} -_* (n+2)_* \cdot_* x^{m*} \cdot_* y^{(n+1)*} = (m+1)_* \cdot_* x^{m*} y^{n*}.$$

From here,

$$n+1 = m+1, n+2 = 0 \qquad \text{or} \qquad n = m = -2.$$

Example 2.5.20 *We consider the equation*

$$\sin_* y \cdot_* d_* x +_* x \cdot_* \cos_* y \cdot_* d_* y = 0_*.$$

Here

$$P(x,y) \quad = \quad \sin_* y,$$

$$Q(x,y) \quad = \quad x \cdot_* \cos_* y.$$

We have

$$\partial_* /_* \partial_* y P(x,y) \quad = \quad \cos_* y,$$

$$\partial_* Q /_* \partial_* x Q(x,y) \quad = \quad \cos_* y.$$

Consequently,

$$\partial_* /_* \partial_* y P(x,y) = \partial_* /_* \partial_* x Q(x,y).$$

Therefore, the considered equation is an exact equation. Let $\phi \in \mathscr{C}_*^1(\mathbb{R}_*)$. We will prove that

$$\mu(x,y) = \phi(x \cdot_* \sin_* y)$$

is a multiplicative integrating factor for the considered equation. We have

$$\mu(x,y) \cdot_* P(x,y) \quad = \quad \sin_* y \cdot_* \phi(x \cdot_* \sin_* y),$$

$$\mu(x,y) \cdot_* Q(x,y) \quad = \quad x \cdot_* \cos_* y \cdot_* \phi(x \cdot_* \sin_* y).$$

We have

$$\partial_* /_* \partial_* y (\mu(x,y) \cdot_* P(x,y)) \quad = \quad \cos_* y \cdot_* \phi(x \cdot_* \sin_* y)$$

$$+_* x \cdot_* \sin_* y \cdot_* \phi^*(x \cdot_* \sin_* y),$$

$$\partial_*/{}_*\partial_* x(\mu(x,y)\cdot_* Q(x,y)) \ = \ \cos_* y \cdot_* \phi(x\cdot_* \sin_* y)$$

$$+_* x\cdot_* \sin_* y\cdot_* \phi^*(x\cdot_* \sin_* y),$$

i.e.,

$$\partial_*/{}_*\partial_* y(\mu(x,y)\cdot_* P(x,y)) = \partial_*/{}_*\partial_* x(\mu(x,y)\cdot_* Q(x,y)).$$

Therefore

$$\phi(x\cdot_* \sin_* y) = c$$

is a solution to the considered equation.

Example 2.5.21 *Now we consider the equation*

$$(y+_* e^x)\cdot_* d_* x +_* x\cdot_* d_* y = 0_*.$$

Here

$$P(x,y) \ = \ y+_* e^x,$$

$$Q(x,y) \ = \ x.$$

Then

$$\partial_*/{}_*\partial_* y P(x,y) \ = \ \partial_*/{}_*\partial_* x Q(x,y)$$

$$= \ e.$$

Therefore, the considered equation is an exact equation. Let $\phi \in \mathscr{C}_*^1(\mathbb{R}_*)$. We will show that the function $\mu(x,y) = \phi(x\cdot_* y +_* e^x)$ is a multiplicative integrating factor for the considered equation. Really,

$$\partial_*/{}_*\partial_* y(\mu(x,y)\cdot_* P(x,y)) \ = \ \partial_*/{}_*\partial_* y((y+_* e^x)\cdot_* \phi(x\cdot_* y+_* e^x))$$

$$= \ \phi(x\cdot_* y+_* e^x)$$

$$+_* x\cdot_* (y+_* e^x)\cdot_* \phi^*(x\cdot_* y+_* e^x),$$

$$\partial_* /_* \partial_* x(\mu(x,y) \cdot_* Q(x,y)) = \partial_* /_* \partial_* x(x \cdot_* \phi(x \cdot_* y +_* e^x))$$

$$= \phi(x \cdot_* y +_* e^x)$$

$$+_* x \cdot_* (y +_* e^x) \cdot_* \phi^*(x \cdot_* y +_* e^x).$$

Therefore,

$$\phi(x \cdot_* y +_* e^x) = c$$

is a solution to the considered equation.

Example 2.5.22 *We consider the equation*

$$(2_* \cdot_* x +_* y) \cdot_* d_* x +_* (x +_* 2_* \cdot_* y) \cdot_* d_* y = 0_*.$$

Here

$$P(x,y) = 2_* \cdot_* x +_* y,$$

$$Q(x,y) = x +_* 2_* \cdot_* y.$$

Then

$$\partial_* /_* \partial_* y P(x,y) = \partial_* /_* \partial_* x Q(x,y)$$

$$= 1.$$

Therefore, the considered equation is an exact equation. Let $\phi \in \mathscr{C}_*^1(\mathbb{R}_*)$. Now we will show that

$$\mu(x,y) = \phi(x^{2*} +_* x \cdot_* y +_* y^{2*})$$

is a multiplicative integrating factor for the considered equation. Indeed,

$$\partial_* /_* \partial_* y(\mu(x,y) \cdot_* P(x,y)) = \partial_* /_* \partial_* y(\phi(x^{2*} +_* x \cdot_* y +_* y^{2*})$$

$$\cdot_*(2_* \cdot_* x +_* y))$$

$$= (x +_* 2_* \cdot_* y) \cdot_* (2_* \cdot_* x +_* y)$$

$$\cdot_* \phi^*(x^{2*} +_* x \cdot_* y +_* y^{2*})$$

$$+_* \phi(x^{2*} +_* x \cdot_* y +_* y^{2*}),$$

$$\partial_*/_* \partial_* x(\mu(x,y) \cdot_* Q(x,y)) = \partial_*/_* \partial_* x(\phi(x^{2*} +_* x \cdot_* y +_* y^{2*})$$

$$\cdot_*(x +_* 2_* \cdot_* y))$$

$$= (x +_* 2_* \cdot_* y) \cdot_* (2_* \cdot_* x +_* y)$$

$$\cdot_* \phi^*(x^{2*} +_* x \cdot_* y +_* y^{2*})$$

$$+_* \phi(x^{2*} +_* x \cdot_* y +_* y^{2*}).$$

Consequently,

$$\phi(x^{2*} +_* x \cdot_* y +_* y^{2*}) = c$$

is a solution of the considered equation.

Example 2.5.23 *We consider the equation*

$$(y -_* \sin_* x) \cdot_* d_* x +_* x \cdot_* d_* y = 0_*.$$

Here

$$P(x,y) = y -_* \sin_* x,$$

$$Q(x,y) = x.$$

Then

$$\partial_*/_* \partial_* y P(x,y) = \partial_*/_* \partial_* x Q(x,y)$$

$$= 1.$$

Therefore, the considered equation is an exact equation. We will prove that

$$\mu_1(x,y) \;=\; (x\cdot_* y +_* \cos_* x)^{2*},$$

$$\mu_2(x,y) \;=\; (x\cdot_* y +_* \cos_* x)$$

are multiplicative integrating factors of the considered equation. Indeed,

$$\partial_*/_*\partial_* y(\mu_1(x,y)\cdot_* P(x,y)) \;=\; \partial_*/_*\partial_* y\left((x\cdot_* y +_* \cos_* x)^{2*}\cdot_* (y-_* \sin_* x)\right)$$

$$=\; 2_*\cdot_* (x\cdot_* y +_* \cos_* x)\cdot_* x\cdot_* (y-_* \sin_* x)$$

$$+_* (x\cdot_* y +_* \cos_* x)^{2*},$$

$$\partial_*/_*\partial_* x(\mu_1(x,y)\cdot_* Q(x,y)) \;=\; \partial_*/_*\partial_* x\left((x\cdot_* y +_* \cos_* x)^{2*}\cdot_* x\right)$$

$$=\; 2_*\cdot_* (x\cdot_* y +_* \cos_* x)\cdot_* x\cdot_* (y-_* \sin_* x)$$

$$+_* (x\cdot_* y +_* \cos_* x)^{2*},$$

$$\partial_*/_*\partial_* y(\mu_2(x,y)\cdot_* P(x,y)) \;=\; \partial_*/_*\partial_* y((x\cdot_* y +_* \cos_* x)\cdot_* (y-_* \sin_* x))$$

$$=\; x\cdot_* (y-_* \sin_* x) +_* (x\cdot_* y +_* \cos_* x),$$

$$\partial_*/_*\partial_* x(\mu_2(x,y)\cdot_* Q(x,y)) \;=\; \partial/_*\partial x((x\cdot_* y +_* \cos_* x)\cdot_* x)$$

$$=\; x\cdot_* y +_* \cos_* x +_* (y-_* \sin_* x)\cdot_* x.$$

Also, we note that

$$\mu_1(x,y)/_*\mu_2(x,y) \neq \text{const.}$$

Therefore

$$x\cdot_* y +_* \cos_* x = c$$

is a solution to the considered equation.

Example 2.5.24 *We consider the equation*

$$(x +_* x^{4*} +_* 2_* \cdot_* x^{2*} \cdot_* y^{2*} +_* y^{4*}) +_* y \cdot_* y^* = 0_*$$

if it admits an integrating factor $\mu = \mu(x^{2*} +_* y^{2*})$. *We can rewrite this equation in the following form:*

$$(x +_* x^{4*} +_* 2_* \cdot_* x^{2*} \cdot_* y^{2*} +_* y^{4*}) \cdot_* d_* x +_* y \cdot_* d_* y = 0_*.$$

Here

$$P(x,y) \quad = \quad x +_* x^{4*} +_* 2_* \cdot_* x^{2*} \cdot_* y^{2*} +_* y^{4*},$$

$$Q(x,y) \quad = \quad y.$$

We have that

$$\partial_* /_* \partial_* y P(x,y) \quad = \quad 4_* \cdot_* x^{2*} \cdot_* y +_* 4_* \cdot_* y^{3*},$$

$$\partial_* \partial_* x Q(x,y) \quad = \quad 1.$$

Hence,

$$\partial_* /_* \partial y P(x,y) \neq \partial_* /_* \partial x Q(x,y)$$

and the considered equation is not an exact equation. Let $v(x,y) = x^{2*} +_* y^{2*}$. *Then*

$$\partial_* /_* \partial_* x \mu(x,y) \quad = \quad 2_* \cdot_* x \cdot_* (d_* \mu /_* d_* v),$$

$$\partial_* /_* \partial_* y \mu(x,y) \quad = \quad 2_* \cdot_* y \cdot_* (d_* \mu /_* d_* v),$$

$$\partial_* /_* \partial_* y (\mu(x,y) \cdot_* P(x,y)) \quad = \quad \partial_* /_* \partial_* x (\mu(x,y) \cdot_* Q(x,y))$$

and

$$P(x,y) \cdot_* (\partial_* /_* \partial_* y \mu(x,y)) +_* \partial_* /_* \partial_* y P(x,y) \cdot_* \mu(x,y)$$

$$= \quad (\partial_* /_* \partial_* x \mu(x,y)) \cdot_* Q(x,y)$$

$$+_*\mu(x,y)\cdot_*(\partial_*/_*\partial_*xQ(x,y)),$$

and

$$(x+_*x^{4*}+_*2_*\cdot_*x^{2*}\cdot_*y^{2*}+_*y^{4*})\cdot_*2_*y\cdot_*(d_*\mu/_*d_*v)$$

$$+_*\mu(4_*\cdot_*x^{2*}\cdot_*y+_*4_*\cdot_*y^{3*})$$

$$=\quad 2_*\cdot_*x\cdot_*y\cdot_*(d_*\mu/_*d_*v),$$

and

$$2_*\cdot_*y\cdot_*(x^{4*}+_*2_*\cdot_*x^{2*}\cdot_*y^{2*}+_*y^{4*})\cdot_*(d_*\mu/_*d_*v)$$

$$=\quad -_*4_*\cdot_*y\cdot_*\mu(x^{2*}+_*y^{2*}),$$

and

$$2_*\cdot_*y\cdot_*(x^{2*}+_*y^{2*})^{2*}\cdot_*(d_*\mu/_*d_*v)=-_*4_*\cdot_*y\cdot_*\mu(x^{2*}+_*y^{2*}),$$

and

$$d_*\mu/_*d_*v=-_*2_*\cdot_*(\mu_*/_*(x^{2*}+_*y^{2*})),$$

and

$$d_*\mu/_*d_*v=-_*2_*\cdot_*(\mu/_*v),$$

and

$$d_*\mu/_*\mu=-_*2_*\cdot_*(d_*v/_*v),$$

and

$$\log_*|\mu|_*=c+_*\log_*(e/_*v^{2*})$$

or

$$\mu(x,y)=c/_*(x^{2*}+_*y^{2*})^{2*}.$$

We take

$$\mu(x,y)=e/_*(x^{2*}+_*y^{2*})^{2*}.$$

We multiplicative multiply the given equation by $\mu(x,y)$ and we find

$$(x +_* x^{4*} +_* 2_* \cdot_* x^{2*} \cdot_* y^{2*} +_* y^{4*})/_* (x^{2*} +_* y^{2*})^{2*} \cdot_* d_* x$$

$$+_* y/_* (x^{2*} +_* y^{2*})^{2*} \cdot_* d_* y$$

$$= 0_*$$

or

$$\left(x/_* (x^{2*} +_* y^{2*})^{2*} +_* e\right) \cdot_* d_* x +_* y/_* (x^{2*} +_* y^{2*})^{2*} \cdot_* d_* y = 0_*,$$

which is an exact equation. There exists a function $u(x,y)$, such that

$$u_x^*(x,y) \;=\; x/_* (x^{2*} +_* y^{2*})^{2*} +_* e,$$

$$u_y^*(x,y) \;=\; y/_* (x^{2*} +_* y^{2*})^{2*}.$$

We multiplicative integrate the second equation with respect to y and we find

$$u(x,y) \;=\; \int_* y/_* (x^{2*} +_* y^{2*})^{2*} \cdot_* d_* y +_* g(x)$$

$$=\; -_*(e/_*(2_* \cdot_* (x^{2*} +_* y^{2*}))) +_* g(x),$$

which we multiplicative differentiate with respect to x, and we become

$$u_x^*(x,y) \;=\; x/_* (x^{2*} +_* y^{2*})^{2*} +_* g^*(x)$$

$$=\; x/_* (x^{2*} +_* y^{2*})^{2*} +_* e$$

or

$$g^*(x) = e.$$

Hence,

$$g(x) = x +_* g(0_*)$$

and

$$u(x,y) = -_*(e/_* 2_* \cdot_* (x^{2*} +_* y^{2*})) +_* x +_* g(0_*).$$

Consequently,

$$x -_* (e/_*2_* \cdot_* (x^{2*} +_* y^{2*})) = c,$$

or

$$2_* \cdot_* (x -_* c) \cdot_* (x^{2*} +_* y^{2*}) = e$$

are the solutions of the considered equation.

Exercise 2.5.25 *Solve the equation*

$$(x -_* y^{2*}) +_* 2_* x \cdot_* y \cdot) *y^* = 0_*$$

if it admits an integrating factor $\mu = \mu(x)$.

Answer

$$y^{2*} +_* x \cdot_* \log_* |x|_* = c \cdot_* x.$$

Exercise 2.5.26 *Solve the equation*

$$y +_* (y^{2*} -_* x) \cdot_* y^* = 0_*$$

if it admits an integrating factor $\mu = \mu(y)$.

Answer

$$y^{2*} +_* x = c \cdot_* x.$$

Exercise 2.5.27 *Solve the equation*

$$(3_* \cdot_* x \cdot_* y +_* y^{2*}) +_* (3_* \cdot_* x \cdot_* y +_* x^{2*}) \cdot_* y^* = 0_*$$

if it admits an integrating factor $\mu = \mu(x +_* y)$.

Answer

$$x^{3*} \cdot_* y +_* 2_* \cdot_* x^{2*} \cdot_* y^{2*} +_* x \cdot_* y^{2*} = c,$$

$$x +_* y = c.$$

Exercise 2.5.28 *Solve the equation*

$$y +_* x \cdot_* (e -_* 3_* \cdot_* x^{2*} \cdot_* y^{2*}) \cdot_* y^* = 0_*$$

if it admits an integrating factor $\mu(x, y) = \mu(x \cdot_* y)$.

Answer

$$y^{6*} \cdot_* e^{e/*(x^{2*}+_*y^{2*})} = c.$$

Theorem 2.5.29 *Let* $P(x,y), Q(x,y) \in \mathscr{C}^1_*(S)$ *be multiplicative homogeneous functions of degree k. Then*

$$e/_*(x \cdot_* P(x,y) +_* y \cdot_* Q(x,y))$$

is a multiplicative integrating factor of the equation (2.21) provided

$$x \cdot_* P +_* y \cdot_* Q \neq 0_*$$

in S.

Proof 2.5.30 *Since* $P(x,y)$ *and* $Q(x,y)$ *are multiplicative homogeneous functions of degree k, using the multiplicative Lagrange identity, we have that*

$$k_* \cdot_* P(x,y) \;=\; x \cdot_* P_x^*(x,y) +_* y \cdot_* P_y^*(x,y)$$

$$k_* \cdot_* Q(x,y) \;=\; x \cdot_* Q_x^*(x,y) +_* y \cdot_* Q_y^*(x,y),$$

from where

$$y \cdot_* P_y^*(x,y) \;=\; k_* \cdot_* P(x,y) -_* x \cdot_* P_x^*(x,y)$$

$$\hspace{6cm}(2.38)$$

$$y \cdot_* Q_y^*(x,y) \;=\; k_* \cdot_* Q(x,y) -_* x \cdot_* Q_x(x,y).$$

Now we multiply the equation (2.21) by the function

$$e/_*(x \cdot_* P(x,y) +_* y \cdot_* Q(x,y))$$

and we get

$$(P(x,y)/_*(x \cdot_* P(x,y) +_* y \cdot_* Q(x,y))) \cdot_* d_* x$$

$$+_* (Q(x,y)/_*(x \cdot_* P(x,y) +_* y \cdot_* Q(x,y))) \cdot_* d_* y = 0_*.$$

Then, using (2.38),

$$\partial_*/_* \partial_* y \, (P(x,y)/_*(x \cdot_* P(x,y) +_* y \cdot_* Q(x,y)))$$

$$= \; (P_y^*(x,y) \cdot_* (x \cdot_* P(x,y) +_* y \cdot_* Q(x,y)) \hspace{2.5cm}(2.39)$$

$$-_* P(x,y)(x \cdot_* P_y^*(x,y) +_* Q(x,y) +_* y \cdot_* Q_y^*(x,y)))$$

$$/_*(x\cdot_* P(x,y) +_* y \cdot_* Q(x,y))^{2_*}$$

$$= (y\cdot_* P_y^*(x,y) \cdot_* Q(x,y) -_* y\cdot_* P(x,y)\cdot_* Q_y^*(x,y)$$

$$-_* P(x,y)\cdot_* Q(x,y))$$

$$/_*(x\cdot_* P(x,y) +_* y\cdot_* Q(x,y))^{2_*}$$

$$= (Q(x,y)\cdot_* (k_* \cdot_* P(x,y) -_* x\cdot_* P_x^*(x,y))$$

$$-_* P(x,y)\cdot_* (k_*\cdot_* Q(x,y) -_* x\cdot_* Q_x^*(x,y)) -_* P(x,y)\cdot_* Q(x,y))$$

$$/_*(x\cdot_* P(x,y) +_* y\cdot_* Q(x,y))^{2_*}$$

$$= (k_* \cdot_* Q(x,y)\cdot_* P(x,y) -_* x\cdot_* Q(x,y)\cdot_* P_x(x,y) -_* k_* \cdot_* Q(x,y)\cdot_* P(x,y)$$

$$+_* x\cdot_* P(x,y)\cdot_* Q_x^*(x,y) -_* P(x,y)\cdot_* Q(x,y))$$

$$/_*(x\cdot_* P(x,y) +_* y\cdot_* Q(x,y))^{2_*}$$

$$= (x\cdot_* P(x,y)\cdot_* Q_x^*(x,y) -_* x\cdot_* Q(x,y)\cdot_* P_x(x,y) -_* P(x,y)\cdot_* Q(x,y))$$

$$/_*(x\cdot_* P(x,y) +_* y\cdot_* Q(x,y))^{2_*}$$

and

$$\partial_*\partial_* x(Q(x,y)/_* x\cdot_* P(x,y) +_* y\cdot_* Q(x,y))$$

$$= (Q_x^*(x,y)\cdot_* (x\cdot_* P(x,y) +_* y\cdot_* Q(x,y))$$

$$-_* Q(x,y)\cdot_* (P(x,y) +_* x\cdot_* P_x^*(x,y) +_* y\cdot_* Q_x^*(x,y)))$$

$$/_*(x\cdot_* P(x,y) +_* y\cdot_* Q(x,y))^{2_*}$$

$$= \; (x \cdot_* P(x,y) \cdot_* Q_x(x,y) -_* x \cdot_* P_x^*(x,y) \cdot_* Q(x,y) -_* P(x,y) \cdot_* Q(x,y))$$

$$/_* (x \cdot_* P(x,y) +_* y \cdot_* Q(x,y))^{2*}.$$

Hence and (2.39) we get that

$$\partial_* /_* \partial_* y \, (P(x,y) /_* x \cdot_* P(x,y) +_* y \cdot_* Q(x,y))$$

$$= \; \partial_* /_* \partial_* x \, (Q(x,y) /_* x \cdot_* P(x,y) +_* y \cdot_* Q(x,y)).$$

This completes the proof.

2.6 Advanced Practical Problems

Problem 2.6.1 *Let* $S = \{(x,y) \in \mathbb{R}^2 : |x|_* < 1_*, |y|_* < 3_*\}$. *Find the solutions of the equation*

$$y^* = 3_* \cdot_* y^{\frac{2}{3}*}, \qquad in \qquad S.$$

Answer

$$y = (x +_* c)^{3*}.$$

$$y = 0_*.$$

Problem 2.6.2 *Let* $S = \{(x,y) \in \mathbb{R}_*^2 : |x|_* < 10_*, |y|_* < 10_*\}$. *Find the solutions of the following equations*

1.
$$(y^{2*} +_* e)^{\frac{1}{2}*} \cdot_* d_* x = x \cdot_* y \cdot_* d_* y.$$

2.
$$(x^{2*} -_* e) \cdot_* y^* +_* 2_* \cdot_* x \cdot_* y^{2*} = 0_*.$$

3.
$$x \cdot_* y^* +_* y = y^{2*}.$$

4.
$$2_* \cdot_* x^{2*} \cdot_* y \cdot_* y^* +_* y^{2*} = 2_*.$$

Answer

1. $\log_* |x|_* = c +_* (y^{2*} +_* e)^{\frac{1}{2}*}, \, x = 0_*.$
2. $\cdot_* \left(\log_* |x^{2*} -_* e|_* +_* c \right) = e, \, y = 0_*.$
3. $y \cdot_* (e -_* c \cdot_* x) = e, \, y = 0_*.$
4. $y^{2*} -_* 2_* = c \cdot_* e^{e/_* x}.$

Problem 2.6.3 *Let* $S = \{(x,y) \in \mathbb{R}^2 : |x|_* < 1_*, |y|_* < 2_*\}$. *Find the solutions of the equation*

$$(x +_* 2_* \cdot_* y) \cdot_* y^* = e.$$

Answer $y = -_* e -_* (x/_* 2_*) +_* c \cdot_* e^y.$

Problem 2.6.4 *Let* $D = \{(x,y) \in \mathbb{R}^2_* : x \neq 0_*$ or $y \neq 0_*\}$. *Prove that the following functions are multiplicative homogeneous. Find the degree of multiplicative homogeneity.*

1.

$$f(x,y) = \cos_* (x^{2*} +_* y^{2*})/_* (2_* \cdot_* x \cdot_* y) +_* \sin_* (x/_* y).$$

2.

$$f(x,y) = x^{4*} +_* 3_* \cdot_* x^{3*} \cdot_* y +_* 7_* \cdot_* x^{2*} \cdot_* y^{2*} -_* y^{4*}.$$

3.

$$f(x,y) = (x +_* y)/_* (x^{5*} +_* y^{5*}).$$

Answer

1. $0_*.$
2. $4_*.$
3. $-_* 4_*.$

Problem 2.6.5 *Let* $D = \mathbb{R}^2_*$. *Check if the function*

$$f(x,y) = 2_* +_* x +_* y^{2*} +_* x \cdot_* y^{3*} +_* 3_* \cdot_* x^{2*} \cdot_* y$$

is a multiplicative homogeneous function.

Answer *No.*

Problem 2.6.6 *Let $D = \mathbb{R}^2_*$. Solve the equation*

$$2_* \cdot_* x^{3*} \cdot_* y^* = y \cdot_* (2_* \cdot_* x^{2*} -_* y^{2*}).$$

Answer

$$x = \pm_* y \cdot_* (\log_*(c \cdot_* x))^{\frac{1}{2}*}.$$
$$y = 0_*$$

.

Problem 2.6.7 *Solve the equation*

$$y^* = 2_* \cdot_* ((y +_* 2_*)/_* (x +_* y -_* 1_*))^{2*}.$$

Answer

$$y +_* 2_* = c \cdot_* e^{-*2*} \arctan_*((y +_* 2_*)/_* (x -_* 3_*)).$$

Problem 2.6.8 *Check that the following equations are exact and find their solutions.*

1.

$$3_* \cdot_* x^{2*} \cdot_* (e +_* \log_* y) \cdot_* d_* x -_* (2_* \cdot_* y -_* (x^{3*})/_* y) \cdot_* d_* y = 0_*.$$

2.

$$(x/_* \sin_* y +_* 2_*) \cdot_* d_* x$$

$$+_* ((x^{2*} +_* e) \cdot_* \cos_* y)/_* (\cos(2_* \cdot_* y) -_* e) \cdot_* d_* y = 0_*.$$

3.

$$\cos_* y \cdot_* d_* x -_* x \cdot_* \sin_* y \cdot_* d_* x = 0_*.$$

Answer

1.
$$x^{3*} +_* x^{3*} \cdot_* \log_* y -_* y^{2*} = c.$$

2.
$$x^{2*} +_* c = 2_* \cdot_* (c -_* 2_* \cdot_* x) \cdot_* \sin_* y.$$

3.
$$x \cdot_* \cos_* y = c.$$

Problem 2.6.9 *Prove that the function* $\mu(x) = e/_* x^{2*}$ *is a multiplicative integrating factor of the equation*

$$(x^{2*} -_* (\sin_* y)^{2*}) \cdot_* d_* x +_* x \cdot_* \sin_* (2_* \cdot_* y) \cdot_* d_* y = 0_*.$$

Problem 2.6.10 *Solve the equation*

$$(4_* \cdot_* x \cdot_* y +_* 3_* \cdot_* y^{3*}) +_* (2_* \cdot_* x^{2*} +_* 5_* \cdot_* x \cdot_* y^{3*}) \cdot_* y^* = 0_*$$

if it admits an integrating factor $\mu = \mu(x^{2*} \cdot_* y).$

Answer
$$x^{4*} \cdot_* y^{2*} +_* x^{3*} \cdot_* y^{5*} = c.$$

3

First-Order Multiplicative Linear Differential Equations

Let $J \subset \mathbb{R}_*$.

3.1 Definition, General Solutions

Definition 3.1.1 *The equations*

$$y^* +_* a(x) \cdot_* y = b(x), \quad x \in J, \tag{3.1}$$

where $a, b : J \to \mathbb{R}_*$, $a, b \in \mathscr{C}(J)$, will be called first-order multiplicative linear differential equations.

The corresponding multiplicative homogeneous equation is

$$y^* +_* a(x) \cdot_* y = 0_*, \quad x \in J. \tag{3.2}$$

The solution of the equation (3.2) is given by

$$y(x) = C_1 \cdot_* e^{-* \int_* a(x) \cdot_* d_* x}, \quad x \in J.$$

Here C_1 is a multiplicative constant. Now we will search a solution of the equation (3.1) in the form

$$y(x) = c(x) \cdot_* e^{-* \int_* a(x) \cdot_* d_* x}, \quad x \in J, \tag{3.3}$$

where $c : J \to \mathbb{R}_*$, $c \in \mathscr{C}_*^1(J)$. We have

$$\begin{aligned} y^*(x) &= c^*(x) \cdot_* e^{-* \int_* a(x) \cdot_* d_* x} \\ &\quad -_* a(x) \cdot_* c(x) \cdot_* e^{-* \int_* a(x) \cdot_* d_* x}, \quad x \in J. \end{aligned} \tag{3.4}$$

DOI: 10.1201/9781003393344-3

Substituting (3.3) and (3.4) into (3.1), we find

$$c^*(x) \cdot_* e^{-* \int_* a(x) \cdot_* d_* x} -_* a(x) \cdot_* c(x) \cdot_* e^{-* \int_* a(x) \cdot_* d_* x}$$

$$+_* a(x) \cdot_* c(x) \cdot_* e^{-* \int_* a(x) \cdot_* d_* x} = b(x), \quad x \in J$$

or

$$c^*(x) \cdot_* e^{-* \int_* a(x) \cdot_* d_* x} = b(x), \quad x \in J,$$

or

$$c(x) = b(x) \cdot_* e^{\int_* a(x) \cdot_* d_* x} +_* C_2, \quad x \in J.$$

Here C_2 is a multiplicative constant. Consequently,

$$y(x) \;=\; e^{-* \int_* a(x) \cdot_* dx} \cdot_* \left(C_2 +_* \int_* b(x) \cdot_* e^{\int_* a(x) \cdot_* d_* x} \cdot_* d_* x \right),$$

(3.5)

$x \in J$, are the solutions of the equation (3.1). Another way to get the formula (3.5) is as follows. We multiplicative multiply the equation (3.1) by the function

$$e^{\int_* a(x) \cdot_* d_* x}, \quad x \in J,$$

and we get

$$y^* \cdot_* e^{\int_* a(x) \cdot_* d_* x} +_* a(x) \cdot_* y(x) \cdot_* e^{\int_* a(x) \cdot_* d_* x} = b(x) \cdot_* e^{\int_* a(x) \cdot_* d_* x}, \quad x \in J.$$

Observe that

$$d_*/_* d_* x \left(y(x) \cdot_* e^{\int_* a(x) \cdot_* d_* x} \right)$$

$$= \; y^*(x) \cdot_* e^{\int_* a(x) \cdot_* d_* x} +_* a(x) \cdot_* y(x) \cdot_* e^{\int_* a(x) \cdot_* d_* x},$$

$x \in J$. Therefore,

$$d_*/_* d_* x \left(y(x) \cdot_* e^{\int_* a(x) \cdot_* d_* x} \right) = b(x) \cdot_* e^{\int_* a(x) \cdot_* d_* x}, \quad x \in J,$$

whereupon

$$y(x) \cdot_* e^{\int_* a(x) \cdot_* d_* x} = C_3 +_* \int_* b(x) \cdot_* e^{\int_* a(x) \cdot_* d_* x} d_* x, \quad x \in J.$$

Here C_3 is a multiplicative constant. Now, multiplicative multiplying both sides of the last equation by the function

$$e^{-*\int_* a(x)\cdot_* d_* x}, \quad x \in J,$$

we find the formula (3.5).

Example 3.1.2 *We consider the equation*

$$x \cdot_* y^* -_* e^2 \cdot_* y = e^2 \cdot_* x^{4*}, \quad x \in \mathbb{R}_*,$$

which we can rewrite in the form

$$y^* -_* (e^2/_* x) \cdot_* y = e^2 \cdot_* x^{3*}, \quad x \in \mathbb{R}_*.$$

Here

$$a(x) = -_*(e^2/_* x),$$

$$b(x) = e^2 \cdot_* x^{3*}, \quad x \in \mathbb{R}_*.$$

Then

$$\int_* a(x) \cdot_* d_* x = -_* e^2 \cdot_* \int_* d_* x/_* x$$

$$= -_* e^2 \cdot_* \log_* |x|_*$$

$$= -_* \log_*(x^{2*}), \quad x \in \mathbb{R}_*,$$

and

$$\int_* b(x) \cdot_* e^{\int_* a(x) \cdot_* d_* x} \cdot_* d_* x = e^2 \cdot_* \int_* x^{3*} \cdot_* e^{-*\log_*(x^{2*})} \cdot_* d_* x$$

$$= e^2 \cdot_* \int_* x^{3*} \cdot_* (e/_* x^{2*}) \cdot_* d_* x$$

$$= e^2 \cdot_* \int_* x \cdot_* d_* x$$

$$= x^{2_*}, \quad x \in \mathbb{R}_*.$$

Consequently,

$$y(x) = e^{\log_*(x^{2_*})} \cdot_* (C +_* x^{2_*})$$

$$= x^{2_*} \cdot_* (C +_* x^{2_*}), \quad x \in \mathbb{R}_*.$$

Example 3.1.3 *Now we consider the equation*

$$(e^2 \cdot_* x +_* e) \cdot_* y^* = e^4 \cdot_* x +_* e^2 \cdot_* y, \quad x > e^{-\frac{1}{2}}.$$

We can rewrite this equation in the form

$$y^* -_* (e^2 /_* (e^2 \cdot_* x +_* e)) \cdot_* y = (e^4 \cdot_* x) /_* (e^2 \cdot_* x +_* e), \quad x > e^{-\frac{1}{2}}.$$

Here

$$a(x) = -_* \frac{e^2}{/_*} -_* e^2 /_* (e^2 \cdot_* x +_* e),$$

$$b(x) = (e^4 \cdot_* x) /_* (e^2 \cdot_* x +_* e), \quad x > e^{-\frac{1}{2}}.$$

Then

$$\int_* a(x) \cdot_* d_* x = -_* e^2 \cdot_* \int_* (e /_* (e^2 \cdot_* x +_* e)) \cdot_* d_* x$$

$$= -_* \int_* ((d_* (e^2 \cdot_* x + *e)) /_* (e^2 \cdot_* x +_* e))$$

$$= -_* \log_* (e^2 \cdot_* x +_* e), \quad x > e^{-\frac{1}{2}}$$

and

$$\int_* b(x) \cdot_* e^{\int_* a(x) \cdot_* d_* x} \cdot_* d_* x = \int_* ((e^4 \cdot_* x) /_* (e^2 \cdot_* x +_* e))$$

$$\cdot_* e^{-_* \log_* (e^2 \cdot_* x +_* e)} \cdot_* d_* x$$

$$= \int_* ((e^4 \cdot_* x)/_*(e^2 \cdot_* x +_* e)^{2*}) \cdot_* d_* x$$

$$= e^2 \cdot_* \int_* (x/_*(e^2 \cdot_* x +_* e)^{2*})$$

$$\cdot_* d_* (e^2 \cdot_* x +_* e)$$

$$= -_* e^2 \cdot_* \int_* x \cdot_* d_*(e/_*(e^2 \cdot_* x +_* e))$$

$$= -_* ((e^2 \cdot_* x)/_*(e^2 \cdot_* x +_* e))$$

$$+_* e^2 \cdot_* \int_* (e/_*(e^2 \cdot_* x +_* e)) \cdot_* d_* x$$

$$= -_* ((e^2 \cdot_* x)/_*(e^2 \cdot_* x +_* e))$$

$$+_* \log_*(e^2 \cdot_* x +_* e), \quad x > e^{-\frac{1}{2}}.$$

Consequently,

$$y(x) = e^{\log_*(e^2 \cdot_* x +_* e)} \cdot_* \left(C -_* ((e^2 \cdot_* x)/_*(e^2 \cdot_* x +_* e)) \right.$$

$$\left. +_* \log_*(e^2 \cdot_* x +_* e) \right)$$

$$= (e^2 \cdot_* x +_* e) \cdot_* \left(C -_* ((e^2 \cdot_* x)/_*(e^2 \cdot_* x +_* e)) \right.$$

$$\left. +_* \log_*(e^2 \cdot_* x +_* e) \right), \quad x > e^{-\frac{1}{2}},$$

are the solutions of the considered equation. Here C is a multiplicative constant.

Example 3.1.4 *Now we consider the equation*

$$(x \cdot_* y +_* e^x) \cdot_* d_* x -_* x \cdot_* d_* y = 0_*, \quad x > 0_*,$$

which we can rewrite in the form

$$x \cdot_* y +_* e^x -_* x \cdot_* y^* = 0_*, \quad x > 0_*,$$

or

$$x \cdot_* y^* -_* x \cdot_* y = e^x, \quad x > 0_*,$$

or

$$y^* -_* y = (e/_* x) \cdot_* e^x, \quad x > 0_*.$$

Here

$$a(x) \;=\; e^{-1},$$

$$b(x) \;=\; (e/_* x) \cdot_* e^x, \quad x > 0_*.$$

Then

$$\int_* a(x) \cdot_* d_* x \;=\; -_* \int_* d_* x$$

$$=\; -_* x, \quad x > 0_*,$$

and

$$\int_* b(x) \cdot_* e^{\int_* a(x) \cdot_* d_* x} \cdot_* d_* x \;=\; \int_* (e/_* x) \cdot_* e^x \cdot_* e^{-_* x} \cdot_* d_* x$$

$$=\; \int_* (e/_* x) \cdot_* d_* x$$

$$=\; \log_* x, \quad x > 0_*.$$

Consequently,

$$y(x) \;=\; e^x \cdot_* (C +_* \log_* x), \quad x > 0_*,$$

are the solutions of the considered equation. Here C is a multiplicative constant.

Exercise 3.1.5 *Solve the equations*

$$1.\; x^{2*} \cdot_* y^* +_* x \cdot_* y +_* e = 0_*,\, x > 0_*.$$

2. $y = x \cdot_* (y^* -_* x \cdot_* \cos_* x)$, $x \in \mathbb{R}_*$.

3. $e^2 \cdot_* x \cdot_* (x^{2*} +_* y) \cdot_* d_* x = d_* y$, $x \in \mathbb{R}_*$.

Answer

1. $x \cdot_* y = C -_* \log_* x$, $x > 0_*$.

2. $y = x \cdot_* (C +_* \sin_* x)$, $x \in \mathbb{R}_*$.

3. $y = C \cdot_* e^{x^{2*}} -_* x^{2*} -_* e$, $x \in \mathbb{R}_*$.

Here C is a multiplicative constant. The equation

$$(x \cdot_* f(y) +_* g(y)) \cdot_* y^* = h(y)$$

may not be integrable as it is. Sometimes, if the roles of x and y are interchanged, then this equation can be rewritten as

$$h(y) \cdot_* x^* = x \cdot_* f(y) +_* g(y),$$

which is a first-order multiplicative linear differential equation in x.

Example 3.1.6 *We consider the equation*

$$\left(e^2 \cdot_* e^y -_* x\right) \cdot_* y^* = e.$$

This equation we can rewrite in the form

$$x^* = -_* x +_* e^2 \cdot_* e^y$$

or

$$x^* +_* x = e^2 \cdot_* e^y.$$

Here

$$a(y) \;=\; e,$$

$$b(y) \;=\; e^2 \cdot_* e^y.$$

Then

$$\int_* a(y) \cdot_* d_* y \;=\; \int_* d_* y$$

$$= \; y,$$

and

$$\int_* b(y) \cdot_* e^{\int_* a(y) \cdot_* d_*y} \cdot_* d_*y \;=\; e^2 \cdot_* \int_* e^y \cdot_* e^y \cdot_* d_*y$$

$$= \; e^2 \cdot_* \int_* e^{e^2 \cdot_* y} \cdot_* d_*y$$

$$= \; e^{e^2 \cdot_* y}.$$

Consequently,

$$x = e^{-*y}\left(C +_* e^{e^2 \cdot_* y}\right)$$

are the solutions of the considered equation. Here C is a multiplicative constant.

Example 3.1.7 *We consider the equation*

$$(x +_* y^{2*}) \cdot_* d_*y = y \cdot_* d_*x, \quad y > 0_*.$$

We can rewrite this equation in the following form

$$y \cdot_* x^* = x +_* y^{2*},$$

or

$$x^* = (e/_*y) \cdot_* x +_* y.$$

Here

$$a(y) \;=\; -_*(e/_*y),$$

$$b(y) \;=\; y.$$

Then

$$\int_* a(y) \cdot_* d_*y \;=\; -_* \int_* (e/_*y) \cdot_* d_*y$$

$$= \; -_* \log_* y,$$

and

$$\int_* b(y) \cdot_* e^{\int_* a(y) \cdot_* d_*y} \cdot_* d_*y \;=\; \int_* y \cdot_* e^{-_* \log_* y} \cdot_* d_*y$$

$$= \int_* d_* \cdot_* y$$

$$= y.$$

Consequently,

$$x = e^{\log_* y} \cdot_* (C +_* y)$$

$$= y \cdot_* (C +_* y)$$

are the solutions of the considered equation. Here C is a multiplicative constant.

Example 3.1.8 *We consider the equation*

$$((\sin_* y)^{2*} +_* x \cdot_* \cot_* y) \cdot_* y^* = e, \quad \sin_* y > 0_*.$$

We can rewrite this equation in the form

$$x^* = x \cdot_* \cot_* y +_* (\sin_* y)^{2*}, \quad \sin_* y > 0_*,$$

or

$$x^* -_* x \cdot_* \cot_* y = (\sin_* y)^{2*}, \quad \sin_* y > 0_*.$$

Here

$$a(y) = -_* \cot_* y,$$

$$b(y) = (\sin_* y)^{2*}, \quad \sin_* y > 0_*.$$

Then

$$\int_* a(y) \cdot_* d_* y = -_* \int_* (\cos_* y /_* \sin_* y) \cdot_* d_* y$$

$$= -_* \log_* (\sin_* y), \quad \sin_* y > 0_*,$$

and

$$\int_* b(y) \cdot_* e^{\int_* a(y) \cdot_* d_* y} \cdot_* d_* y = \int_* (\sin_* y)^{2*} \cdot_* e^{-_* \log_* (\sin_* y)} \cdot_* d_* y$$

$$= \int_* (\sin_* y)^{2*} \cdot_* (e/_* \sin_* y) \cdot_* d_* y$$

$$= \int_* \sin_* y \cdot_* d_* y$$

$$= -_* \cos_* y, \quad \sin_* y > 0_*.$$

Consequently,

$$x = e^{\log_*(\sin_* y)} \cdot_* (C -_* \cos_* y)$$

$$= \sin_* y \cdot_* (C -_* \cos_* y), \quad \sin_* y > 0_*.$$

Here C is a multiplicative constant.

Exercise 3.1.9 *Solve the equations*

1. $(e^2 \cdot_* x +_* y) \cdot_* d_* y = y \cdot_* d_* x +_* e^4 \cdot_* \log_* y \cdot_* d_* y.$
2. $y^* \cdot_* (e^3 \cdot_* x -_* y^{2*}) = y.$
3. $(e -_* e^2 \cdot_* x \cdot_* y) \cdot_* y^* = y \cdot_* (y -_* e).$

Answer

1. $x = e^2 \cdot_* \log_* y -_* y +_* e +_* C \cdot_* y^{2*}.$
2. $x = C \cdot_* y^{3*} +_* y^{2*}, \ y = 0_*.$
3. $(y -_* e)^{2*} \cdot_* x = y -_* \log_*(C \cdot_* y), \ y = 0_*, \ y = e.$

3.2   The Multiplicative Bernoulli Equation

Now, we consider the equation

$$y^* +_* a(x) \cdot_* y = b(x) \cdot_* y^{n*}, \qquad n \neq 0, 1, \qquad (3.6)$$

where $a, b : J \to \mathbb{R}_*, \ a, b \in \mathcal{C}(J)$.

Definition 3.2.1 *The equation (3.6) is called the multiplicative Bernoulli equation.*

If $n > 0$, then $y = 0_*$ is a solution to the equation (3.6). If $y \neq 0_*$, then (3.6) can be rewritten as follows:

$$y^{(-n)*} \cdot_* y^* +_* a(x) \cdot_* y^{(1-n)*} = b(x).$$

We set

$$v = y^{(1-n)*}.$$

Then

$$v^* = (1 - n)_* \cdot_* y^{(-n)*} \cdot_* y^*$$

or

$$y^{(-n)*} \cdot_* y^* = (e/_*(1 - n)_*) \cdot_* v^*.$$

From here, the equation (3.6) takes the form

$$(e/_*(1 - n)_*) \cdot_* v^* +_* a(x) \cdot_* v = b(x)$$

or

$$v^* +_* (1 - n)_* \cdot_* a(x) \cdot_* v = (1 - n)_* \cdot_* b(x),$$

which is a first-order multiplicative linear differential equation.

Example 3.2.2 *We consider the equation*

$$y^* +_* e^2 \cdot_* y = y^{2*} \cdot_* e^x, \quad x \in \mathbb{R}_*.$$

We note that $y = 0_*$ is a solution of the considered equation. Let $y \neq 0_*$. Then we can rewrite the given equation in the following way:

$$(e/_*y^{2*}) \cdot_* y^* +_* (e^2/_*y) = e^x, \quad x \in \mathbb{R}_*.$$

We set

$$v = e/_*y, \quad x \in \mathbb{R}_*.$$

Then

$$v^* = -_*(y^*/_*y^{2*}), \quad x \in \mathbb{R}_*,$$

or

$$y^*/_*y^{2*} = -_*v^*, \quad x \in \mathbb{R}_*.$$

Therefore,

$$-_*v^*(x) +_* e^2 \cdot_* v = e^x, \quad x \in \mathbb{R}_*,$$

or

$$v^* -_* e^2 \cdot_* v = -_* e^x, \quad x \in \mathbb{R}_*.$$

Here

$$a(x) \;=\; -_* e^2,$$

$$b(x) \;=\; -_* e^x, \quad x \in \mathbb{R}_*.$$

Then

$$\int_* a(x) \cdot_* d_* x \;=\; -_* e^2 \cdot_* \int_* d_* x$$

$$=\; -_* e^2 \cdot_* x, \quad x \in \mathbb{R}_*,$$

and

$$\int_* b(x) \cdot_* e^{\int_* a(x) \cdot_* d_* x} d_* x \;=\; -_* \int_* e^x \cdot_* e^{-_* e^2 \cdot_* x} \cdot_* d_* x$$

$$=\; -_* \int_* e^{-_* x} \cdot_* d_* x$$

$$=\; e^{-_* x}, \quad x \in \mathbb{R}_*.$$

Consequently,

$$v(x) = e^{e^2 \cdot_* x} \cdot_* (c +_* e^{-_* x}), \quad x \in \mathbb{R}_*,$$

whereupon, if $y \neq 0_*$,

$$y(x) = e/_* (e^{e^2 \cdot_* x} \cdot_* (c +_* e^{-_* x})), \quad x \in \mathbb{R}_*.$$

Here c is a multiplicative constant.

Example 3.2.3 *We consider the equation*

$$(x +_* e)(y^* +_* y^{2*}) = -_* y, \quad x > e^{-1}.$$

We can rewrite it in the following way

$$y^* +_* y^{2*} = -_* (e/_* (x +_* e)) \cdot_* y, \quad x > e^{-1},$$

or

$$y^* +_* (e/_*(x+_* e)) \cdot_* y = -_* y^{2*}, \quad x > e^{-1}.$$

We note that $y = 0_*$ *is a solution. Let* $y \neq 0_*$. *Then*

$$(e/_* y^{2*}) \cdot_* y^* +_* (e/_*(x+_* e)) \cdot_* (e/_* y) = e^{-1}, \quad x > e^{-1}.$$

We set

$$v = e/_* y.$$

Then

$$v^* = -_*(y^*/_* y^{2*}),$$

or

$$y^*/_* y^{2*} = -_* v^*.$$

Hence,

$$-_* v^* +_* (e/_*(x+_* e)) \cdot_* v = e^{-1},$$

or

$$v^* -_* (e/_*(x+_* e)) \cdot_* v = e.$$

Here

$$a(x) \;=\; -_*(e/_*(x+_* e)),$$

$$b(x) \;=\; e, \quad x > e^{-1}.$$

Then

$$\int_* a(x) \cdot_* d_* x \;=\; -_* \int_* (e/_*(x+_* e)) \cdot_* d_* x$$

$$=\; -_* \log_*(x+_* e), \quad x > e^{-1},$$

and

$$\int_* b(x) \cdot_* e^{\int_* a(x) \cdot_* d_* x} \cdot_* d_* x \;=\; \int_* e^{-_* \log_*(x+_* e)} \cdot_* d_* x$$

$$=\; \int_* (e/_*(x+_* e)) \cdot_* d_* x$$

$$= \log_*(x+_*e), \quad x > e^{-1}.$$

Consequently,

$$v(x) = e^{-*\log_*(x+_*e)} \cdot_* (c+_*\log_*(x+_*e))$$

$$= (e/_*(x+_*e)) \cdot_* (c+_*\log_*(x+_*e)), \quad x > e^{-1},$$

whereupon, if $y \neq 0_*$,

$$y(x) = (x+_*e)/_*(c+_*\log_*(x+_*e)), \quad x > e^{-1}.$$

Here c is a multiplicative constant.

Example 3.2.4 *Now we consider the equation*

$$y^* \cdot_* x^{3*} \cdot_* \sin_* y = x \cdot_* y^* -_* e^2 \cdot_* y, \quad y > 0_*.$$

We can rewrite it in the following form

$$y^* \cdot_* (x^{3*} \cdot_* \sin_* y -_* x) = e^{-2} \cdot_*, \quad y > 0_*,$$

or

$$e^2 \cdot_* y \cdot_* x^* = x -_* x^{3*} \cdot_* \sin_* y, \quad y > 0_*,$$

or

$$x^* = (e/_*(e^2 \cdot_* y)) \cdot_* x -_* x^{3*} \cdot_* (\sin_* y/_*(e^2 \cdot_*)y), \quad y > 0_*,$$

or

$$x^* -_* (e/_*(e^2 \cdot_* y)) \cdot_* x = -_* x^{3*} \cdot_* (\sin_* y/_*(e^2 \cdot_* y)), \quad y > 0_*.$$

We note that $x = 0_*$ *is a solution. Let* $x \neq 0_*$*. Then*

$$(e/_*x^{3*})/_*x^* -_* (e/_*(e^2 \cdot_* y)) \cdot_* (e/_*x^{2*}) = -_*(\sin_* y/_*(e^2 \cdot_* y)), \quad y > 0_*.$$

We set $v = e/_*x^{2*}$*. Hence,*

$$v^* = e^{-2} \cdot_* (x^*/_*x^{3*}),$$

or

$$x^*/_*x^{3*} = -_*(v^{*\cdot}_*e^2).$$

Therefore,

$$-_*(v^*/_*e^2) -_* (e/_*(e^2 \cdot _*y)) \cdot_* v = -_*(\sin_* y/_*(e^2 \cdot _*y)), \quad y > 0_*,$$

or

$$v^* +_* (e/_*y) \cdot_* v = \sin_* y/_*y, \quad y > 0_*.$$

Here

$$a(y) = e/_*y,$$

$$b(y) = \sin_* y/_*y, \quad y > 0_*.$$

Then

$$\int_* a(y) \cdot_* d_*y = \int_* e/_*y \cdot_* d_*y$$

$$= \log_* y, \qquad y > 0_*,$$

and

$$\int_* b(y) \cdot_* e^{\int_* a(y) \cdot_* d_*y} \cdot_* dy = \int_* (\sin_* y/_*y) \cdot_* e^{\log_* y} \cdot_* d_*y$$

$$= \int_* (\sin_* y/_*y) \cdot_* y \cdot_* d_*y$$

$$= \int_* \sin_* y \cdot_* d_*y$$

$$= -_* \cos_* y, \quad y > 0_*.$$

Consequently,

$$v(y) = e^{-_* \log_* y} \cdot_* (c -_* \log_* y)$$

$$= (e/_*y) \cdot_* (c -_* \log_* y), \quad y > 0_*,$$

whereupon, if $x \neq 0_*$,

$$x^{2*}(y) = y/_*(c -_* \cos_* y), \quad y > 0_*.$$

Example 3.2.5 *We consider the equation*

$$e^2 \cdot_* y^* -_* (x/_* y) = (x \cdot_* y)/_* (x^{2*} -_* e), \quad y > 0_*, \quad x > e.$$

This equation we can rewrite in the form

$$e^2 \cdot_* y \cdot_* y^* -_* x = (x/_* (x^{2*} -_* e)) \cdot_* y^{2*}, \quad y > 0_*, \quad x > e,$$

or

$$e^2 \cdot_* y \cdot_* y^* -_* (x/_* (x^{2*} -_* e)) \cdot_* y^{2*} = x, \quad y > 0_*, \quad x > e.$$

We set

$$v = y^{2*}.$$

Then

$$v^* = e^2 \cdot_* y \cdot_* y^*, \quad y > 0_*, \quad x > e.$$

Hence,

$$v^* -_* (x/_* (x^{2*} -_* e)) \cdot_* v = x, \quad y > 0_*, \quad x > e.$$

Here

$$a(x) = -_*(x/_*(x^{2*} -_* e)),$$

$$b(x) = x, \quad x > e.$$

Then

$$\int_* a(x) \cdot_* d_*x = -_* \int_* (x/_*(x^{2*} -_* e)) \cdot_* d_*x$$

$$= e^{-\frac{1}{2}} \cdot_* \int_* d_*(x^{2*} -_* e)/_*(x^{2*} -_* e)$$

$$= \log_* e/_*(x^{2*} -_* e)^{\frac{1}{2}*}, \quad x > e,$$

and

$$\int_* b(x) \cdot_* e^{\int_* a(x) \cdot_* d_*x} \cdot_* d_*x = \int_* x \cdot_* e^{\log_*(e/_*(x^{2*} -_* e)^{\frac{1}{2}*})} \cdot_* d_*x$$

$$= \int_* (x/_*(x^{2*} -_* e)^{\frac{1}{2}*}) \cdot_* d_* x$$

$$= e^{\frac{1}{2}} \cdot_* \int_* (d_*(x^{2*} -_* e)/_*(x^{2*} -_* e)^{\frac{1}{2}*})$$

$$= (x^{2*} -_* e)^{\frac{1}{2}*}, \quad x > e.$$

Consequently,

$$v(x) = e^{-_* \log_*(e/_*(x^{2*} -_* e))} \cdot_* (c +_* (x^{2*} -_* e)^{\frac{1}{2}*})$$

$$= (x^{2*} -_* e)^{\frac{1}{2}*} \cdot_* (c +_* (x^{2*} -_* e)^{\frac{1}{2}*}), \quad x > e,$$

whereupon

$$(y(x))^2 = (x^{2*} -_* e)^{\frac{1}{2}*} \cdot_* (c +_* (x^{2*} -_* e)^{\frac{1}{2}*}), \quad x > e.$$

Exercise 3.2.6 *Solve the equations*

1. $y^* = y^{4*} \cdot_* \cos_* x +_* y \tan_* x,\ x \in \mathbb{R}_*.$
2. $x \cdot_* y^{2*} \cdot_* y^* = x^{2*} +_* y^{3*},\ x \in \mathbb{R}_*.$
3. $x \cdot_* y \cdot_* d_* y = (y^{2*} +_* x) \cdot_* d \cdot_* x,\ x \in \mathbb{R}_*.$

Answer

1. $(y(x))^{-*3*} = c \cdot_* (\cos_* x)^{3*} -_* e^3 \cdot_*$
   $\sin_* x \cdot_* (\cos_* x)^{2*},\ y = 0_*,\ x \in \mathbb{R}_*.$
2. $(y(x))^{3*} = c \cdot_* x^{3*} -_* e^3 \cdot_* x^{2*},\ x \in \mathbb{R}_*.$
3. $(y(x))^{2*} = c \cdot_* x^{2*} -_* e^2 \cdot_* x,\ x \in \mathbb{R}_*,\ x = 0_*.$

3.3 The Multiplicative Riccati Equation

Definition 3.3.1 *The equation*

$$y^* = a(x) \cdot_* y^{2*} +_* b(x) \cdot_* y +_* c(x), \quad x \in J, \tag{3.7}$$

where $a, b, v : J \to \mathbb{R}_*$, $a, b, c \in \mathscr{C}(J)$, is called the multiplicative Riccati equation.

If y_1 is a solution of the multiplicative Riccati equation (3.7), then the substitution

$$y(x) = y_1(x) +_* e/_* v(x), \quad x \in J, \tag{3.8}$$

converts it into a first-order multiplicative linear differential equation. Really,

$$y^*(x) = y_1^*(x) -_* (v^*(x)/_*(v(x))^{2*}), \quad x \in J, \tag{3.9}$$

$$y_1^*(x) = a(x) \cdot_* (y_1(x))^{2*} +_* b(x) \cdot_* y_1(x) +_* c(x), \quad x \in J. \tag{3.10}$$

We put (3.8) and (3.9) in the equation (3.7) and using (3.10), we get

$$y_1^*(x) -_* (v^*(x)/_*(v(x))^{2*}) \quad = \quad a(x) \cdot_* (y_1(x) +_* (e/_* v(x)))^{2*}$$

$$+_* b(x) \cdot_* (y_1(x) +_* (e/_* v(x))) +_* c(x),$$

$x \in J$, or

$$y_1^*(x) -_* (v^*(x)/_*(v(x))^{2*}) = a(x) \cdot_* \left((y_1(x))^{2*} +_* e^2 \cdot_* y_1(x) \right.$$

$$\left. \cdot_* (e/_* v(x)) +_* (e/_*(v(x))^{2*}) \right)$$

$$+ * b(x) \cdot_* (y_1(x) +_* (e/_* v(x))) +_* c(x), \quad x \in J,$$

or 1

$$y_1^*(x) -_* (v^*(x)/_*(v(x))^{2*}) = a(x) \cdot_* (y_1(x))^{2*}$$
$$+_* e^2 \cdot_* a(x) \cdot_* y_1(x) \cdot_* (e/_* v(x))$$

$$+_* a(x) \cdot_* (e/_*(v(x))^{2*}) +_* b(x) \cdot_* y_1(x)$$

$$+_* b(x) \cdot_* (e/_* v(x)) +_* c(x), \quad x \in J,$$

or

$$\left(y_1^*(x) -_* a(x) \cdot_* (y_1(x))^{2*} -_* b(x) \cdot_* y_1(x) -_* c(x) \right)$$

$$-_* (v^*(x)/_*(v(x))^{2*})$$

$$= \quad e^2 \cdot_* a(x) \cdot_* y_1(x) \cdot_* (e/_* v(x))$$

$$+_* a(x) \cdot_* (e/_* (v(x))^{2*}) +_* b(x) \cdot_* (e/_* v(x)), \quad x \in J,$$

or

$$-_* (v^*(x)/_* (v(x))^{2*} \quad = \quad \left(e^2 \cdot_* a(x) \cdot_* y_1(x) +_* b(x) \right) \cdot_* (e/_* v(x))$$

$$+_* a(x) \cdot_* (e/_* (v(x))^{2*}), \quad x \in J,$$

or

$$v^*(x) +_* \left(e^2 \cdot_* a(x) \cdot_* y_1(x) +_* b(x) \right) \cdot_* v(x) = -_* a(x), \quad x \in J.$$

Example 3.3.2 *We consider the equation*

$$x^{2*} \cdot_* y^* +_* x \cdot_* y +_* x^{2*} \cdot_* y^{2*} = e^4, \quad x > 0_*.$$

We will show that

$$y_1(x) = e^2/_* x, \quad x > 0_*,$$

is its solution. Really,

$$y_1^*(x) = -_* (e^2/_* x^{2*}), \quad x > 0_*,$$

and

$$x^{2*} \cdot_* y_1^*(x) +_* x \cdot_* y_1(x) +_* x^{2*} \cdot_* (y_1(x))^{2*}$$

$$= \quad x^{2*} \cdot_* \left(-_* (e^2/_* x^{2*}) \right)$$

$$+_* x \cdot_* (e^2/_* x) +_* x^{2*} \cdot_* (e^4/_* x^{2*})$$

$$= \quad -_* e^2 +_* e^2 +_* e^4$$

$$= \quad e^4, \quad x > 0_*.$$

Now, we will find the general solution of the considered equation. For this aim we make the substitution

$$y(x) = (e^2/_* x) +_* v(x), \quad x > 0_*.$$

Then

$$y^*(x) = -_*(e^2/_*x^{2*}) -_* (v^*(x)/_*(v(x))^{2*}), \quad x > 0_*,$$

and the considered equation takes the form

$$x^{2*} \cdot_* \left(-_*(e^2/_*x^{2*}) -_* (v^*(x)/_*(v(x))^{2*})\right)$$

$$+_*x \cdot_* \left(e^2/_*x +_* (e/_*v(x))\right)$$

$$+_*x^{2*} \cdot_* \left(\,^{(}e^2/_*x) +_* (e/_*v(x))\right)^{2*}$$

$$= \; e^4, \quad x > 0_*,$$

or

$$-_*e^2 -_* x^{2*} \cdot_* (v^*(x)/_*(v(x))^{2*}) +_* e^2 +_* (x/_*v(x))$$

$$+_*x^{2*} \cdot_* \left((e^4/_*x^{2*}) +_* (e^4/_*(x \cdot_* v(x)))\right.$$

$$\left. +_*(e/_*(v(x))^{2*})\right)$$

$$= \; e^4, \quad x > 0_*,$$

or

$$-_*x^{2*} \cdot_* (v^*(x)/_*(v(x))^{2*}) +_* (x/_*v(x)) +_* e^4$$

$$+_*((e^4 \cdot_* x)/_*v(x)) +_* (x^{2*}/_*(v(x))^{2*})$$

$$= \; e^4, \quad x > 0_*,$$

or

$$-_*x^{2*} \cdot_* (v^*(x)/_*(v(x))^{2*}) +_* ((e^5 \cdot_* x)/_*v(x))$$

$$+_*(x^{2*}/_*(v(x))^{2*})$$

$$=\ 0_*, \quad x > 0_*,$$

or

$$-_*x^{2*} \cdot_* v^*(x) +_* e^5 \cdot_* x \cdot_* v(x) +_* x^{2*} = 0_*, \quad x > 0_*,$$

or

$$v^*(x) -_* (e^5/_*x) \cdot_* v = e, \quad x > 0_*.$$

Here

$$a(x)\ =\ -_*(e^5/_*x),$$

$$b(x)\ =\ 1, \quad x > 0_*.$$

Then

$$\int_* a(x) \cdot_* d_*x\ =\ -_*e^5 \cdot_* \int_* (e/_*x) \cdot_* d_*x$$

$$=\ -_*e^5 \cdot_* \log_* x$$

$$=\ \log_*(x^{5*}), \quad x > 0_*,$$

and

$$\int_* b(x) \cdot_* e^{\int_* a(x) \cdot_* d_*x} \cdot_* d_*x\ =\ \int_* e^{-_*\log_*(e/_*x^{5*})} \cdot_* d_*x$$

$$=\ \int_* x^{5*} \cdot_* d_*x$$

$$=\ e^{\frac{1}{6}} \cdot_* x^{6*}, \quad x > 0_*.$$

Consequently,

$$v(x)\ =\ e^{-_*\log_*(e/_*x^{5*})} \cdot_* \left(c +_* e^{\frac{1}{6}} \cdot_* x^{6*}\right)$$

$$=\ x^{5*} \cdot_* \left(c +_* e^{\frac{1}{6}} \cdot_* x^{6*}\right), \quad x > 0_*.$$

Hence,

$$y(x) = (e^2/{}_*x) +{}_* x^{5*} \cdot{}_* \left(c +{}_* e^{\frac{1}{6}} \cdot{}_* x^{6*}\right), \quad x > 0_*.$$

Here c is a multiplicative constant.

Example 3.3.3 *Now, we consider the equation*

$$e^3 \cdot{}_* y^* +{}_* y^{2*} +{}_* (e^2/{}_*x^{2*}) = 0_*, \quad x > 0_*.$$

We will show that

$$y_1(x) = e/{}_*x, \quad x > 0_*,$$

is its solution. Really,

$$y_1^*(x) = -{}_*(e/{}_*x^{2*}), \quad x > 0_*,$$

and

$$e^3 \cdot{}_* \left(-{}_*(e/{}_*x^{2*})\right) +{}_* (e/{}_*x^{2*}) +{}_* (e^2/{}_*x^{2*} = 0_*,$$

$x > 0_*.$ *Now we will find the general solution of the considered equation. We set*

$$\begin{aligned} y(x) &= y_1(x) +{}_* (e/{}_*v(x)) \\[2mm] &= (e/{}_*x) +{}_* (e/{}_*v(x)) \quad x > 0_*. \end{aligned}$$

Then

$$y^*(x) = -{}_*(e/{}_*x^{2*}) -{}_* (v^*(x)/{}_*(v(x))^{2*}), \quad x > 0_*.$$

Therefore, the considered equation takes the form

$$\begin{aligned} e^3 \cdot{}_* \left(-{}_*(e/{}_*x^{2*}) -{}_* (v^*(x)/{}_*(v(x))^{2*})\right) & \\[2mm] +{}_* ((e/{}_*x) +{}_* (e/{}_*v(x)))^{2*} +{}_* (e^2/{}_*x^{2*}) & \\[2mm] = 0, \quad x > 0_*. \end{aligned}$$

or

$$-_*(e^3/_*x^{2*}) -_* e^3 \cdot_* (v^*(x)/_*(v(x))^{2*}) +_* (e/_*x^{2*})$$

$$+_*e^2 \cdot_* (e/_*(x \cdot_* v(x))) +_* (e/_*(v(x))^{2*}) +_* (e^2/_*x^{2*})$$

$$= 0_*, \quad x > 0_*,$$

or

$$-_*e^3 \cdot_* (v^*(x)/_*(v(x))^{2*}) +_* (e^2/_*(x \cdot_* v(x))) +_* (e/_*(v(x))^{2*}) = 0_*,$$

$x > 0_*,$

or

$$v^*(x) -_* (e^2/_*(e^3 \cdot_* x)) \cdot_* v(x) = e^{\frac{1}{3}}, \quad x > 0_*,$$

which is a first-order multiplicative linear differential equation. Here

$$a(x) = -_*(e^2/_*(e^3 \cdot_* x)),$$

$$b(x) = e^{\frac{1}{3}}, \quad x > 0_*.$$

Then

$$\int_* a(x) \cdot_* d_*x = -_*e^{\frac{2}{3}} \cdot_* \int_* (d_*x/_*x)$$

$$= \log_* x^{e^{-\frac{2}{3}}}, \quad x > 0_*,$$

and

$$\int_* b(x) \cdot_* e^{\int_* a(x) \cdot_* d_*x} \cdot_* d_*x = e^{\frac{1}{3}} \cdot_* \int_* e^{\log_* x^{e^{-\frac{2}{3}}}} \cdot_* d_*x$$

$$= e^{\frac{1}{3}} \cdot_* \int_* x^{e^{-\frac{2}{3}}} \cdot_* d_*x$$

$$= x^{e^{\frac{1}{3}}}, \quad x > 0_*.$$

Consequently,

$$v(x) \;=\; e^{-*\log_* x^{e^{-\frac{2}{3}}}} \cdot_* \left(c +_* x^{e^{\frac{1}{3}}}\right)$$

$$=\; x^{e^{\frac{2}{3}}} \cdot_* \left(c +_* x^{e^{\frac{1}{3}}}\right)$$

$$=\; c \cdot_* x^{e^{\frac{2}{3}}} +_* x, \quad x > 0_*.$$

Hence,

$$y(x) = (e/_* x) +_* (e/_* (c \cdot_* x^{e^{\frac{2}{3}}} +_* x)), \quad x > 0_*.$$

Here c is a multiplicative constant.

Example 3.3.4 *We consider the equation*

$$x \cdot_* y^* -_* (e^2 \cdot_* x +_* e) \cdot_* y +_* y^{2*} = -_* x^{2*}, \qquad x > 0_*.$$

We will show that

$$y_1(x) = x, \quad x > 0_*$$

is its solution. Really,

$$y_1^*(x) = e, \quad x > 0_*,$$

and

$$x \cdot_* y_1^*(x) -_* (e^2 \cdot_* x +_* e) \cdot_* y_1(x) +_* (y_1(x))^{2*}$$

$$=\; x -_* (e^2 \cdot_* x +_* e) \cdot_* x +_* x^{2\cdot 8}$$

$$=\; -_* x^{2*}, \quad x > 0_*.$$

Now, we will find its general solution. We set

$$y(x) \;=\; y_1(x) +_* (e/_* v(x))$$

$$=\; x +_* (e/_* v(x)), \quad x > 0_*.$$

Then

$$y^*(x) = e -_* (v^*(x)/_*(v(x))^{2*}), \quad x > 0_*,$$

and the considered equation takes the form

$$x \cdot_* \left(e -_* (v^*(x)/_*(v(x))^{2*}) \right)$$

$$-_*(e^2 \cdot_* x +_* e) \cdot_* (x +_* (e/_* v(x)))$$

$$+_* (x +_* (e/_* v(x)))^{2*}$$

$$= -_* x^{2*}, \quad x > -0_*,$$

or

$$x -_* x \cdot_* (v^*(x)/_*(v(x))^{2*}) -_* e^2 \cdot_* x^{2*} -_* x$$

$$-_*(e^2 \cdot_* x +_* e) \cdot_* (e/_* v(x)) +_* x^{2*}$$

$$+_*((e^2 \cdot_* x)/_* v(x)) +_* (e/_*(v(x))^{2*})$$

$$= -_* x^{2*}, \quad x > 0_*,$$

or

$$-_* x \cdot_* (v^*(x)/_*(v(x))^{2*}) -_* (e^2 \cdot_* x +_* e) \cdot_* (e/_* v(x))$$

$$+_*((e^2 \cdot_* x)/_* v(x)) +_* (e/_*(v(x))^{2*})$$

$$= 0_*, \quad x > 0_*,$$

or

$$-_* x \cdot_* (v^*(x)/_*(v(x))^{2*}) -_* (e/_* v(x)) +_* (e/_*(v(x))^{2*}) = 0_*,$$

$$x > 0_*,$$

or

$$-_* x \cdot_* v^*(x) -_* v(x) +_* e = 0_*, \quad x > 0_*,$$

or

$$v^*(x) +_* (e/_*x) \cdot_* v(x) = 1/_*x, \quad x > 0_*,$$

which is a first-order multiplicative linear differential equation. Here

$$a(x) \;=\; e/_*x,$$

$$b(x) \;=\; e/_*x, \quad x > 0_*.$$

Then

$$\int_* a(x) \cdot_* d_*x \;=\; \int_* (e/_*x) \cdot_* d_*x$$

$$=\; \log_* x, \quad x > 0_*,$$

and

$$\int_* b(x) \cdot_* e^{\int_* a(x) \cdot_* d_*x} \cdot_* d_*x \;=\; \int_* (e/_*x) \cdot_* e^{\log_* x} \cdot_* d_*x$$

$$=\; \int_* (e/_*x) \cdot_* x \cdot_* d_*x$$

$$=\; \int_* d_*x$$

$$=\; x, \quad x > 0_*.$$

Consequently,

$$v(x) \;=\; e^{-_* \log_* x} \cdot_* (c +_* x)$$

$$=\; (e/_*x) \cdot_* (c +_* x), \quad x > 0_*,$$

and

$$y(x) = x +_* (x/_*(c +_* x)), \quad x > 0_*.$$

Exercise 3.3.5 *Find the general solution of the following multiplicative Riccati equations.*

1. $y^* +_* y^{2*} -_* \cos_* x -_* (\sin_* x)^{2*} = 0_*$, $y_1(x) = \sin_* x$, $x \in \mathbb{R}_*$.

2. $y^* -_* e^2 \cdot_* x \cdot_* y +_* y^{2*} = e^5 -_* x^{2*}$, $y_1(x) = x +_* e^2$, $x \in \mathbb{R}_*$.

3. $y^* +_* e^2 \cdot_* y \cdot_* e^x -_* y^{2*} = e^{e^2 \cdot_* x} +_* e^x$, $y_1(x) = e^x$, $x \in \mathbb{R}_*$.

Answer

1.

$$y(x) = \sin_* x +_* e^{-_* e^2 \cdot_* \cos_* x} \cdot_* \left(c +_* \int_* e^{e^2 \cdot_* \cos_* x} \cdot_* d_* x \right), \quad x \in \mathbb{R}_*.$$

2.

$$y(x) = x +_* e^2 +_* (e^4 /_* (c \cdot_* e^{e^4 \cdot_* x} -_* e)), \quad x \in \mathbb{R}_*.$$

3.

$$y(x) = e^x -_* (e /_* (x +_* c)), \quad x \in \mathbb{R}_*.$$

3.4 Applications

In many problems, the terms a and b in the equation (3.1) are specified by different formulas in different intervals. This is often the case when (3.1) is considered as an input-output relation, i.e., the functions a and b are some inputs and the solution y is an output corresponding to the inputs a and b. Usually, the solution y is not defined at some certain points, so that it is not continuous throughout the interval of interest. Let

$$a(x) = \begin{cases} a_1(x), & x_0 \leq x \leq x_1, \\ a_2(x), & x_1 \leq x \leq x_2, \\ \cdots \\ a_k(x), & x_{k-1} \leq x \leq x_k, \end{cases}$$

$$b(x) = \begin{cases} b_1(x), & x_0 \leq x \leq x_1, \\[2mm] b_2(x), & x_1 \leq x \leq x_2, \\[2mm] \cdots \\[2mm] b_k(x), & x_{k-1} \leq x \leq x_k, \end{cases}$$

where a_1 and b_1 are positive continuous functions in $[x_0, x_1)$, a_l and b_l, $l = 2, \ldots, k$, are positive continuous functions in $(x_{l-1}, x_l]$. Then solution y of the equation (3.1) can be written in the form

$$y(x) = \begin{cases} y_1(x) = e^{-*\int_* a_1(x)\cdot_* d_* x} \cdot_* \left(c_1 +_* \int_* b_1(x) \cdot_* e^{\int_* a_1(x)\cdot_* d_* x} \cdot_* dx \right), \\[2mm] x_0 \leq x < x_1, \\[4mm] y_2(x) = e^{-*\int_* a_2(x)\cdot_* d_* x} \cdot_* \left(c_2 +_* \int_* b_2(x) \cdot_* e^{\int_* a_2(x)\cdot_* d_* x} \cdot_* d_* x \right), \\[2mm] x_1 < x \leq x_2, \\[4mm] \vdots \\[4mm] y_k(x) = e^{-*\int_* a_k(x)\cdot_* d_* x} \cdot_* \left(c_k +_* \int_* b_k(x) \cdot_* e^{\int_* a_k(x)\cdot_* d_* x} \cdot_* d_* x \right), \\[2mm] x_{k-1} < x \leq x_k. \end{cases}$$

If the limits

$$\lim_{x \to x_1^{-*}} y_1(x), \qquad \lim_{x \to x_1^{+*}} y_2(x),$$

$$\vdots$$

$$\lim_{x \to x_{k-1}^{-*}} y_{k-1}(x), \qquad \lim_{x \to x_k^{+*}} y_k(x),$$

exist, then the relations

$$\lim_{x \to x_1^{-*}} y_1(x) = \lim_{x \to x_1^{+*}} y_2(x),$$

$$\vdots$$

$$\lim_{x \to x_{k-1}^{-*}} y_{k-1}(x) \;=\; \lim_{x \to x_{k-1}^{+*}} y_k(x)$$

determine the multiplicative constants c_l, $1 \le l \le k$, so that the solution y is continuous on $[x_0, x_k]$.

Example 3.4.1 *We consider the equation*

$$y^* +_* e^2 \cdot_* y = \begin{cases} e & 1 \le x \le e \\ \\ 1 & x > e. \end{cases}$$

1. $x \in [1, e]$. Then we have the equation

$$y^* +_* e^2 \cdot_* y = e.$$

Here

$$a(x) \;=\; e^2,$$

$$b(x) \;=\; e, \quad x \in [1, e].$$

Then

$$\int_* a(x) \cdot_* d_* x \;=\; e^2 \cdot_* \int_* d_* x$$

$$=\; e^2 \cdot_* x, \quad x \in [1, e],$$

and

$$\int_* b(x) \cdot_* e^{\int_* a(x) \cdot_* d_* x} \cdot_* d_* x \;=\; \int_* e^{e^2 \cdot_* x} \cdot_* d_* x$$

$$=\; e^{\frac{1}{2}} \cdot_* e^{e^2 \cdot_* x}, \quad x \in [a, e].$$

Consequently,

$$y_1(x) \;=\; e^{-_* e^2 \cdot_* x} \cdot_* \left(c_1 +_* e^{\frac{1}{2}} \cdot_* e^{e^2 \cdot_* x} \right)$$

$$=\; c_1 \cdot_* e^{-_* e^2 \cdot_* x} +_* e^{\frac{1}{2}}, \quad x \in [1, e].$$

Here c_1 is a multiplicative constant.

2. $x > e$. *Then we have the equation*

$$y^* +_* e^2 \cdot_* y = 0_*, \quad x > e,$$

or

$$y^* = -_* e^2 \cdot_* y, \quad x > e,$$

or

$$d_* y /_* y = -_* e^2 \cdot_* d_* x, \quad x > e,$$

whereupon

$$\log_* |y|_* = -_* e^2 \cdot_* x +_* c_2, \quad x > e,$$

or

$$y_2(x) = c_2 \cdot_* e^{-_* e^2 \cdot_* x}, \quad x > e.$$

Here c_2 is a multiplicative constant. Hence,

$$\lim_{x \to e^{-*}} y_1(x) = c_1 \cdot_* e^{-_* e^2} +_* e^{\frac{1}{2}},$$

$$\lim_{x \to e^{+*}} y_2(x) = c_2 \cdot_* e^{-_* e^2}$$

and

$$\lim_{x \to e^{-*}} y_1(x) = \lim_{x \to e^{+*}} y_2(x),$$

whereupon

$$c_1 \cdot_* e^{-_* e^2} +_* e^{\frac{1}{2}} = c_2 \cdot_* e^{-_* e^2},$$

and

$$c_2 = c_1 +_* e^{\frac{1}{2}} \cdot_* e^{e^2}.$$

Therefore,

$$y(x) = \begin{cases} y_1(x) = c_1 \cdot_* e^{-_* e^2 \cdot_* x} +_* e^{\frac{1}{2}}, & 1 \le x \le e, \\[2mm] y_2(x) = \left(c_1 +_* e^{\frac{1}{2}} \cdot_* e^{e^2} \right) \cdot_* e^{-_* e^2 \cdot_* x}, & x > e, \end{cases}$$

is a continuous solution of the considered equation.

Example 3.4.2 *Now we consider the equation*

$$y^* +_* a(x) \cdot_* y = x, \qquad x \in [1, e^4],$$

where

$$a(x) = \begin{cases} e & 1 \le x < e \\ e/_*x & e < x \le e^4. \end{cases}$$

1. $x \in [1, e)$. Then we have the equation

$$y^* +_* y = x, \quad x \in [1, e).$$

Here

$$a(x) \;=\; 1,$$

$$b(x) \;=\; x, \quad x \in [1, e).$$

Then

$$\int_* a(x) \cdot_* d_* x \;=\; \int_* d_* x$$

$$=\; x, \quad x \in [1, e),$$

and

$$\int_* b(x) \cdot_* e^{\int_* a(x) \cdot_* d_* x} \cdot_* d_* x \;=\; \int_* x \cdot_* e^x \cdot_* d_* x$$

$$=\; x \cdot_* e^x -_* \int_* e^x \cdot_* d_* x$$

$$=\; (x -_* e) \cdot_* e^x, \quad x \in [1, e).$$

Consequently,

$$y_1(x) \;=\; e^{-_*x} \cdot_* (c_1 +_* (x -_* e) \cdot_* e^x)$$

$$=\; c_1 \cdot_* e^{-_*x} +_* x -_* e, \quad x \in [1, e).$$

Here c_1 is a multiplicative constant.

2. $x \in (e, e^4]$. *Then we have the equation*

$$y^* +_* (e/_* x) \cdot_* y = x, \quad x \in [e, e^4].$$

Here

$$a(x) = e/_* x,$$

$$b(x) = x, \quad x \in [e, e^4].$$

Then

$$\int_* a(x) \cdot_* d_* x = \int_* (e/_* x) \cdot_* d_* x$$

$$= \log_* x, \quad x \in [e, e^4],$$

and

$$\int_* b(x) \cdot_* e^{\int_* a(x) \cdot_* d_* x} \cdot_* d_* x = \int_* x \cdot_* e^{\log_* x} \cdot_* d_* x$$

$$= \int_* x^{2*} \cdot_* d_* x$$

$$= e^{\frac{1}{3}} \cdot_* x^{3*}, \quad x \in [e, e^4].$$

Consequently,

$$y_2(x) = e^{-_* \log_* x} \cdot_* \left(c_2 +_* e^{\frac{1}{3}} \cdot_* x^{3*} \right)$$

$$= c_2 \cdot_* (e/_* x) +_* e^{\frac{1}{3}} \cdot_* x^{2*}, \quad x \in [e, e^4].$$

Here c_2 is a multiplicative constant. Hence,

$$\lim_{x \to e^{-*}} y_1(x) = c_1 \cdot_* e^{-*e},$$

$$\lim_{x \to e^{+*}} y_2(x) = c_2 +_* e^{\frac{1}{3}}.$$

Therefore,

$$\lim_{x \to e^{-*}} y_1(x) = \lim_{x \to e^{+*}} y_2(x)$$

and

$$c_1 \cdot_* e^{-*e} = c_2 +_* e^{\frac{1}{3}},$$

and

$$c_2 = c_1 \cdot_* e^{-*e} -_* e^{\frac{1}{3}}.$$

Consequently,

$$y(x) = \begin{cases} c_1 \cdot_* e^{-*x} +_* x -_* e, & 1 \le x < e, \\ \left(c_1 \cdot_* e^{-*e} -_* e^{\frac{1}{3}} \right) \cdot_* (e/_*x) +_* e^{\frac{1}{3}} \cdot_* x^{2*}, \\ e \le x \le e^4, \end{cases}$$

is a continuous solution of the considered equation in $[1, e^4]$.

Example 3.4.3 *Now we consider the equation (3.1) for*

$$a(x) = \begin{cases} e & 1 \le x < e \\ e^{-1} & e \le x < e^2 \\ (e/_*x) & e^2 \le x, \end{cases}$$

$$b(x) = \begin{cases} x & 0 \le x < 3 \\ 2x & 3 \le x. \end{cases}$$

1. $x \in [1, e)$. Then we have the equation

$$y^* +_* y = x, \quad x \in [1, e).$$

Here

$$a(x) = 1,$$

$$b(x) = x, \quad x \in [1, e).$$

Then

$$\int_* a(x) \cdot_* d_*x = x, \quad x \in [1, e),$$

and

$$\int_* b(x) \cdot_* e^{\int_* a(x) \cdot_* d_* x} \cdot_* d_* x = \int_* x \cdot_* e^x \cdot_* d_* x$$

$$= (x -_* e) \cdot_* e^x, \quad x \in [1, e).$$

Consequently,

$$y_1(x) = e^{-_* x} \cdot_* (c_1 +_* (x -_* e) \cdot_* e^x)$$

$$= c_1 \cdot_* e^{-_* x} +_* x -_* e, \quad x \in [1, e).$$

Here c_1 is a multiplicative constant.

2. $x \in [e, e^2)$. Then we have the equation

$$y^* -_* y = x, \quad x \in [e, e^2).$$

Here

$$a(x) = e^{-1},$$

$$b(x) = x, \quad x \in [e, e^2).$$

Then

$$\int_* a(x) \cdot_* d_* x = -_* \int_* d_* x$$

$$= -_* x, \quad x \in [e, e^2),$$

and

$$\int_* b(x) \cdot_* e^{\int_* a(x) \cdot_* d_* x} \cdot_* d_* x = \int_* x \cdot_* e^{-_* x} \cdot_* d_* x$$

$$= -_* x \cdot_* e^{-_* x} +_* \int_* e^{-_* x} \cdot_* d_* x$$

$$= -_* (x +_* e) \cdot_* e^{-_* x}, \quad x \in [e, e^2).$$

Consequently,

$$y_2(x) \;=\; e^x \cdot_* \left(c_2 -_* (x +_* e) \cdot_* e^{-_*x} \right)$$

$$=\; c_2 \cdot_* e^x -_* (x +_* e), \quad x \in [e, e^2).$$

Here c_2 is a multiplicative constant.

3. $x \in [e^2, e^3)$. Then we have the equation

$$y^* +_* (e/_*x) \cdot_* y = x, \quad x \in [e^2, e^3).$$

Here

$$a(x) \;=\; (e/_*x),$$

$$b(x) \;=\; x, \quad x \in [e^2, e^3).$$

Then

$$\int_* a(x) \cdot_* d_*x \;=\; \int_* (e/_*x) \cdot_* d_*x$$

$$=\; \log_* x, \quad x \in [e^2, e^3),$$

and

$$\int_* b(x) \cdot_* e^{\int_* a(x) \cdot_* d_*x} \cdot_* d_*x \;=\; \int_* x \cdot_* e^{\log_* x} \cdot_* d_*x$$

$$=\; \int_* x^{2*} \cdot_* d_*x$$

$$=\; e^{\frac{1}{3}} \cdot_* x^{3*}, \quad x \in [e^2, e^3).$$

Consequently,

$$y_3(x) \;=\; e^{-_* \log_* x} \cdot_* \cdot_* \left(c_3 +_* e^{\frac{1}{3}} \cdot_* x^{3*} \right)$$

$$=\; c_3 \cdot_* (e/_*x) +_* e^{\frac{1}{3}} \cdot_* x^{2*}, \quad x \in [e^2, e^3).$$

Here c_3 is a multiplicative constant.

4. $x \in [e^3, \infty)$. Then we have the equation

$$y^* +_* (e/_* x) \cdot_* y = e^2 \cdot_* x, \quad x \in [e^3, \infty).$$

Here

$$a(x) \quad = \quad e/_* x,$$

$$b(x) \quad = \quad e^2 \cdot_* x, \quad x \in [e^3, \infty).$$

Then

$$\int_* a(x) \cdot_* d_* x \quad = \quad \int_* (e/_* x) \cdot_* d_* x$$

$$= \quad \log_* x, \quad x \in [e^3, \infty),$$

and

$$\int_* b(x) \cdot_* e^{\int_* a(x) \cdot_* d_* x} \cdot_* dx \quad = \quad e^2 \cdot_* \int_* x \cdot_* e^{\log_* x} \cdot_* d_* x$$

$$= \quad e^2 \cdot_* \int_* x^{2*} \cdot_* d_* x$$

$$= \quad e^{\frac{2}{3}} \cdot_* x^{3*}, \quad x \in [e^3, \infty).$$

Consequently,

$$y_4(x) \quad = \quad e^{-_* \log_* x} \cdot_* \left(c_4 +_* e^{\frac{2}{3}} \cdot_* x^{3*} \right)$$

$$= \quad c_4 \cdot_* (e/_* x) +_* e^{\frac{2}{3}} \cdot_* x^{2*}, \quad x \in [e^3, \infty).$$

Here c_4 is a multiplicative constant.

Hence,

$$\lim_{x \to e^{-_*}} y_1(x) \quad = \quad c_1 \cdot_* e^{-_* e},$$

$$\lim_{x \to e^{+_*}} y_2(x) \quad = \quad c_2 \cdot_* e -_* e^2,$$

$$\lim_{x \to e^{2-}*} y_2(x) \;=\; c_2 \cdot_* e^{e^2} -_* e^3,$$

$$\lim_{x \to e^{2+}*} y_3(x) \;=\; c_3 \cdot_* e^{\frac{1}{2}} +_* e^{\frac{4}{3}},$$

$$\lim_{x \to e^{3-}*} y_3(x) \;=\; c_3 \cdot_* e^{\frac{1}{3}} +_* e^3,$$

$$\lim_{x \to e^{3+}*} y_4(x) \;=\; c_4 \cdot_* e^{\frac{1}{3}} +_* e^6.$$

Therefore,

$$\left\{ \begin{array}{l} \displaystyle \lim_{x \to e^-*} y_1(x) = \lim_{x \to e^+*} y_2(x), \\[1.5em] \displaystyle \lim_{x \to e^{2-}*} y_2(x) = \lim_{x \to e^{2+}*} y_3(x), \\[1.5em] \displaystyle \lim_{x \to e^{3-}*} y_3(x) = \lim_{x \to e^{3+}*} y_4(x), \end{array} \right.$$

whereupon

$$\left\{ \begin{array}{l} c_1 \cdot_* e^{-_* e} = c_2 -_* e^2 \\[1.5em] c_2 \cdot_* e^{e^2} -_* e^3 = c_3 \cdot_* e^{\frac{1}{2}} +_* e^{\frac{4}{3}} \\[1.5em] c_3 \cdot_* e^{\frac{1}{3}} +_* e^3 = c_4 \cdot_* e^{\frac{1}{3}} +_* e^6, \end{array} \right.$$

and

$$\left\{ \begin{array}{l} c_2 = c_1 \cdot_* e^{-_* e^2} +_* e^2 \cdot_* e^{-_* e} \\[1.5em] c_3 = e^2 \cdot_* c_2 \cdot_* e^{e^2} -_* e^{\frac{26}{3}} = e^2 \cdot_* c_1 +_* e^{4e} -_* e^{\frac{26}{3}} \\[1.5em] c_3 = c_3 -_* e^9 = e^2 \cdot_* c_1 +_* e^{4e - \frac{53}{3}}. \end{array} \right.$$

Consequently,

$$y(x) = \begin{cases} c_1 \cdot_* e^{-_*x} +_* x -_* e & 1 \le x < e \\[2mm] \left(c_1 \cdot_* e^{e^{-2}} +_* e^{2e^{-1}} \right) \cdot_* e^x -_* (x +_* e) & e \le x < e^2 \\[2mm] \left(e^2 \cdot_* c_1 +_* e^{4e-\frac{26}{3}} \right) \cdot_* (e/_*x) +_* e^{\frac{1}{3}} \cdot_* x^{2_*} & \\[2mm] e^2 \le x < e^3 & \\[2mm] \left(e^2 \cdot_* c_1 +_* e^{4e-\frac{53}{3}} \right) \cdot_* (e/_*x) & \\[2mm] +_* e^{\frac{2}{3}} \cdot_* x^{2_*} \quad e^3 \le x \end{cases}$$

is a continuous solution of the considered equation in $[1, \infty)$.

Exercise 3.4.4 *Find a continuous solution in* $[e, e^4]$ *of the equation*

$$y^* -_* (e^4/_*x) \cdot_* y = \begin{cases} -_* e^2 \cdot_* x -_* (e^4/_*x) & x \in [e, e^2) \\[2mm] x^{2_*} & x \in (e^2, e^4]. \end{cases}$$

Answer

$$y(x) = \begin{cases} c_1 \cdot_* x^{4_*} +_* x^{2_*} +_* e & x \in [e, e^2) \\[2mm] \left(c_1 +_* e^{\frac{13}{16}} \right) \cdot_* x^{4_*} -_* x^{3_*} & e^2 < x \le e^4. \end{cases}$$

Here c_1 *is a multiplicative constant.*

Exercise 3.4.5 *Find a continuous solution in* $[1, \infty)$ *of the equation*

$$y^* +_* a(x) \cdot_* y = 0_*, \quad x \in [1, \infty),$$

where

$$a(x) = \begin{cases} e^2 & 1 \le x < e \\[2mm] e & x \ge e. \end{cases}$$

Answer

$$y(x) = \begin{cases} c_1 \cdot_* e^{-_* e^2 \cdot_* x} & 1 \leq x < e \\ c_1 \cdot_* e^{-_* x -_* e} & x \geq e. \end{cases}$$

Here c_1 is a multiplicative constant.

Exercise 3.4.6 *Find a continuous solution in $[1, e^3]$ of the equation*

$$y^* +_* a(x) \cdot_* y = 0_*, \quad x \in [1, e^3],$$

where

$$a(x) = \begin{cases} e^{-1} & 1 \leq x < e^2 \\ e^2 & e^2 \leq x \leq e^3. \end{cases}$$

Answer

$$y(x) = \begin{cases} c_1 \cdot_* e^x & 1 \leq x < e^2 \\ c_1 \cdot_* e^{e^{-2} \cdot_* x +_* e^6} & e^2 \leq x \leq e^3. \end{cases}$$

Here c_1 is a multiplicative constant.

Exercise 3.4.7 *Find a continuous solution in $[1, \infty)$ of the equation*

$$y^* +_* y = \begin{cases} 1 & x \in [1, e) \\ x & x \geq e. \end{cases}$$

Answer

$$y(x) = \begin{cases} c_1 \cdot_* e^{-_* x} +_* e & x \in [1, e) \\ (c_1 +_* e^e) \cdot_* e^{-_* x} +_* x -_* e & x \geq e. \end{cases}$$

Here c_1 is a multiplicative constant.

3.5 Multiplicative Initial Value Problems

If we consider the equation (3.1) subject to the initial condition

$$y(x_0) = y_0, \tag{3.11}$$

where $x_0 \in J$, $y_0 \in \mathbb{R}_*$, then its solution is given by the expression

$$y(x) = e^{-* \int_{*x_0}^{x} a(s) \cdot_* d_* s} \cdot_* \left(y_0 +_* \int_{*x_0}^{x} b(s) \cdot_* e^{\int_{*x_0}^{s} a(\tau) \cdot_* d_* \tau} \cdot_* d_* s \right),$$

$x \in J$.

Example 3.5.1 *We consider the Cauchy problem*

$$y^* +_* e^2 \cdot_* y \;\; = \;\; x, \qquad x > 1,$$

$$y(1) \;\; = \;\; e.$$

Here

$$a(x) \;\; = \;\; 2,$$

$$b(x) \;\; = \;\; x, \quad x > 1.$$

Then

$$\int_{*1}^{x} a(s) \cdot_* d_* s \;\; = \;\; e^2 \cdot_* \int_{*1}^{x} d_* s$$

$$= \;\; e^2 \cdot_* x, \quad x > 1,$$

and

$$\int_{*1}^{x} b(s) \cdot_* e^{\int_{*1}^{s} a(\tau) \cdot_* d_* \tau} \cdot_* d_* s \;\; = \;\; \int_{*1}^{x} s \cdot_* e^{e^2 \cdot_* s} \cdot_* d_* s$$

$$= \;\; e^{\frac{1}{2}} \cdot_* s \cdot_* e^{e^2 \cdot_* s} \Big|_{s=1}^{s=x}$$

$$-_* e^{\frac{1}{2}} \cdot_* \int_1^x e^{e^2 \cdot_* s} \cdot_* d_* s$$

$$= e^{\frac{1}{2}} \cdot_* x \cdot_* e^{e^2 \cdot_* x} -_* e^{\frac{1}{4}} \cdot_* e^{e^2 \cdot_* s} \Big|_{s=1}^{s=x}$$

$$= e^{\frac{1}{2}} \cdot_* x \cdot_* e^{e^2 \cdot_* x}$$

$$-_* e^{\frac{1}{4}} \cdot_* e^{e^2 \cdot_* x} +_* e^{\frac{1}{4}}, \quad x > 1,$$

and

$$y(x) \;=\; e^{-_* e^2 \cdot_* x} \cdot_* \left(e +_* e^{\frac{1}{2}} \cdot_* x \cdot_* e^{e^2 \cdot_* x} \right.$$

$$\left. -_* e^{\frac{1}{4}} \cdot_* e^{e^2 \cdot_* x} +_* e^{\frac{1}{4}} \right)$$

$$= e^{-_* e^2 \cdot_* x} \cdot_* \left(e^{\frac{5}{4}} +_* e^{\frac{1}{2}} \cdot_* x \cdot_* e^{e^2 \cdot_* x} -_* e^{\frac{1}{4}} \cdot_* e^{e^2 \cdot_* x} \right)$$

$$= e^{\frac{5}{4}} \cdot_* e^{-_* e^2 \cdot_* x} +_* e^{\frac{1}{2}} \cdot_* x -_* e^{\frac{1}{4}}, \quad x > 1.$$

Example 3.5.2 *We consider the Cauchy problem*

$$y^* +_* (\sin_* x) \cdot_* y \;=\; \sin_* x, \quad x > 1,$$

$$y(1) \;=\; e.$$

Here

$$a(x) \;=\; \sin_* x,$$

$$b(x) \;=\; \sin_* x, \quad x > 1.$$

Then

$$\int_{*1}^s a(\tau) \cdot_* d_* \tau \;=\; \int_{*0}^s \sin_* \tau \cdot_* d_* \tau$$

$$= \;-_* \cos_* \tau \Big|_{\tau=1}^{\tau=s}$$

$$= -_* \cos_* s + e, \quad s \in (1, x],$$

$$\int_{*1}^{x} a(s) \cdot_* d_* s = -_* \cos_* x +_* e, \quad x > 1,$$

and

$$\int_{*1}^{x} b(s) \cdot_* e^{\int_{*1}^{s} a(\tau) \cdot_* d_* \tau} \cdot_* ds = \int_{*1}^{x} \sin_* s \cdot_* e^{-_* \cos_* s +_* e} \cdot_* d_* s$$

$$= e^{e} \cdot_* \int_{*1}^{x} e^{-_* \cos_* s} \cdot_* d(-_* \cos_* s)$$

$$= e^{e - _* \cos_* s} \Big|_{s=1}^{s=x}$$

$$= e^{e - _* \cos_* x} -_* e, \quad x > 1,$$

and

$$y(x) = e^{\cos_* x - _* e} \cdot_* \left(e +_* e^{e - _* \cos_* x} -_* e \right)$$

$$= e, \quad x > 1.$$

Example 3.5.3 *We consider the Cauchy problem*

$$y^* +_* y = x^{2*} -_* e, \qquad x > e,$$

$$y(e) = e^2.$$

Here

$$a(x) = 1,$$

$$b(x) = x^{2*} -_* e, \quad x > e.$$

Then

$$\int_{*e}^{s} a(\tau) \cdot_* d_* \tau = \int_{*e}^{s} d_* \tau$$

$$= s -_* e, \quad s \in (1, x],$$

$$\int_{*e}^{x} a(s) \cdot_* d_* s = x -_* e, \quad x > e.$$

and

$$\int_{*e}^{x} b(s) \cdot_* e^{\int_{*e}^{s} a(\tau) \cdot_* d_* \tau} \cdot_* ds$$

$$= \int_{*e}^{x} (s^{2*} -_* e) \cdot_* e^{s -_* e} \cdot_* d_* s$$

$$= (s^{2*} -_* e) \cdot_* e^{s -_* e} \Big|_{s=e}^{s=x} -_* e^2 \int_{*e}^{x} s \cdot_* e^{s -_* e} \cdot_* d_* s$$

$$= (x^{2*} -_* e) \cdot_* e^{x -_* e} -_* e^2 \cdot_* s \cdot_* e^{s -_* e} \Big|_{s=e}^{s=x}$$

$$\quad +_* e^2 \cdot_* \int_{*e}^{x} e^{s -_* e} \cdot_* d_* s$$

$$= (x^{2*} -_* e) \cdot_* e^{x -_* e} -_* e^2 \cdot_* x \cdot_* e^{x -_* e} +_* e^2 +_* e^2 \cdot_* e^{s -_* e} \Big|_{s=e}^{s=x}$$

$$= (x^{2*} -_* e^2 \cdot_* x -_* e) \cdot_* e^{x -_* e} +_* e^2 +_* e^2) \cdot_* e^{x -_* e} -_* e^2$$

$$= (x^{2*} -_* e^2 \cdot_* x +_* e) \cdot_* e^{x -_* e}$$

$$= (x -_* e) \cdot_* e^2 \cdot_* e^{x -_* e}, \quad x > e,$$

and

$$y(x) = e^{-_* x +_* e} \cdot_* \left(e^2 +_* (x -_* e)^{2*} \cdot_* e^{x -_* e} \right)$$

$$= e^2 \cdot_* e^{-_* x +_* e} +_* (x -_* e)^{2*}, \quad x > e.$$

Exercise 3.5.4 *Solve the Cauchy problems*

1.

$$y^* -_* y \;=\; \cos_* x, \quad x > 1,$$

$$y(1) \;=\; 1.$$

2.

$$y^* -_* e^2 \cdot_* x \cdot_* y \;=\; x^{3*}, \quad x > 1,$$

$$y(1) \;=\; e.$$

3.

$$x \cdot_* d_* x \;=\; (x^{2*} -_* e^2 \cdot_* y +_* e) \cdot_* d_* y, \quad y > 1,$$

$$x(1) \;=\; 1.$$

Answer

1. $y(x) = e^{\frac{1}{2}} \cdot_* e^x +_* ((\sin_* x -_* \cos_* x)/_* e^2), \; x > 1.$

2. $y(x) = e^{\frac{3}{2}} \cdot_* e^{x^{2*}} -_* e^{\frac{1}{2}} \cdot_* (x^{2*} +_* e), \; x > 1.$

3. $(x(y))^{2*} = e^2 \cdot_* y, \; y > 1.$

Exercise 3.5.5 *Find solution y of the equation*

$$x^{3*} \cdot_* y^* -_* \sin_* y = e$$

for which

$$\lim_{x \to \infty} y(x) = e^{5\pi}.$$

Solution 3.5.6 *Firstly, we will find the general solution of the given equation. We have*

$$x^{3*} \cdot_* y^* \;=\; e +_* \sin_* y$$

$$=\; (\sin_* (y/_* e^2))^{2*} +_* (\cos_* (y/_* e^2))^{2*}$$

$$+_* e^2 \cdot_* \sin_* (y/_* e^2) \cdot_* \cos_* (y/_* e^2)$$

$$= \left(\sin_* (y/_* e^2) +_* \cos_* (y/_* e^2) \right)^{2*}$$

$$= \left(\cos_* (y/_* e^2) \right)^{2*} \cdot_* \left(e +_* \tan_* (y/_* e^2) \right)^{2*},$$

or

$$x^{3*} \cdot_* d_* y = \left(\cos_* (y/_* e^2) \right)^{2*} \cdot_* \left(e +_* \tan_* (y/_* e^2) \right)^* d_* x,$$

or

$$d_* y /_* \left((\cos_* (y/_* e^2))^{2*} \cdot_* \left(e +_* \tan_* (y/_* e^2) \right)^{2*} \right) = d_* x /_* (x^{3*}),$$

whereupon

$$\int_* d_* y /_* \left((\cos_* (y/_* e^2))^{2*} \cdot_* \left(e +_* \tan_* (y/_* e^2) \right)^{2*} \right) = \int_* d_* x /_* x^{3*} +_* c,$$

or

$$e^2 \cdot_* \int_* (d_* (y/_* e^2)) /_* \left((\cos_* (y/_* e^2))^{2*} \cdot_* \left(e +_* \tan_* (y/_* e^2) \right)^{2*} \right)$$

$$= -_* (e /_* (e^2 \cdot_* x^{2*})) +_* c,$$

or

$$e^2 \cdot_* \int_* (d_* (\tan_* (y/_* e^2))) /_* \left((e +_* \tan_* (y/_* e^2))^{2*} \right) = -_* (e /_* (e^2 \cdot_* x^{2*})) +_* c,$$

or

$$-_* (e^2 /_* (e +_* \tan_* (y/_* e^2))) = -_* (e /_* (e^2 \cdot_* x^{2*})) +_* c,$$

or

$$e^2 /_* (e +_* \tan_) *(y/_* e^2) = e /_* (e^2 \cdot_* x^{2*}) +_* c.$$

Hence, using

$$\lim_{x \to \infty} y(x) = e^{5\pi},$$

we get

$$\lim_{x \to \infty} (e^2 /_* (e +_* \tan_* (y/_* e^2))) = \lim_{x \to \infty} (e /_* (e^2 \cdot_* x^{2*})) +_* c,$$

or

$$0_* = c.$$

Consequently,

$$e^2/_*(e +_* \tan_*(y/_*e^2)) = e/_*(e^2 \cdot_* x^{2*}),$$

or

$$e^4 \cdot_* x^{2*} = e +_* \tan_*(y/_*e^2).$$

Exercise 3.5.7 *Find solution y of the equation*

$$y^* \cdot_* \sin_*(e^2 \cdot_* x) = e^2 \cdot_* (y +_* \cos_* x), \qquad \tan_* x > 1,$$

which is bounded when $x \to e^{\frac{\pi}{2}}$.

Answer

$$y(x) = ((\sin_* x -_* e)/_*\cos_* x), \quad x \in \mathbb{R}_*.$$

Remark 3.5.8 *We note that*

$$y(x) = c \cdot_* e^{-* \int_* a(x) \cdot_* d_* x}, \quad x \in \mathbb{R}_*,$$

where c is a multiplicative constant, is the general solution to the equation (3.2). We consider

$$v(x) = e^{-* \int_* a(x) \cdot_* d_* x} \cdot_* \int_* b(x) \cdot_* e^{\int_* a(x) \cdot_* d_* x} \cdot_* d_* x,$$

$x \in J$. Then

$$
\begin{aligned}
v^*(x) &= -_*a(x) \cdot_* e^{-* \int_* a(x) \cdot_* d_* x} \cdot_* \int_* b(x) \cdot_* e^{\int_* a(x) \cdot_* d_* x} \cdot_* d_* x \\[4pt]
&\quad +e^{-* \int_* a(x) \cdot_* d_* x} \cdot_* b(x) \cdot_* e^{\int_* a(x) \cdot_* d_* x} \\[6pt]
&= -_*a(x) \cdot_* e^{-* \int_* a(x) \cdot_* d_* x} \cdot_* \int_* b(x) \cdot_* e^{\int_* a(x) \cdot_* d_* x} \cdot_* d_* x +_* b(x) \\[6pt]
&= -_*a(x) \cdot_* v(x) +_* b(x), \quad x \in J.
\end{aligned}
$$

i.e., v is a particular solution of the equation (3.1). Therefore the general solution of the equation (3.1) is obtained by adding a particular solution of (3.1) to the general solution of the equation (3.2).

Exercise 3.5.9 *Solve the equation*

$$y(x) = x +_* \lambda \cdot_* \int_{*1}^x \cdot_* e^{x-*s} \cdot_* y(s) \cdot_* d_* s,$$

where λ is a multiplicative constant.

Solution 3.5.10 *We note that*

$$y(1) = 1,$$

and

$$
\begin{aligned}
y^*(x) &= e +_* \lambda \cdot_* y(x) +_* \lambda \cdot_* e^x \cdot_* \int_{*1}^x e^{-*s} \cdot_* y(s) \cdot_* d_* s \\[2mm]
&= e -_* x +_* \lambda \cdot_* y(x) +_* x \\[2mm]
&\quad +_* \lambda \cdot_* e^x \cdot_* \int_{*0}^x e^{-*s} \cdot_* y(s) \cdot_* d_* s \\[2mm]
&= e -_* x +_* \lambda \cdot_* y(x) +_* y(x) \\[2mm]
&= (\lambda +_* e) \cdot_* y(x) +_* e -_* x, \quad x > 1.
\end{aligned}
$$

Consequently, y is a solution to the Cauchy problem

$$y^*(x) -_* (\lambda +_* e) \cdot_* y(x) = e -_* x, \qquad x > 1,$$

$$y(1) = 1.$$

Here

$$a(x) = -_*(\lambda +_* e),$$

$$b(x) = e -_* x, \quad x > 1.$$

1. Let $\lambda \neq e^{-1}$. Then

$$\int_* a(x) \cdot_* d_* x = -_*(\lambda +_* e) \cdot_* \int_* d_* x$$

$$= \; -_*(\lambda +_* e) \cdot_* x, \quad x > 1,$$

and

$$\int_* b(x) \cdot_* e^{\int_* a(x) \cdot_* d_* x} \cdot_* d_* x = \int_* (e -_* x) \cdot_* e^{-_*(\lambda +_* e) \cdot_* x} \cdot_* d_* x$$

$$= \; (e/_*(\lambda +_* e)) \cdot_* \int_* (x -_* e) \cdot_* e^{-_*(\lambda +_* e) \cdot_* x} \cdot_* d_* (-_*(\lambda +_* e) \cdot_* x)$$

$$= \; (e/_*(\lambda +_* e)) \cdot_* (x -_* e) \cdot_* e^{-_*(\lambda +_* e) \cdot_* x}$$

$$-_*(e/_*(\lambda +_* e)) \cdot_* \int_* e^{-_*(\lambda +_* e) \cdot_* x} \cdot_* d_* x$$

$$= \; (e/_*(\lambda +_* e)) \cdot_* (x -_* e) \cdot_* e^{-_*(\lambda +_* e) \cdot_* x}$$

$$+_*(e/_*(\lambda +_* e)^{2*}) \cdot_* e^{-_*(\lambda +_* e) \cdot_* x}, \quad x > 1.$$

Then

$$y(x) \; = \; e^{(\lambda +_* e) \cdot_* x} \cdot_* \left(c +_* ((x -_* e)/_*(\lambda +_* e)) \cdot_* e^{-_*(\lambda +_* e) \cdot_* x} \right.$$

$$\left. +_*(e/_*(\lambda +_* e)^{2*}) \cdot_* e^{-_*(\lambda +_* e) \cdot_* x} \right)$$

$$= \; c \cdot_* e^{(\lambda +_* e) \cdot_* x} +_* ((x -_* e)/_*(\lambda +_* e))$$

$$+_*(e/_*(\lambda +_* e)^{2*}), \quad x > 1,$$

and

$$y(1) \; = \; c -_* (e/_*(\lambda +_* e)) +_* (e/_*(\lambda +_* e)^{2*})$$

$$= \; 1,$$

from where

$$c \; = \; (e/_*(\lambda +_* e)) -_* (e/_*(\lambda +_* e)^{2*})$$

$$= \ ((\lambda +_* e -_* e)/_*(\lambda +_* e)^{2*})$$

$$= \ (\lambda /_*(\lambda +_* e)^{2*}).$$

Therefore

$$y(x) \ = \ (\lambda /_*(\lambda +_* e)^{2*}) \cdot_* e^{(\lambda +_* e)\cdot_* x}$$

$$+_* ((x -_* e)/_*(\lambda +_* e)) +_* (e/_*(\lambda +_* e)^{2*}).$$

2. *Let* $\lambda = e^{-1}$. *Then* y *satisfies the Cauchy problem*

$$y^* \ = \ e -_* x, \quad x > 1,$$

$$y(1) \ = \ 1.$$

We have

$$d_* y = (e -_* x) \cdot_* d_* x$$

and

$$\int_* d_* y = \int_* (e -_* x) \cdot_* d_* x +_* c,$$

and

$$y(x) \ = \ -_*((e -_* x)^{2*}/_* e^2) +_* c, \quad x > 1,$$

$$y(1) \ = \ -_*(e/_* 2) +_* c$$

$$= \ 1,$$

whereupon

$$c = e^{\frac{1}{2}}.$$

Therefore

$$y(x) = -_* e^{\frac{1}{2}} \cdot_* (e -_* x)^{2*} +_* e^{\frac{1}{2}}, \quad x \geq 1.$$

Exercise 3.5.11 *Let $f \in \mathscr{C}_*^1(\mathbb{R}_*)$ be a multiplicative ω-periodic function. Prove that the equation*

$$y^* = f(x), \quad x \in \mathbb{R}_*,$$

has a multiplicative ω-periodic solution if and only if

$$\int_{*1}^{\omega} f(s) \cdot_* d_*s = 1.$$

Solution 3.5.12 *We note that for every x_0, $y_0 \in \mathbb{R}_*$ we have that*

$$y(x) = \int_{*x_0}^{x} f(s) \cdot_* d_*s +_* y_0, \quad x \in \mathbb{R}_*,$$

is a solution of the considered equation for which $y(x_0) = y_0$. Then y is multiplicative ω-periodic solution of the considered equation if and only if

$$y(x +_* \omega) = y(x), \quad x \in \mathbb{R}_*,$$

if and only if

$$\int_{*x_0}^{x+_*\omega} f(s) \cdot_* d_*s = \int_{*x_0}^{x} f(s) \cdot_* d_*s +_* y_0$$

if and only if

$$\int_{*x}^{x+_*\omega} f(s) \cdot_* d_*s = 1, \quad x \in \mathbb{R}_*,$$

if and only if

$$\int_{*x}^{1} f(s) \cdot_* d_*s +_* \int_{*1}^{\omega} f(s) \cdot_* d_*s$$

$$+_* \int_{*\omega}^{x+_*\omega} f(s) \cdot_* d_*s$$

$$= 1, \quad x \in \mathbb{R}_*,$$

if and only if

$$\int_{*x}^{1} f(s) \cdot_* d_*s +_* \int_{*1}^{\omega} f(s) \cdot_* d_*s$$

$$+ \int_{*1}^{x} f(s) \cdot_* d_*s$$

$$= 1, \quad x \in \mathbb{R}_*,$$

if and only if

$$\int_{*1}^{\omega} f(s) \cdot_* d_*s = 1.$$

Exercise 3.5.13 *Let $f \in \mathscr{C}_*^1(\mathbb{R}_*)$ be a multiplicative ω-periodic func-tion. Prove that if the equation*

$$y^* = f(x), \quad x \in \mathbb{R}_*,$$

has a multiplicative bounded solution, then it has a multiplicative ω-periodic solution.

Solution 3.5.14 *Let y be a multiplicative bounded solution of the con-sidered equation. Then*

$$y(x) = \int_{*x_0}^{x} f(s) \cdot_* d_*s +_* y_0, \quad x \in \mathbb{R}_*,$$

for some $x_0, y_0 \in \mathbb{R}_*$. Hence,

$$-_*\infty < y(\pm_*\infty) < +_*\infty.$$

Therefore

$$-_*\infty < \int_{*x_0}^{\infty} f(s) \cdot_* d_*s < \infty$$

if and only if

$$-_*\infty < \int_{*x_0}^{x_0+_*\omega} f(s) \cdot_* d_*s +_* \int_{*x_0+_*\omega}^{x_0+_*e^2\omega} f(s) \cdot_* d_*s +_* \cdots < -_*\infty$$

if and only if

$$-_*\infty < \int_{*1}^{\omega} f(s) \cdot_* d_*s +_* \int_{*1}^{\omega} f(s) \cdot_* d_*s +_* \cdots < \infty$$

if and only if

$$\int_{*1}^{\omega} f(s) \cdot_* d_*s = 1.$$

From here and from the previous problem, it follows that the considered equation has a multiplicative ω-periodic solution.

Exercise 3.5.15 *Let a be a multiplicative constant, $f(x) \to b$ as $x \to 1$. Prove that there exists a unique solution of the equation*

$$x \cdot_* y^* +_* a \cdot_* y = f(x), \quad x \in \mathbb{R}_*,$$

which is multiplicative bounded when $x \to 1$ and find the limit of this solution when $x \to 1$.

Solution 3.5.16 *The given equation we can rewrite in the form*

$$y^* +_* (e/_* x) \cdot_* a \cdot_* y = (f(x)/_* x), \quad x \in \mathbb{R}_*.$$

Then its solution for which

$$y(x_0) = y_0, \quad x_0, \quad y_0 \in \mathbb{R}_*,$$

is given by the expression

$$y(x) = e^{-*\int_{*x_0}^{x}(a/_*s)\cdot_*d_*s}$$

$$\cdot_* \left(y_0 +_* \int_{*x_0}^{x} (f(s)/_*s) \cdot_* e^{\int_{*x_0}^{s}(a/_*\tau)\cdot_*d_*\tau} \cdot_* d_*s \right)$$

$$= e^{-*a\cdot_*\log_*|x/_*x_0|_*}$$

$$\cdot_* \left(y_0 +_* \int_{*x_0}^{x} (f(s)/_*s) \cdot_* e^{e^a \cdot_* \log_*|s/_*x_0|_*} \cdot_* d \cdot_* s \right)$$

$$= |x_0/_*x|_*^{a_*}$$

$$\cdot_* \left(y_0 +_* \int_{*x_0}^{x} (f(s)/_*s) \cdot_* |s/_*x_0|_*^{a_*} \cdot_* d_*s \right)$$

$$= y_0 \cdot_* |x_0/_*x|_*^{a_*} +_* ((|x_0|_*^{a_*})/_*|x|_*^{a_*})$$

$$\cdot_* \int_{*x_0}^{x} (f(s)/_*s) \cdot_* (|s|_*^{a_*}/_*|x_0|_*^{a_*}) \cdot_* d_*s$$

$$= y_0 \cdot_* |x_0/_*x|_*^{a_*}$$

$$+_* (e/_* |x|_*^{a_*}) \cdot_* \int_{*x_0}^x (f(s)/_* s) \cdot_* |s|_*^{a_*} \cdot_* d_* s,$$

$x \in \mathbb{R}_*$. *Let* y *be a multiplicative bounded solution of the considered equation when* $x \to 1$. *Therefore,*

$$y(x) = (e/_* |x|_*^{a_*}) \cdot_* \int_{*1}^x (f(s)/_* s) \cdot_* |s|_*^{a_*} \cdot_* d_* s, \quad x \in \mathbb{R}_*.$$

1. Let $x > 1$. *Then*

$$y(x) = (e/_* x^{a_*}) \cdot_* \int_{*1}^x f(s) \cdot_* s^{(a-1)_*} \cdot_* d_* s, \quad x \in \mathbb{R}_*.$$

Hence,

$$
\begin{aligned}
\lim_{x \to 1} y(x) &= \lim_{x \to 1} (e/_* x^{a_*}) \cdot_* \int_{*1}^x f(s) \cdot_* s^{(a-1)_*} \cdot_* d_* s \\
&= \lim_{x \to 1} ((f(x) \cdot_* x^{(a-1)_*}) /_* (e^a \cdot_* x^{(a-1)_*})) \\
&= \lim_{x \to 1} (f(x)/_* e^a) \\
&= e^{\frac{b}{a}}.
\end{aligned}
$$

2. $x < 1$. *Then*

$$
\begin{aligned}
y(x) &= ((e^{-1})^{a_*} \cdot_* \int_{*1}^x (f(s)/_* s) \cdot_* s^{a_*} \cdot_* d_* s) \\
&\quad /_* (e^{-1})^{a_*} \cdot_* x^{a_*} \\
&= (\int_{*1}^x (f(s)/_* s) \cdot_* s^{a_*} \cdot_* d_* s) /_* x^{a_*} \\
&= (\int_{*10}^x f(s) \cdot_* s^{(a-1)_*} \cdot_* d_* s) /_* x^{a_*}, \quad x \in \mathbb{R}_*,
\end{aligned}
$$

whereupon, as in the previous case,

$$\lim_{x \to 1} y(x) = e^{\frac{b}{a}}.$$

Exercise 3.5.17 *Let $a < 0_*$, $f(x) \to e^b$ as $x \to 1$. Prove that every so-lution y of the equation*

$$x \cdot_* y^* +_* e^a \cdot_* y = f(x), \quad x \in R_*,$$

there exists $\lim\limits_{x \to 1} y(x)$. Find this limit.

Solution 3.5.18 *As in the previous exercise, we have that every solu-tion of the considered equation for which $y(x_0) = y_0$ can be expressed in the following way:*

$$y(x) \;=\; y_0 \cdot_* \left(|x_0|^{a_*} /_* |x|^{a_*} \right)$$

$$+_* (e /_* |x|^{a_*}) \cdot_* \int_{*x_0}^{x} (f(s) /_* s) \cdot_* |s|^{a_*} \cdot_* d_* s, \quad x \in R_*.$$

Hence,

$$\lim_{x \to 1} y(x) \;=\; \lim_{x \to 1} \left(y_0 \cdot_* \left(|x_0|^{a_*}_* /_* |x|^{a_*}_* \right) \right.$$

$$\left. +_* (e /_* |x|^{a_*}_*) \cdot_* \int_{*x_0}^{x} (f(s) /_* s) \cdot_* |s|^{a_*}_* \cdot_* d_* s \right)$$

$$= \; \lim_{x \to 1} (e /_* |x|^{a_*}) \cdot_* \int_{*x_0}^{x} (f(s) /_* s) \cdot_* |s|^{a_*}_* \cdot_* d_* s$$

$$= \; e^{\frac{b}{a}}.$$

3.6   Some Multiplicative Nonlinear Differential Equations

Here we consider the equation

$$y^* = f(x, y), \quad x \in J. \tag{3.12}$$

Often, it is possible to introduce a new set of multiplicative variables

$$u \;=\; \phi(x, y),$$

$$v \;=\; \psi(x, y), \tag{3.13}$$

which convert the equation (3.12) into a form that can be solved easily. We assume that

$$\partial_*(u,v)/_*\partial_*(x,y) = \left| \begin{array}{cc} \partial_*u/_*\partial_*x & \partial_*v/_*\partial_*x \\ \partial_*u/_*\partial_*y & \partial_*v/_*\partial_*y \end{array} \right|_* |_* \neq 1$$

over a region in $\mathbb{R}^2_*$. This implies that there is a multiplicative functional relationship between u and v. If (u_0, v_0) is the image of (x_0, y_0) under the transformation (3.13), then it can be uniquely solved for x and y in a multiplicative neighborhood of the point (x_0, y_0). This leads to the multiplicative inverse transformation

$$x = x(u,v),$$

$$y = y(u,v).$$

(3.14)

Also, we have

$$y^* = d_*y/_*d_*x$$

$$= ((y_u^* \cdot_* d_*u +_* y_v^* \cdot_* d_*v)/_*(x_u^* \cdot_* d_*u +_* x_v^* \cdot_* d_*v))$$

$$= ((y_u^* +_* y_v^* \cdot_* (d_*v/_*d_*u))/_*(x_u^* +_* x_v^* \cdot_* (d_*v/_*d_*u))).$$

(3.15)

The relations (3.14) and (3.15) can be used to convert the equation (3.12) in the terms of u and v, which hopefully can be solved explicitly. Finally, replacement of u and v in the terms of x and y, by using (3.13), leads to an implicit solution of the equation (3.12).

Example 3.6.1 *We will use the transformation*

$$u = x^{3*},$$

$$v = y^{2*}$$

to solve the equation

$$e^3 \cdot_* x^{5*} -_* y \cdot_* (y^{2*} -_* x^{3*}) \cdot_* y^* = 1, \qquad x > 1, y > 1.$$

We have

$$x = u^{\frac{1}{3}*},$$

$$y = v^{\frac{1}{2}}*,$$

and

$$y^* = d_*y/_*d_*x$$

$$= ((y_u^* \cdot_* d_*u +_* y_v^* \cdot_* d_*v)/_*(x_u^* \cdot_* d_*u +_* x_v^* \cdot_* d_*v))$$

$$= ((e_*(e^2 \cdot_* v^{\frac{1}{2}}*)) \cdot_* d_*v)/_*(e^{\frac{1}{3}} \cdot_* (e/_*u^{\frac{2}{3}}*) \cdot_* d_*u)$$

$$= e^{\frac{3}{2}} \cdot_* (u^{\frac{2}{3}}*/_*v^{\frac{1}{2}}*) \cdot_* (d_*v/_*d_*u).$$

Then the given equation takes the form

$$e^3 \cdot_* \left(u^{\frac{1}{3}}*\right)^{5*} -_* v^{\frac{1}{2}}* \cdot_* (v -_* u) \cdot_* e^{\frac{3}{2}}$$

$$\cdot_*(u^{\frac{2}{3}}*/_*(v^{\frac{1}{2}}*)) \cdot_* (d_*v/_*d_*u)$$

$$= 1,$$

or

$$e^3 \cdot_* u^{\frac{5}{3}}* -_* e^{\frac{3}{2}} \cdot_* (v -_* u) \cdot_* u^{\frac{2}{3}}* \cdot_* (d_*v/_*d_*u) = 1$$

or

$$e^3 \cdot_* u^{\frac{2}{3}}* \cdot_* \left(u -_* e^{\frac{1}{2}} \cdot_* (v -_* u) \cdot_* (d_*v/_*d_*u)\right) = 1,$$

whereupon

$$u = 1$$

and

$$u -_* e^{\frac{1}{2}} \cdot_* (v -_* u) \cdot_* (d_*v/_*d_*u) = 1.$$

From $u = 1$, we get $x = 1$, which is not a solution to the considered equation. Now, we consider

$$u -_* e^{\frac{1}{2}} \cdot_* (v -_* u) \cdot_* (d_*v/_*d_*u) = 1,$$

whereupon

$$e^{\frac{1}{2}} \cdot_* (v -_* u) \cdot_* (d_* v/_* d_* u) = 1$$

or

$$d_* v/_* d_* u = e^2 \cdot_* (u/_* (v -_* u))$$

or

$$d_* v/_* d_* u = e^2 \cdot_* (e/_* ((v/_* u) -_* e)).$$

We set

$$z = v/_* u.$$

Then

$$v = zu,$$

$$v^* = z^* \cdot_* u +_* z.$$

Therefore

$$z^* \cdot_* u +_* z = (e^2/_* (z -_* e))$$

and

$$
\begin{aligned}
z \cdot_* u^* &= (e^2/_* (z -_* e)) -_* z \\[1ex]
&= ((e^2 -_* z^{2*} +_* z)/_* (z -_* e)) \\[1ex]
&= -_* ((z^{2*} -_* z -_* e^2)/_* (z -_* e)) \\[1ex]
&= -_* (((z +_* e) \cdot_* (z -_* e^2))/_* (z -_* e)),
\end{aligned}
$$

and

$$
\begin{aligned}
u \cdot_* d_* z &= -_* (((z +_* e) \cdot_* (z -_* e^2))/_* (z -_* e)) \cdot_* d_* u \\[1ex]
& \qquad \cdot_* \left| (((z +_* e) \cdot_* (z -_* e^2))/_* (z -_* e)) \right|_* \\[1ex]
&\neq 1,
\end{aligned}
$$

and

$$((z-_*e)/_*((z+_*e)\cdot_*(z-_*e^2)))\cdot_*d_*z = -_*(d_*u/_*u),$$

and

$$\int_*((z-_*e)/_*((z+_*e)\cdot_*(z-_*e^2)))\cdot_*d_*z = -_*\int_*(d_*u/_*u)+_*c,$$

and

$$e^{\frac{2}{3}}\cdot_*\int_*(d_*z/_*(z+_*e))+_*e^{\frac{1}{3}}\cdot_*\int_*(d_*z/_*(z-_*e^2)) = -_*\log_* u+_*c,$$

and

$$\log_*|z+_*e|^{\frac{2}{3}}*+_*\log_*|z-_*e^2|^{\frac{1}{3}}* = -_*\log_* u+_*c,$$

and

$$\log_*\left(|z+_*e|^{\frac{2}{3}}*\cdot_*|z-_*e^2|^{\frac{1}{3}}*\cdot_*u\right) = c,$$

and

$$(z+_*e)^{\frac{2}{3}}*\cdot_*(z-_*e^2)^{\frac{1}{3}}*\cdot_*u = c,$$

and

$$((v/_*u)+_*e)^{\frac{2}{3}}*\cdot_*\left((v/_*u)-_*e^2\right)^{\frac{1}{3}}*\cdot_*u = c,$$

and

$$(v+_*u)^{\frac{2}{3}}*\cdot_*(v-_*e^2\cdot_*u)^{\frac{1}{3}}* = c,$$

and

$$(y^{2*}+_*x^{3*})^{\frac{2}{3}}*\cdot_*(y^{2*}-_*e^2\cdot_*x^{3*})^{\frac{1}{3}}* = c,$$

and

$$(y^{2*}+_*x^{3*})^{2*}\cdot_*(y^{2*}-_*e^2\cdot_*x^{3*}) = c.$$

Here c is a multiplicative constant. We note that $z = e^{-1}$ and $z = e^2$ are included in the last equality for $c = 1$.

Example 3.6.2 *Now, we will use the transformation*

$$u = x^{2*}-_*y^{3*},$$

$$v = x+_*e^2\cdot_*y^{2*}$$

so that to solve the equation

$$(x^{4*} +_* e^4 \cdot_* x \cdot_* y^{2*} +_* y^{3*})$$

$$-_* (e^3 \cdot_* x \cdot_* y^{2*} +_* e^{10} \cdot_* y^{4*} -_* e^4 \cdot_* x^{2*} \cdot_* y) \cdot_* y^*$$

$$= 1.$$

We have

$$d_* u = e^2 \cdot_* x \cdot_* d_* x -_* e^3 \cdot_* y^{2*} \cdot_* d_* y,$$

$$d_* v = d_* x +_* e^4 \cdot_* y \cdot_* d_* y.$$

Also,

$$(x^{2*} +_* e^4 \cdot_* x \cdot_* y^{2*} +_* y^{3*})$$

$$-_* (e^3 \cdot_* x \cdot_* y^{2*} +_* e^{10} \cdot_* y^{4*} -_* e^4 \cdot_* x^{2*} \cdot_* y) \cdot_* y^*$$

$$= 1$$

and

$$(x^{2*} +_* e^4 \cdot_* x \cdot_* y^{2*} +_* y^{3*}) \cdot_* d_* x$$

$$-_* (e^3 \cdot_* x \cdot_* y^{2*} +_* e^{10} \cdot_* y^{4*} -_* e^4 \cdot_* x^{2*} \cdot_* y) \cdot_* d_* y$$

$$= 1,$$

and

$$(e^2 \cdot_* x^{2*} +_* e^4 \cdot_* x \cdot_* y^{2*} -_* x^{2*} +_* y^{3*}) \cdot_* d_* x$$

$$-_* (e^3 \cdot_* x \cdot_* y^{2*} +_* e^6 \cdot_* y^{4*} -_* e^4 \cdot_* y \cdot_* x^{2*} +_* e^4 \cdot_* y^{4*}) \cdot_* d_* y$$

$$= 1,$$

and

$$e^2 \cdot_* x \cdot_* (x +_* e^2 \cdot_* y^{2*}) \cdot_* d_*x -_* e^3 \cdot_* y^{2*} \cdot_* (x +_* e^2 \cdot_* y^{2*}) \cdot_* d_*y$$

$$= (x^{2*} -_* y^{3*}) \cdot_* d_*x +_* e^4 \cdot_* y \cdot_* (x^{2*} -_* y^{3*}) \cdot_* d_*y,$$

and

$$((e^2 \cdot_* x \cdot_* d_*x -_* e^3 \cdot_* y^{2*} \cdot_* d_*y) /_* (x^{2*} -_* y^{3*}))$$

$$= ((d_*x +_* e^4 \cdot_* y \cdot_* d_*y) /_* (x +_* e^2 \cdot_* y^{2*})),$$

and

$$d_*u /_* u = d_*v /_* v,$$

and

$$\int_* (d_*u /_* u) = \int_* (d_*v /_* v) +_* c,$$

and

$$\log_* |u|_* = \log_* |v|_* +_* c,$$

and

$$u = c \cdot_* v,$$

and

$$x^{2*} -_* y^{3*} = c \cdot_* (x +_* e^2 \cdot_* y^{2*}).$$

Here c is a multiplicative constant.

Example 3.6.3 *Now, we will use the transformation*

$$x = r \cdot_* \cos_* \phi,$$

$$y = r \cdot_* \sin_* \phi, \quad \phi \in [1, e^{2\pi}], \quad r > 1.$$

We have

$$r = (x^{2*} +_* y^{2*})^{\frac{1}{2}*},$$

$$\phi = \arctan_* (y /_* x),$$

and

$$x_r^* = \cos_* \phi,$$

$$x_\phi^* = -_* r \cdot_* \sin_* \phi,$$

$$y_r^* = \sin_* \phi,$$

$$y_\phi^* = r \cdot_* \cos_* \phi,$$

and

$$x +_* e^2 \cdot_* y = r \cdot_* \cos_* \phi +_* e^2 \cdot_* r \sin_* \phi$$

$$= r \cdot_* (\cos_* \phi +_* e^2 \cdot_* \sin_* \phi),$$

and

$$y -_* e^2 \cdot_* x = r \cdot_* \sin_* \phi -_* e^2 \cdot_* r \cdot_* \cos_* \phi$$

$$= r \cdot_* (\sin_* \phi -_* e^2 \cdot_* \cos_* \phi),$$

and

$$y^* = d_* y /_* d_* x$$

$$= ((y_r^* \cdot_* d_* r +_* y_\phi^* \cdot_* d_* \phi) /_* (x_r^* \cdot_* d_* r +_* x_\phi^* \cdot_* d_* \phi))$$

$$= ((\sin_* \phi \cdot_* d_* r +_* r \cdot_* \cos_* \phi \cdot_* d_* \phi))$$

$$/_* (\cos_* \phi \cdot_* d_* r -_* r \cdot_* \sin_* \phi \cdot_* d_* \phi)$$

$$= ((\sin_* \phi +_* r \cdot_* \cos_* \phi \cdot_* \phi^*) /_* (\cos_* \phi -_* r \cdot_* \sin_* \phi \cdot_* \phi^*)),$$

$$\phi^* = d_* \phi /_* d_* r.$$

Then the given equation takes the form

$$r \cdot_* (\cos_* \phi +_* e^2 \cdot_* \sin_* \phi) +_* r \cdot_* (\sin_* \phi -_* e^2 \cdot_* \cos_* \phi)$$

$$/_* ((\sin_* \phi +_* r \cdot_* \cos_* \phi \cdot_* \phi^*) /_* (\cos_* \phi -_* r \cdot_* \sin_* \phi \cdot_* \phi^*))$$

$$= \ 1,$$

or

$$r \cdot_* (\cos_* \phi +_* e^2 \cdot_* \sin_* \phi) \cdot_* (\cos_* \phi -_* r \cdot_* \sin_* \phi \cdot_* \phi^*)$$

$$+_* r \cdot_* (\sin_* \phi -_* e^2 \cdot_* \cos_* \phi) \cdot_* (\sin_* \phi +_* r \cdot_* \cos_* \phi \cdot_* \phi^*)$$

$$= \ 1,$$

or

$$r \cdot_* ((\cos_* \phi)^{2*} +_* e^2 \cdot_* \sin_* \phi \cdot_* \cos_* \phi$$

$$-_* r \cdot_* \sin_* \phi \cdot_* \cos_* \phi \cdot_* \phi^* -_* e^2 \cdot_* r \cdot_* (\sin_* \phi)^{2*} \cdot_* \phi^*)$$

$$+_* r \cdot_* ((\sin_* \phi)^{2*} +_* r \cdot_* \sin_* \phi \cdot_* \cos_* \phi \cdot_* \phi^*$$

$$-_* e^2 \cdot_* \sin_* \phi \cdot_* \cos_* \phi -_* e^2 \cdot_* r \cdot_* (\cos_* \phi)^{2*} \cdot_* \phi^*)$$

$$= \ 1,$$

and

$$r \cdot_* ((\sin_* \phi)^{2*} +_* (\cos_* \phi)^{2*}) -_* e^2 \cdot_* r^{2*}$$

$$\cdot_* ((\sin_* \phi)^{2*} +_* (\cos_* \phi)^{2*}) \cdot_* \phi^*$$

$$= \ 1 \quad \Longrightarrow,$$

and

$$r -_* e^2 \cdot_* r^{2*} \cdot_* \phi^* = 1,$$

and

$$\phi^* = e/_*(e2 \cdot_* r),$$

and

$$d_*\phi = (e/_*(e^2 \cdot_* r)) \cdot_* d_*r,$$

and

$$\int_* d_*\phi = e^{\frac{1}{2}} \cdot_* \int_* (d_*r/_*r) +_* c,$$

and

$$\phi = \log_*(r^{\frac{1}{2}}_*) +_* c,$$

and

$$e^\phi = r^{\frac{1}{2}}_* \cdot_* c,$$

and

$$r^{2*} = c \cdot_* e^{e^4 \cdot_* \phi},$$

and

$$x^{2*} +_* y^{2*} = c \cdot_* e^{e^4 \cdot_* \arctan_*(y/_*x)}.$$

Here c is a multiplicative constant.

Exercise 3.6.4 *Use the transformation*

$$u \;=\; x^{3*},$$

$$v \;=\; y \cdot_* e^y$$

to solve the equation

$$e^3 \cdot_* x^{2*} \cdot_* y \cdot_* e^y +_* x^{3*} \cdot_* e^y \cdot_* (y +_* e) \cdot_* y^* = 1.$$

Answer

$$x^{3*} \cdot_* y \cdot_* e^y = c.$$

Exercise 3.6.5 *Use the transformation*

$$u = x^{2*},$$
$$v = y^{2*}$$

to solve the equation

$$(e^2 \cdot_* x^{2*} +_* e^3 \cdot_* y^{2*} -_* e^7) \cdot_* x -_* (e^3 \cdot_* x^{2*} +_* e^2 \cdot_* y^{2*} -_* e^8) \cdot_* y \cdot_* y^* = 1.$$

Answer

$$(x^{2*} -_* y^{2*} -_* e)^{5*} = c \cdot_* (x^{2*} +_* y^{2*} -_* e^3).$$

Exercise 3.6.6 *Use the transformation*

$$y = x^{n*} \cdot_* v$$

for appropriate $n \in \mathbb{R}$ to solve the equation

$$x^{3*} \cdot_* y^* = e^2 \cdot_* y \cdot_* (x^{2*} -_* y).$$

Answer

$$y \cdot_* (e^2 \cdot_* \log_* |x|_* +_* c) = x^{2*},$$

$y = 0.$

3.7   Advanced Practical Problems

Problem 3.7.1 *Solve the equations*

1. $(x \cdot_* y^* -_* e) \cdot_* \log_* x = e^2 \cdot_* y.$
2. $x \cdot_* y^* +_* (x +_* e) \cdot_* y = e^3 \cdot_* x^{2*} \cdot_* e^{-*x}.$
3. $y^* -_* (\cot_* x) \cdot_* y = e^2 \cdot_* x \cdot_* \sin_* x.$

Answer

1. $y = C \cdot_* (\log_* x)^{2*} -_* \log_* x.$
2. $x \cdot_* y = (x^{3*} +_* C) \cdot_* e^{-*x}.$
3. $y = C \cdot_* \sin_* x +_* x^{2*} \cdot_* \sin_* x.$

Here C is a multiplicative constant.

Problem 3.7.2 *Solve the equations*

1. $\left(e^{-*y} -_* x\right) \cdot_* y^* = e.$

2. $y^{2*} -_* e +_* e^2 \cdot_* \left(x -_* y \cdot_* (e +_* y)^{2*}\right) \cdot_* y^* = 1.$

3. $e +_* y^{2*} = ((e/_* \tan_* y) -_* x) \cdot_* y^*,\ \tan_* y \neq 1.$

Answer

1. $x \cdot_* e^y = y +_* C.$

2. $x \cdot_* (y -_* e) = (y^{2*} +_* C) \cdot_* (y +_* e).$

3. $x = (e/_* \tan_* y) -_* e +_* C \cdot_* e^{-*(e/_* \tan_* y)}.$

Here C is a multiplicative constant.

Problem 3.7.3 *Solve the equations*

1. $x \cdot_* y^* -_* e^2 \cdot_* x^{2*} \cdot_* y^{\frac{1}{2}*} = e^4 \cdot_* y.$

2. $x \cdot_* y^* +_* e^2 \cdot_* y +_* x^{5*} \cdot_* y^{3*} \cdot_* e^x = 1.$

3. $(e^2 \cdot_* x^{2*} \cdot_* y \cdot_* \log_* y -_* x) \cdot_* y^* = y.$

Answer

1. $y = x^{4*} \cdot_* (\log_*(c \cdot_* x))^{2*},\ y = 1.$

2. $y^{-*2*} = x^{4*} \cdot_* (e^2 \cdot_* e^x +_* c),\ y = 1.$

3. $x \cdot_* y \cdot_* (c -_* (\log_* y)^{2*}) = e.$

Here c is a multiplicative constant.

Problem 3.7.4 *Find the general solution of the following Riccati equations.*

1. $y^* = e +_* x^{2*} -_* e^2 \cdot_* x \cdot_* y +_* y^{2*},\ y_1(x) = x.$

2. $(e -_* x^{2*}) \cdot_* y^* +_* y^{2*} -_* e = 1,\ y_1(x) = x.$

3. $y^* +_* e^{-*x} \cdot_* y^{2*} -_* y -_* e^x = 1,\ y_1(x) = e^x.$

Answer

1. $y = x +_* (e/_*(c -_* x)).$

2. $y = x +_* (e/_*(e -_* x^{2*})) \cdot_* (c +_* x).$

3. $y = e^x +_* e^x \cdot_* \left(c -_* e^{\frac{1}{2}} \cdot_* e^{-*e^2 \cdot_* x}\right).$

Here c is a multiplicative constant.

Problem 3.7.5 *Find a continuous solution in $[1,\infty)$ of the equation*

$$y^* +_* a(x) \cdot_* y = 1,$$

where

$$a(x) = \begin{cases} x & 1 \le x < e \\ e^2 & x \ge e. \end{cases}$$

Answer

$$y(x) = \begin{cases} c_1 \cdot_* e^{-_*(x^{2_*}/_*e^2)} & 1 \le x < e \\ c_1 \cdot_* e^{-_*e^2 \cdot_* x + _* e^{\frac{3}{2}}} & x \ge e. \end{cases}$$

Here c_1 is a multiplicative constant.

Problem 3.7.6 *Find a continuous solution in $[1,\infty)$ of the equation*

$$y^* -_* e^2 \cdot_* y = \begin{cases} e^3 & x \in [1,e^2) \\ e^{-1} & x \ge e^2. \end{cases}$$

Answer

$$y(x) = \begin{cases} c_1 \cdot_* e^{e^2 \cdot_* x} -_* e^{\frac{3}{2}} & x \in [1,e^2) \\ \left(c_1 -_* e^2 \cdot_* e^{-_*e^4}\right) \cdot_* e^{e^2 \cdot_* x} +_* e^{\frac{1}{2}} & x \ge e^2. \end{cases}$$

Here c_1 is a multiplicative constant.

Problem 3.7.7 *Find a continuous solution in $[1,\infty)$ of the equation*

$$y^* +_* a(x) \cdot_* y = e,$$

where

$$a(x) = \begin{cases} e^{-1} & x \in [1,e) \\ e^{-3} & x \ge e. \end{cases}$$

Answer

$$y(x) = \begin{cases} c_1 \cdot_* e^x -_* e & x \in [1, e) \\ \left(c_1 \cdot_* e^{-_* e^2} -_* e^{\frac{2}{3}} \cdot_* e^{-_* e^3} \right) \cdot_* e^{e^3 \cdot_* x} -_* e^{\frac{1}{3}} & x \geq e. \end{cases}$$

Here c_1 is a multiplicative constant.

Problem 3.7.8 *Find a continuous solution in $[1, \infty)$ of the equation*

$$y^* +_* a(x) \cdot_* y = x,$$

where

$$a(x) = \begin{cases} x & 1 \leq x < e \\ e^{-1} & x \geq e. \end{cases}$$

Answer

$$y(x) = \begin{cases} c_1 \cdot_* e^{-_*(x^{2_*}/_* e^2)} +_* e & x \in [1, e) \\ \left(c_1 \cdot_* e^{-_* e^{\frac{3}{2}}} \cdot_* e^3 \cdot_* e^{-_* e} \right) \cdot_* e^x -_* x -_* e & x \geq e. \end{cases}$$

Here c_1 is a multiplicative constant.

Problem 3.7.9 *Solve the Cauchy problem*

1.

$$y^* +_* \tan_* x \cdot_* y \;=\; e/_* \cos_* x, x > 1,$$

$$y(1) \;=\; 1,$$

2.

$$y^* +_* y \;=\; x^{3_*}, x > 1,$$

$$y(1) \;=\; 1,$$

3.

$$y^* +_* e^2 \cdot_* y \;=\; e^x, x > 0,$$

$$y(1) \;=\; e.$$

Answer

1. $y(x) = \sin_* x,.$
2. $y(x) = -_* e^6 \cdot_* e^{-_* x} +_* x^{3*} +_* e^3 \cdot_* x^{2*} -_* e^6 \cdot_* x +_* e^6.$
3. $y(x) = e^{\frac{2}{3}} \cdot_* e^{-_* e^2 \cdot_* x} +_* e^{\frac{1}{3}} \cdot_* e^x.$

Problem 3.7.10 *Use the transformation*

$$y = v \cdot_* x^{n*}$$

for appropriate $n \in \mathbb{R}$ to solve the equation

$$y^* = ((y -_* x \cdot_* y^{2*})/_* (x +_* x^{2*} \cdot_*)y).$$

Answer

$$y \cdot_* e^{x \cdot_* y} = c \cdot_* x.$$

4

Second-Order Linear Multiplicative Differential Equations

Suppose that $J \subseteq \mathbb{R}_*$.

4.1 General Properties

Consider the homogeneous linear second-order MDE

$$a_0(x) \cdot_* y^{**} +_* a_1(x) \cdot_* y^* +_* a_2(x) \cdot_* y = 0_*, \quad x \in J, \tag{4.1}$$

and the nonhomogeneous linear second-order MDE

$$a_0(x) \cdot_* y^{**} +_* a_1(x) \cdot_* y^* +_* a_2(x) \cdot_* y = f(x), \quad x \in J, \tag{4.2}$$

where $a_0, a_1, a_2, f : J \to \mathbb{R}_*$, $a_0, a_1, a_2, f \in \mathscr{C}(J)$, $a_0(x) \neq 0_*$, $x \in J$, are given functions and y is unknown. The equation (4.2) can be rewritten in the following form:

$$e^{\log(f(x))} \; = \; e^{\log(a_0(x)) \log(y^{**}(x))} +_* e^{\log(a_1(x)) \log(y^*(x))}$$

$$+_* e^{\log(a_2(x)) \log(y(x))}$$

$$= \; e^{\log(a_0(x)) \log e^{x(x(\log y)''(x) + (\log y)'(x))}}$$

$$+_* e^{\log(a_1(x)) \log e^{x(\log y)'(x)}}$$

$$+_* e^{\log(a_2(x)) \log(y(x))}$$

DOI: 10.1201/9781003393344-4

$$= e^{\log(a_0(x))}(x(x(\log y)''(x)+(\log y)'(x)))$$

$$+_* e^{\log(a_1(x))}(x(\log y)'(x))$$

$$+_* e^{\log(a_2(x))}\log(y(x))$$

$$= \exp\Big(x^2 \log(a_0(x))(\log y)''(x)$$

$$+x\log(a_1(x))(\log y)'(x)+\log(a_2(x))\log(y(x))\Big)$$

$$= \exp\Big(x^2 \log(a_0(x))(\log y)''(x)$$

$$+x\,(\log(a_0(x))+\log(a_1(x)))\,(\log y)'(x)$$

$$+\log(a_2(x))\log(y(x))\Big)$$

$$= \exp\Big(x^2 \log(a_0(x))(\log y)''(x)+x\log(a_0(x)a_1(x))(\log y)'(x)$$

$$+\log(a_2(x))\log y(x)\Big), \quad x\in J,$$

whereupon

$$x^2 \log(a_0(x))(\log y)''(x)+x\log(a_0(x)a_1(x))(\log y)'(x)$$

$$+\log(a_2(x))\log(y(x))=\log(f(x)), \quad x\in J.$$

Set

$$z(x)=\log(y(x)), \quad x\in J.$$

Thus, we get the equation

$$x^2 \log(a_0(x))z''(x)+x\log(a_0(x)a_1(x))z'(x)$$

$$+\log(a_2(x))z(x)=\log f(x), \quad x\in J. \tag{4.3}$$

Definition 4.1.1 *We say that a function* $y : J \to \mathbb{R}_*$ *is a solution to the equation (4.2) if it satisfies the following conditions.*

    1. $y \in \mathscr{C}_*^2(J)$.

    2.

$$a_0(x) \cdot_* y^{**}(x) +_* a_1(x) \cdot_* y^*(x) +_* a_2(x) \cdot_* y(x) = f(x)$$

    for any $x \in J$.

Example 4.1.2 *Consider the equation*

$$y^{**} +_* y^* -_* e^2 \cdot_* y = 0_*, \quad x \in J.$$

We will prove that

$$y(x) = e^{-2x}, \quad x \in \mathbb{R}_*,$$

is its solution.

First Way. *Note that*

$$y'(x) \;\; = \;\; -2e^{-2x},$$

$$y^*(x) \;\; = \;\; e^{-2x}, \quad x \in \mathbb{R}_*,$$

and

$$(y^*)'(x) \;\; = \;\; -2e^{-2x},$$

$$y^{**}(x) \;\; = \;\; e^{-2x}, \quad x \in \mathbb{R}_*.$$

Then

$$
\begin{aligned}
y^{**}(x) +_* y^*(x) -_* e^2 \cdot_* y(x) \;\; &= \;\; e^{-2x} +_* e^{-2x} -_* e^2 \cdot_* e^{-2x} \\
&= \;\; e^{-4x} -_* e^{\log(e^2)\log(e^{-2x})} \\
&= \;\; e^{-4x} -_* e^{-4x}
\end{aligned}
$$

$$= \frac{e^{-4x}}{e^{-4x}}$$

$$= 1$$

$$= 0_*, \quad x \in \mathbb{R}_*.$$

Thus, y is a solution to the given equation.

Second Way. *Let*

$$v(x) = 2x, \quad x \in \mathbb{R}_*.$$

Then

$$v(x) = \log(y(x)), \quad x \in \mathbb{R}_*.$$

Using (4.3), the given equation can be rewritten in the form

$$x^2 z'' + 2xz' - 2z = 0, \quad x \in \mathbb{R}^*. \tag{4.4}$$

We have

$$v'(x) \;=\; 2,$$

$$v''(x) \;=\; 0, \quad x \in \mathbb{R}_*.$$

Then

$$x^2 v''(x) + 2xv'(x) - 2v(x) \;=\; 4x - 4x$$

$$=\; 0, \quad x \in \mathbb{R}_*.$$

Thus, v is a solution to the equation (4.4) and then y is a solution to the given equation.

Example 4.1.3 *Consider the equation*

$$y^{**} +_* y = e^4 \cdot_* x \cdot_* e^x, \quad x \in \mathbb{R}_*.$$

We will prove that

$$y(x) = \sin_* x +_* \left(e^2 \cdot_* x -_* e^2\right) \cdot_* e^x, \quad x \in \mathbb{R}_*,$$

is its solution.

First Way. *We have*

$$y^*(x) = \cos_* x +_* e^2 \cdot_* e^x +_* \left(e^2 \cdot_* x -_* e^2\right) \cdot_* e^x$$

$$= \cos_* x +_* e^2 \cdot_* x \cdot_* e^x,$$

$$y^{**}(x) = -_* \sin_* x +_* e^2 \cdot_* e^x +_* e^2 \cdot_* x \cdot_* e^x$$

$$= -_* \sin_* x +_* e^2 \cdot_* \left(1_* +_* x\right) \cdot_* e^x, \quad x \in \mathbb{R}_*.$$

Then

$$y^{**}(x) +_* y(x) = \sin_* x +_* \left(e^2 \cdot_* x -_* e^2\right) \cdot_* e^x$$

$$-_* \sin_* x +_* e^2 \cdot_* \left(1_* +_* x\right) \cdot_* e^x$$

$$= e^4 \cdot_* x \cdot_* e^x, \quad x \in \mathbb{R}_*.$$

Second Way. *Here*

$$f(x) = e^4 \cdot_* x \cdot_* e^x$$

$$= e^{\log(e^4)\log x \log(e^x)}$$

$$= e^{4x\log x}, \quad x \in \mathbb{R}_*.$$

Then, using (4.3), the given equation can be written in the form

$$x^2 z'' + xz' + z = 4x\log x, \quad x \in \mathbb{R}_*. \tag{4.5}$$

Next, for y we have the following representation

$$y(x) = e^{\sin(\log x)} +_* \left(e^{2\log x} -_* e^2\right) \cdot_* e^x$$

$$= e^{\sin(\log x)} +_* e^{2\log x - 2} \cdot_* e^x$$

$$= e^{\sin(\log x)} +_* e^{2x(\log x - 1)}$$

$$= e^{\sin(\log x) + 2x(\log x - 1)}, \quad x \in \mathbb{R}_*.$$

Set

$$v(x) = \sin(\log x) + 2x(\log x - 1), \quad x \in \mathbb{R}_*.$$

Then

$$v(x) = \log(y(x)), \quad x \in \mathbb{R}_*,$$

and

$$v'(x) = \frac{\cos(\log x)}{x} + 2(\log x - 1) + 2,$$

$$v''(x) = -\frac{\sin(\log x)}{x^2} - \frac{\cos(\log x)}{x^2} + \frac{2}{x}, \quad x \in \mathbb{R}_*.$$

Hence,

$$
\begin{aligned}
x^2 v''(x) + x v'(x) + v(x) &= x^2 \left(-\frac{\sin(\log x)}{x^2} - \frac{\cos(\log x)}{x^2} + \frac{2}{x} \right) \\
&\quad + x \left(\frac{\cos(\log x)}{x} + 2(\log x - 1) + 2 \right) \\
&\quad + \sin(\log x) + 2x(\log x - 1) \\
&= -\sin(\log x) - \cos(\log x) + 2x \\
&\quad + \cos(\log x) + 2x(\log x - 1) + 2x \\
&\quad + \sin(\log x) + 2x(\log x - 1) \\
&= 4x(\log x - 1) + 4x \\
&= 4x \log x, \quad x \in \mathbb{R}_*.
\end{aligned}
$$

Thus, v satisfies (4.5) and hence, y satisfies the given equation.

Example 4.1.4 *Consider the equation*

$$x^{2*} \cdot_* y^{**} +_* x \cdot_* y^* +_* e^4 \cdot_* y = e^{10} \cdot_* x, \quad x \in \mathbb{R}_*.$$

We will prove that

$$y(x) \;=\; \cos_* \left(e^2 \cdot_* \log_* |x|_*\right) +_* \sin_* \left(e^2 \cdot_* \log_* |x|_*\right)$$

$$+_* e^2 \cdot_* x, \quad x \in \mathbb{R}_*,$$

is its solution.

First Way. *We have*

$$y^*(x) \;=\; -_* \left(\sin_* \left(e^2 \cdot_* \log_* |x|_*\right)\right) \cdot_* \left(e^2 /_* x\right)$$

$$+_* \left(\cos_* \left(e^2 \cdot_* \log_* |x|_*\right)\right) \cdot_* \left(e^2 /_* x\right) +_* e^2,$$

$$y^{**}(x) \;=\; -_* \left(\cos_* \left(e^2 \cdot_* \log_* |x|_*\right)\right) \cdot_* \left(e^2 /_* x\right)^{2*}$$

$$-_* \left(\sin_* \left(e^2 \cdot_* \log_* |x|_*\right)\right) \cdot_* \left(e^2 /_* x\right)^{2*}$$

$$+_* \left(\sin_* \left(e^2 \cdot_* \log_* |x|_*\right)\right) \cdot_* \left(e^2 /_* x^{2*}\right)$$

$$-_* \left(\cos_* \left(e^2 \cdot_* \log_* |x|_*\right)\right) \cdot_* \left(e^2 /_* x^{2*}\right), \quad x \in \mathbb{R}_*.$$

Therefore,

$$x^{2*} \cdot_* y^{**} +_* x \cdot_* y^* +_* e^4$$

$$= x^{2*} \cdot_* \left(-_* \left(\cos_* \left(e^2 \cdot_* \log_* |x|_*\right)\right) \cdot_* \left(e^2 /_* x\right)^{2*} \right.$$

$$-_* \left(\sin_* \left(e^2 \cdot_* \log_* |x|_*\right)\right) \cdot_* \left(e^2 /_* x\right)^{2*}$$

$$+_* \left(\sin_* \left(e^2 \cdot_* \log_* |x|_*\right)\right) \cdot_* \left(e^2 /_* x^{2*}\right)$$

$$-_* \left(\cos_* \left(e^2 \cdot_* \log_* |x|_* \right) \right) \cdot_* \left(e^2 /_* x^{2*} \right) \Big)$$

$$+_* x \cdot_* \left(-_* \left(\cos_* \left(e^2 \cdot_* \log_* |x|_* \right) \right) \cdot_* \left(e^2 /_* x \right)^{2*} \right.$$

$$-_* \left(\sin_* \left(e^2 \cdot_* \log_* |x|_* \right) \right) \cdot_* \left(e^2 /_* x \right)^{2*}$$

$$+_* \left(\sin_* \left(e^2 \cdot_* \log_* |x|_* \right) \right) \cdot_* \left(e^2 /_* x^{2*} \right)$$

$$-_* \left(\cos_* \left(e^2 \cdot_* \log_* |x|_* \right) \right) \cdot_* \left(e^2 /_* x^{2*} \right) \right) +_* e^2 \cdot_* x$$

$$+_* e^4 \cdot_* \left(\cos_* \left(e^2 \cdot_* \log_* |x|_* \right) +_* \sin_* \left(e^2 \cdot_* \log_* |x|_* \right) \right.$$

$$+_* e^2 \cdot_* x \right)$$

$$= \ -_* e^4 \cdot_* \cos_* \left(e^2 \cdot_* \log_* |x|_* \right)$$

$$-_* e^4 \cdot_* \sin_* \left(e^2 \cdot_* \log_* |x|_* \right)$$

$$+_* e^2 \cdot_* \sin_* \left(e^2 \cdot_* \log_* |x|_* \right)$$

$$-_* e^2 \cdot_* \cos_* \left(e^2 \cdot_* \log_* |x|_* \right)$$

$$-_* e^2 \cdot_* \sin_* \left(e^2 \cdot_* \log_* |x|_* \right)$$

$$+_* e^2 \cdot_* \cos_* \left(e^2 \cdot_* \log_* |x|_* \right) +_* e^2 \cdot_* x$$

$$+_* e^4 \cdot_* \cos_* \left(e^2 \cdot_* \log_* |x|_* \right)$$

$$+_* e^4 \cdot_* \sin_* \left(e^2 \cdot_* \log_* |x|_* \right) +_* e^8 \cdot_* x$$

$$= e^{10} \cdot_* x, \quad x \in \mathbb{R}_8.$$

Second Way. *Here*

$$f(x) = e^{10} \cdot_* x$$

$$= e^{10 \log x},$$

$$a_0(x) = x^{2_*}$$

$$= e^{(\log x)^2},$$

$$a_1(x) = x,$$

$$a_2(x) = e^4, \quad x \in \mathbb{R}_*.$$

Then, using (4.3), we can rewrite the given equation in the following form.

$$x^2 (\log x)^2 z'' + x \log x (\log x + 1) z' + 4z = 10 \log x, \quad x \in \mathbb{R}_*. \quad (4.6)$$

For the function y we have the following representation

$$y(x) = \cos_* \left(e^2 \cdot_* \log_* |x|_* \right) +_* \sin_* \left(e^2 \cdot_* \log_* |x|_* \right)$$

$$+_* e^2 \cdot_* x$$

$$= \cos_* \left(e^{2 \log |\log x|} \right) +_* \sin_* \left(e^{2 \log |\log x|} \right) +_* e^{2 \log x}$$

$$= \cos_* \left(e^{\log((\log x)^2)} \right) +_* \sin_* \left(e^{\log((\log x)^2)} \right) +_* e^{2 \log x}$$

$$= e^{\cos(\log((\log x)^2)) + \sin \cos(\log((\log x)^2)) + 2 \log x}, \quad x \in \mathbb{R}_*.$$

Set

$$v(x) = \cos\left(\log\left((\log x)^2\right)\right) + \sin\left(\log\left((\log x)^2\right)\right) + 2 \log x, \quad x \in \mathbb{R}_*.$$

We have

$$v(x) = \log(y(x)), \quad x \in \mathbb{R}_*,$$

and

$$
\begin{aligned}
v'(x) &= -\sin\left(\log\left((\log x)^2\right)\right) \frac{1}{(\log x)^2} \cdot \frac{2\log x}{x} \\
&\quad + \cos\left(\log\left((\log x)^2\right)\right) \frac{1}{(\log x)^2} \cdot \frac{2\log x}{x} + \frac{2}{x} \\
&= -\sin\left(\log\left((\log x)^2\right)\right) \frac{2}{x\log x} \\
&\quad + \cos\left(\log\left((\log x)^2\right)\right) \frac{2}{x\log x} + \frac{2}{x},
\end{aligned}
$$

$$
\begin{aligned}
v''(x) &= -\cos\left(\log\left((\log x)^2\right)\right) \frac{4}{x^2(\log x)^2} \\
&\quad - \sin\left(\log\left((\log x)^2\right)\right) \frac{4}{x^2(\log x)^2} - \frac{2}{x^2} \\
&\quad + 2\sin\left(\log\left((\log x)^2\right)\right) \frac{\log x + 1}{x^2(\log x)^2} \\
&\quad - 2\cos\left(\log\left((\log x)^2\right)\right) \frac{\log x + 1}{x^2(\log x)^2}, \quad x \in \mathbb{R}_*.
\end{aligned}
$$

Hence,

$$x^2(\log x)^2 v''(x) + x\log x(\log x + 1)v'(x) + 4v(x)$$

$$
\begin{aligned}
&= x^2(\log x)^2 \Bigg(-\cos\left(\log\left((\log x)^2\right)\right) \frac{4}{x^2(\log x)^2} \\
&\quad - \sin\left(\log\left((\log x)^2\right)\right) \frac{4}{x^2(\log x)^2} - \frac{2}{x^2} \\
&\quad + 2\sin\left(\log\left((\log x)^2\right)\right) \frac{\log x + 1}{x^2(\log x)^2} \\
&\quad - 2\cos\left(\log\left((\log x)^2\right)\right) \frac{\log x + 1}{x^2(\log x)^2} \Bigg)
\end{aligned}
$$

$$+x\log x(\log x+1)\left(-\sin\left(\log\left((\log x)^2\right)\right)\frac{2}{x\log x}\right.$$

$$+\cos\left(\log\left((\log x)^2\right)\right)\frac{2}{x\log x}+\frac{2}{x}\right)$$

$$+4\cos\left(\log\left((\log x)^2\right)\right)+4\sin\left(\log\left((\log x)^2\right)\right)$$

$$+8\log x$$

$$=\quad-4\cos\left(\log\left((\log x)^2\right)\right)-4\sin\left(\log\left((\log x)^2\right)\right)$$

$$-2(\log x)^2$$

$$+2(\log x+1)\sin\left(\log\left((\log x)^2\right)\right)$$

$$-2(\log x+1)\cos\left(\log\left((\log x)^2\right)\right)$$

$$-2(\log x+1)\sin\left(\log\left((\log x)^2\right)\right)$$

$$+2(\log x+1)\cos\left(\log\left((\log x)^2\right)\right)$$

$$+2\log x(\log x+1)+4\cos\left(\log\left((\log x)^2\right)\right)$$

$$+4\sin\left(\log\left((\log x)^2\right)\right)+8\log x$$

$$=\quad10\log x,\quad x\in\mathbb{R}_*.$$

Thus, v is a solution to the equation (4.6) and then y is a solution to the given equation.

Exercise 4.1.5 *Prove that the function*

$$y(x)=e^x+_*e^{2x}+_*e^{\frac{1}{10}}\cdot_*\sin_*x,\quad x\in\mathbb{R}_*,$$

is a solution to the equation

$$y^{**} -_* e^3 \cdot_* y^* +_* e^2 \cdot_* y = \sin_* x, \quad x \in \mathbb{R}_*.$$

Theorem 4.1.6 *Let y_1, y_2 be solutions to the equations*

$$a_0(x) \cdot_* y^{**} +_* a_1(x) \cdot_* y^* +_* a_2(x) \cdot_* y = f_1(x), \quad x \in J,$$

and

$$a_0(x) \cdot_* y^{**} +_* a_1(x) \cdot_* y^* +_* a_2(x) \cdot_* y = f_2(x), \quad x \in J,$$

respectively, where $f_1, f_2 : J \to \mathbb{R}_*$, $f_1, f_2 \in \mathscr{C}(J)$. Then, for any $c_1, c_2 \in \mathbb{R}_*$, the function $c_1 \cdot_* y_1 +_* c_2 \cdot_* y_2$ is a solution to the equation

$$a_0(x) \cdot_* y^{**} +_* a_1(x) \cdot_* y^* +_* a_2(x) \cdot_* y = c_1 \cdot_* f_1(x) +_* c_2 \cdot_* f_2(x), \quad x \in J. \tag{4.7}$$

Proof 4.1.7 First Way. *We have*

$$a_0(x) \cdot_* y_1^{**}(x) +_* a_1(x) \cdot_* y_1^*(x) +_* a_2(x) \cdot_* y_1(x) = f(x), \quad x \in J,$$

and

$$a_0(x) \cdot_* y_2^{**}(x) +_* a_1(x) \cdot_* y_2^*(x) +_* a_2(x) \cdot_* y_2(x) = f(x), \quad x \in J.$$

Then

$$a_0(x) \cdot_* (c_1 \cdot_* y_1 +_* c_2 \cdot_* y_2)^{**}(x)$$

$$+_* a_1(x) \cdot_* (c_1 \cdot_* y_1 +_* c_2 \cdot_* y_2)^*(x)$$

$$+_* a_2(x) \cdot_* (c_1 \cdot_* y_1(x) +_* c_2 \cdot_* y_2(x))$$

$$= \; a_0(x) \cdot_* (c_1 \cdot_* y_1^{**}(x) +_* c_2 \cdot_* y_2^{**}(x))$$

$$+_* a_1(x) \cdot_* (c_1 \cdot_* y_1^*(x) +_* c_2 \cdot_* y_2^*(x))$$

$$+_* a_2(x) \cdot_* (c_1 \cdot_* y_1(x) +_* c_2 \cdot_* y_2(x))$$

$$= \quad c_1 \cdot_* (a_0(x) \cdot_* y_1^{**}(x) +_* a_1(x) \cdot_* y_1^*(x) +_* a_2(x) \cdot_* y_1(x))$$

$$+_* c_2 \cdot_* (a_0(x) \cdot_* y_2^{**}(x) +_* a_1(x) \cdot_* y_2^*(x) +_* a_2(x) \cdot_* y_2(x))$$

$$= \quad c_1 \cdot_* f_1(x) +_* c_2 \cdot_* f_2(x), \quad x \in J.$$

This completes the proof.

Second Way. *Let*

$$z_1(x) \quad = \quad \log(y_1(x)),$$

$$z_2(x) \quad = \quad \log(y_2(x)), \quad x \in J.$$

Then

$$x^2 \log(a_0(x)) z_1''(x) + x \log(a_0(x) a_1(x)) z_1'(x)$$

$$+ \log(a_2(x)) z_1(x) = \log(f_1(x)), \quad x \in J,$$

and

$$x^2 \log(a_0(x)) z_2''(x) + x \log(a_0(x) a_1(x)) z_2'(x)$$

$$+ \log(a_2(x)) z_2(x) = \log(f_2(x)), \quad x \in J.$$

Hence,

$$x^2 \log(a_0(x)) (\log c_1 z_1 + \log c_2 z_2)''(x)$$

$$+ x \log(a_0(x) a_1(x)) (\log c_1 z_1 + \log c_2 z_2)'(x)$$

$$+ \log(a_2(x)) (\log c_1 z_1(x) + \log c_2 z_2(x))$$

$$= x^2 \log(a_0(x)) \left(\log c_1 z_1''(x) + \log c_2 z_2''(x) \right)$$

$$+ x \log(a_0(x) a_1(x)) \left(\log c_1 z_1'(x) + \log c_2 z_2'(x) \right)$$

$$+ \log(a_2(x)) \left(\log c_1 z_1(x) + \log c_2 z_2(x) \right)$$

$$= \log c_1 \left(x^2 \log(a_0(x)) z_1''(x) + \log(a_0(x) a_1(x)) z_1'(x) \right.$$

$$\left. + \log(a_2(x)) z_1(x) \right)$$

$$+ \log c_2 \left(x^2 \log(a_0(x)) z_2''(x) + \log(a_0(x) a_1(x)) z_2'(x) \right.$$

$$\left. + \log(a_2(x)) z_2(x) \right)$$

$$= \log c_1 \log f_1(x) + \log c_2 \log f_2(x), \quad x \in J,$$

i.e., $\log c_1 z_1 + \log c_2 z_2$ *is a solution to the equation*

$$x^2 \log(a_0(x)) z'' + x \log(a_0(x) a_1(x)) z'$$

$$+ \log(a_2(x)) z(x) = \log c_1 \log f_1(x) + \log c_2 \log f_2(x), \quad x \in J.$$

Because

$$c_1 \cdot_* f_1(x) +_* c_2 \cdot_* f_2(x) = e^{\log c_1 \log(f_1(x))} +_* e^{\log c_2 \log(f_2(x))}$$

$$= e^{\log c_1 \log(f_1(x)) + \log c_2 \log(f_2(x))}, \quad x \in J,$$

we conclude that $c_1 \cdot_* y_1 +_* c_2 \cdot_* y_2$ *is a solution to the equation* (4.7). *This completes the proof.*

Corollary 4.1.8 *Let* y_1, y_2 *be solutions to the homogeneous equation* (4.1). *Then*

$$c_1 \cdot_* y_1 +_* c_2 \cdot_* y_2$$

is a solution to the equation (4.1) *for any* $c_1, c_2 \in \mathbb{R}_*$.

Proof 4.1.9 *We apply Theorem 4.1.6 for*

$$f_1 = 0_* \quad and \quad f_2 = 0_*$$

and we get the desired result. This completes the proof.

Corollary 4.1.10 *Let y be a solution to the equation (4.2). Then $v :$ $J \to \mathbb{R}_*$, $v \in \mathscr{C}^1_*(J)$, is a solution to the equation (4.2) if and only if $y -_* v$ is a solution to the equation (4.1).*

Proof 4.1.11   *1. Let $v : J \to \mathbb{R}_*$, $v \in \mathscr{C}^1_*(J)$, be a solution to the equation (4.2). Now, we apply Theorem 4.1.6 for*

$$c_1 = 1_* \quad and \quad c_2 = -_* 1_*$$

and we get that $y -_* v$ is a solution to the equation (4.1).

    2. Let $v : J \to \mathbb{R}_*$, $Rv \in \mathscr{C}^1_*(J)$, be such that $y -_* v$ is a solution to the equation (4.1). Hence, applying Theorem 4.1.6 for

$$c_1 = c_2 = -_* 1_*,$$

we get that

$$-_* y -_* (y -_* v) = v$$

is a solution to the equation (4.2). This completes the proof.

4.2 Multiplicative Linear Dependence

Definition 4.2.1 *The vector-valued functions $u^j : J \to \mathbb{R}^n_*$, $j \in \{1,\ldots,m\}$, are said to be multiplicative linearly independent on J if the relation*

$$c_1 \cdot_* u^1(x) +_* \cdots +_* c_m \cdot_* u^m(x) = 0_*, \quad x \in J, \tag{4.8}$$

implies that

$$c_1 = \ldots = c_m = 0_*.$$

Conversely, these functions are said to be multiplicative linearly dependent on J.

The relation (4.8) can be rewritten in the form

$$e^0 = 0_*$$

$$= c_1 \cdot_* \begin{pmatrix} u_1^1(x) \\ \vdots \\ u_n^1(x) \end{pmatrix} +_* \cdots +_* c_m \cdot_* \begin{pmatrix} u_1^m(x) \\ \vdots \\ u_n^m(x) \end{pmatrix}$$

$$= \begin{pmatrix} c_1 \cdot_* u_1^1(x) \\ \vdots \\ c_1 \cdot_* u_n^1(x) \end{pmatrix} +_* \cdots +_* \begin{pmatrix} c_m \cdot_* u_1^m(x) \\ \vdots \\ c_m \cdot_8 u_n^m(x) \end{pmatrix}$$

$$= \begin{pmatrix} e^{\log c_1 \log(u_1^1(x))} \\ \vdots \\ e^{\log c_1 \log(u_n^1(x))} \end{pmatrix} +_* \cdots +_* \begin{pmatrix} e^{\log c_m \log(u_1^m(x))} \\ \vdots \\ e^{\log c_m \log(u_n^m(x))} \end{pmatrix}$$

$$= \begin{pmatrix} e^{\log c_1 \log(u_1^1(x))+_*\cdots+_*\log c_m \log(u_1^m(x))} \\ \vdots \\ e^{\log c_1 \log(u_n^1(x))+\cdots+\log c_m \log(u_n^m(x))} \end{pmatrix}, \quad x \in J.$$

Set

$$\log u^j(x) = \begin{pmatrix} \log(u_1^j(x)) \\ \vdots \\ \log(u_n^j(x)) \end{pmatrix}, \quad x \in J, \quad j \in \{1,\ldots,m\}.$$

Then (4.8) takes the form

$$\log c_1 \log(u^1(x)) + \cdots + \log c_m \log(u^m(x)) = 0, \quad x \in J. \qquad (4.9)$$

Example 4.2.2 *Let*

$$u^1(x) = e^x,$$

$$u^2(x) = e^{2x},$$

$$u^3(x) = e^{3x}, \quad x \in \mathbb{R}_*.$$

We have that u^1, u^2, u^3 are linearly independent on $\mathbb{R}_*$. On the other hand, for $c_1, c_2, c_3 \in \mathbb{R}_*$, we have

$$\log c_1 \log(u^1(x)) + \log c_2 \log(u^2(x)) + \log c_3 \log(u^3(x))$$

$$= \log c_1 \log(e^x) + \log c_2 \log(e^{2x}) + \log c_3 \log(e^{3x})$$

$$= \log c_1 x + 2\log c_2 x + 3\log c_3 x$$

$$= (\log c_1 + 2\log c_2 + 3\log c_3)x, \quad x \in \mathbb{R}_*.$$

Take

$$c_1 = e^{-5},$$

$$c_2 = e,$$

$$c_3 = e.$$

Then

$$\log c_1 + 2\log c_2 + 3\log c_3 = \log(e^{-5}) + 2\log e + 3\log e$$

$$= -5 + 2 + 3$$

$$= 0.$$

Therefore, u^1, u^2, u^3 are multiplicative linearly dependent.

Example 4.2.3 *Let*

$$u^1(x) = x,$$

$$u^2(x) = 2x,$$

$$u^3(x) = 3x, \quad x \in \mathbb{R}_*.$$

We have that u^1, u^2, u^3 are linearly dependent on $\mathbb{R}_*$. Let now $c_1, c_2, c_3 \in \mathbb{R}_*$ be such that

$$\log c_1 \log(u^1(x)) + \log c_2 \log(u^2(x)) + \log c_3 \log(u^3(x)) = 0, \quad x \in \mathbb{R}_*.$$

Then

$$
\begin{aligned}
0 &= \log c_1 \log x + \log c x_2 \log(2x) + \log c_3 \log(3x) \\[2mm]
&= \log c_1 \log x + \log c_2 (\log 2 + \log x) + \log c_3 (\log 3 + \log x) \\[2mm]
&= (\log c_1 + \log c_2 + \log c_3) \log x \\[2mm]
&\quad + \log c_2 \log 2 + \log c_3 \log 3, \quad x \in \mathbb{R}_*,
\end{aligned}
$$

whereupon

$$(\log c_1 + \log c_2 + \log c_3) \log x = -(\log c_2 \log 2 + \log c_3 \log 3), \quad x \in \mathbb{R}_*.$$

The last equation holds for any $x \in \mathbb{R}_*$ if and only if

$$
\begin{aligned}
\log c_1 + \log c_2 + \log c_3 &= 0 \\[2mm]
\log c_2 \log 2 + \log c_3 \log 3 &= 0,
\end{aligned}
$$

if and only if

$$
\begin{aligned}
\log c_1 + \log c_2 + \log c_3 &= 0 \\[2mm]
\log c_2 &= -\frac{\log 3}{\log 2} \log c_3.
\end{aligned}
$$

Hence,

$$
\begin{aligned}
0 &= \log c_1 + \log c_2 + \log c_3 \\[2mm]
&= \log c_1 - \frac{\log 3}{\log 2} \log c_3 + \log c_3
\end{aligned}
$$

$$= \log c_1 + \left(1 - \frac{\log 3}{\log 2}\right)\log c_3$$

$$= \log c_1 + \frac{\log 2 - \log 3}{\log 2}\log c_3$$

$$= \log c_1 + \frac{\log \frac{2}{3}}{\log 2}\log c_3,$$

whereupon

$$\log c_1 = -\frac{\log \frac{2}{3}}{\log 2}\log c_3.$$

Take

$$c_3 = e$$

$$\log c_2 = -\frac{\log 3}{\log 2}$$

$$\log c_1 = -\frac{\log \frac{2}{3}}{\log 2},$$

or

$$c_1 = e^{-\frac{\log \frac{2}{3}}{\log 2}}$$

$$c_2 = e^{-\frac{\log 3}{\log 2}}$$

$$c_3 = e.$$

For these values of c_1, c_2, c_3 we have

$$\log c_1 \log(u^1(x)) + \log c_2 \log(u^2(x)) + \log c_3 \log(u^3(x)) = 0, \quad x \in \mathbb{R}_*.$$

Thus, u^1, u^2, u^3 are multiplicative linearly dependent.

Example 4.2.4 *Let*

$$u^1(x) = x,$$

$$u^2(x) \;=\; 2x, \quad x \in \mathbb{R}_*.$$

We have that u^1, u^2 are linearly dependent on $\mathbb{R}_*$. Let now, $c_1, c_2 \in \mathbb{R}_*$ be such that

$$\log c_1 \log(u^1(x)) + \log c_2 \log(u^2(x)) = 0, \quad \in \mathbb{R}_*.$$

Then

$$
\begin{aligned}
0 \;&=\; \log c_1 \log x + \log c_2 \log(2x) \\[2mm]
&=\; \log c_1 \log x + \log c_2 (\log 2 + \log x) \\[2mm]
&=\; (\log c_1 + \log c_2) \log x + \log c_2 \log 2, \quad x \in \mathbb{R}_*,
\end{aligned}
$$

whereupon

$$(\log c_1 + \log c_2) \log x = -\log c_2 \log 2, \quad x \in \mathbb{R}_*.$$

The last equation holds for any $x \in \mathbb{R}_*$ if and only if

$$
\begin{aligned}
\log c_1 + \log c_2 \;&=\; 0 \\[2mm]
\log c_2 \;&=\; 0,
\end{aligned}
$$

or if and only if

$$
\begin{aligned}
\log c_1 \;&=\; 0 \\[2mm]
\log c_2 \;&=\; 0,
\end{aligned}
$$

or if and only if

$$
\begin{aligned}
c_1 \;&=\; 0_* \\[2mm]
c_2 \;&=\; 0_*.
\end{aligned}
$$

Thus, u^1, u^2 are multiplicative linearly independent.

Exercise 4.2.5 *Using the definition, prove that the functions*

$$u^1(x) = x +_* e^2,$$

$$u^2(x) = x -_* e^2, \quad x \in \mathbb{R}_*,$$

are multiplicative linearly independent.

Definition 4.2.6 *For given n vector-valued functions* $u^1, \ldots, u^n : J \to \mathbb{R}_*^n$, *the multiplicative determinant*

$$W_*(u^1, \ldots, u^n)(x) = W_*(x)$$

$$= \det_*\begin{pmatrix} u_1^1(x) & \cdots & u_1^n(x) \\ \vdots & \vdots & \vdots \\ u_n^1((x) & \cdots & u_n^n(x) \end{pmatrix}, \quad x \in J,$$

is said to be multiplicative Wronskian of these functions.

Theorem 4.2.7 *If* $U^1, \ldots, u^n : J \to \mathbb{R}_*^n$ *and*

$$W_*(u^1, \ldots, u^n)(x_0) \neq 0_*$$

for some $x_0 \in J$, *then* $u^1, \ldots, u^n$ *are multiplicative linearly independent on J.*

Proof 4.2.8 *Assume that* $u^1, \ldots, u^n$ *are multiplicative linearly dependent on J. Then there are* $c_1, \ldots, c_n \in \mathbb{R}_*$ *such that*

$$(c_1, \ldots, c_n) \neq (0_*, \ldots, 0_*)$$

and

$$\sum_{j=1_*}^n c_j \cdot_* u^j(x) = 0_*, \quad x \in J. \tag{4.10}$$

We consider the system (4.10) with respect to $c_1, \ldots, c_n$. *By (4.2.8), it follows that the system (4.10) has a nontrivial solution for any* $x \in J$. *Hence,*

$$W_*(u^1, \ldots, u^n)(x) = 0_*, \quad x \in J.$$

In particular, we get

$$W_*(u^1,\ldots,u^n)(x_0) = 0_*,$$

which is a contradiction. Therefore, $u^1,\ldots,u^n$ are multiplicative linearly independent on J. This completes the proof.

Example 4.2.9 *Consider*

$$u^1(x) = \begin{pmatrix} x^{2*} \\ e \end{pmatrix},$$

$$u^2(x) = \begin{pmatrix} x^{3*} \\ x \end{pmatrix}, \quad x \in \mathbb{R}_*.$$

We have

$$
\begin{aligned}
W_*(u^1,\ldots,u^n)(x) &= \det_* \begin{pmatrix} x^{2*} & x^{3*} \\ e & x \end{pmatrix} \\
&= x^{2*} \cdot_* x -_* e \cdot_* x^{3*} \\
&= x^{3*} -_* x^{3*} \\
&= 0_*, \quad x \in \mathbb{R}_*.
\end{aligned}
$$

Let now, $c_1, c_2 \in \mathbb{R}_*$ be such that

$$c_1 \cdot_* u^1(x) +_* c_2 \cdot_* u^2(x) = 0_*, \quad x \in \mathbb{R}_*.$$

Then, applying (4.9), we find

$$\log c_1 \log x^{2*} + \log c_2 \log x^{3*} = 0$$

$$\log c_1 + \log c_2 \log x = 0, \quad x \in \mathbb{R}_*,$$

$$\log c_1 \log e^{(\log x)^2} + \log c_2 \log e^{(\log x)^3} = 0$$

$$\log c_1 + \log c_2 \log x = 0, \quad x \in \mathbb{R}_*,$$

or

$$\log c_1 (\log x)^2 + \log c_2 (\log x)^3 = 0$$

$$\log c_1 + \log c_2 \log x = 0, \quad x \in \mathbb{R}_*,$$

which is possible if and only if

$$\log c_1 = 0$$

$$\log c_2 = 0,$$

or

$$c_1 = 0_*$$

$$c_2 = 0_*.$$

Thus, u^1, u^2 are multiplicative linearly independent.

Remark 4.2.10 *The last example shows that the condition*

$$W_*(u^1, u^2) \neq 0_* \quad on \quad J$$

is not necessary for the multiplicative linear independence of u^1 and u^2 on J and the condition

$$W_*(u^1, u^2) = 0_* \quad on \quad J$$

does not imply that u^1 and u^2 are multiplicative linearly dependent on J.

Definition 4.2.11 *The multiplicative Wronskian of n functions $y_1, \ldots, y_n :$ $J \to \mathbb{R}_*$, which are $(n-1)$ times multiplicative differentiable on J, is defined by the multiplicative determinant*

$$W_*(y_1, \ldots, y_n)(x) = W_*(x)$$

$$= \det_{*} \begin{pmatrix} y_1(x) & \cdots & y_n(x) \\ y_1^*(x) & \cdots & y_n^*(x) \\ \vdots & \vdots & \vdots \\ y_1^{*(n-1)} & \cdots & y_n^{*(n-1)} \end{pmatrix}, \quad x \in J.$$

Theorem 4.2.12 *Let y_1, y_2, y_3 and λ be multiplicative differentiable on J. Then we have the following.*

1. $W_*(y_1, y_2 +_* y_3)(x) = W_*(y_1, y_2)(x) +_* W_*(y_1, y_3)(x)$, $x \in J$.

2. $W_*(\lambda \cdot_* y_1, \lambda \cdot_* y_2) = (\lambda(x))^{2*} \cdot_* W_*(y_1, y_2)(x)$, $x \in J$.

3. $W_*(y_1, \lambda \cdot_* y_1) = \lambda^*(x) \cdot_* (y_1(x))^{2*}$, $x \in J$.

Proof 4.2.13   *1. We have*

$$W_*(y_1, y_2)(x) = \det_{*} \begin{pmatrix} y_1(x) & y_2(x) \\ y_1^*(x) & y_2^*(x) \end{pmatrix}$$

$$= y_1(x) \cdot_* y_2^*(x) -_* y_1^*(x) \cdot_* y_2(x), \quad x \in J,$$

and

$$W_*(y_1, y_3)(x) = \det_{*} \begin{pmatrix} y_1(x) & y_3(x) \\ y_1^*(x) & y_3^*(x) \end{pmatrix}$$

$$= y_1(x) \cdot_* y_3^*(x) -_* y_1^*(x) \cdot_* y_3(x), \quad x \in J.$$

Hence,

$$W_*(y_1, y_2 +_* y_3)(x) = \det_{*} \begin{pmatrix} y_1(x) & y_2(x) +_* y_3(x) \\ y_1^*(x) & y_2^*(x) +_* y_3^*(x) \end{pmatrix}$$

$$= y_1(x) \cdot_* (y_2^*(x) +_* y_3(x))$$

$$-_* y_1^*(x) \cdot_* (y_2(x) +_* y_3(x))$$

$$= \quad y_1(x) \cdot_* y_2^*(x) +_* y_1(x) \cdot_* y_3^*(x)$$

$$-_* y_1^*(x) \cdot_* y_2(x) -_* y_1^*(x) \cdot_* y_3(x)$$

$$= \quad \left(y_1(x) \cdot_* y_2^*(x) -_* y_1^*(x) \cdot_* y_2(x) \right)$$

$$+_* \left(y_1(x) \cdot_* y_3^*(x) -_* y_1^*(x) \cdot_* y_3(x) \right)$$

$$= \quad W_*(y_1, y_2)(x) +_* W_*(y_1, y_3)(x), \quad x \in J.$$

2. We have

$$W_*(\lambda \cdot_* y_1, \lambda \cdot_* y_2)(x) = \det_* \left(\begin{array}{c} \lambda(x) \cdot_* y_1(x) \\ \lambda^*(x) \cdot_* y_1(x) +_* \lambda(x) \cdot_* y_1(x) \end{array} \right.$$

$$\left. \begin{array}{c} \lambda(x) \cdot_* y_2(x) \\ \lambda^*(x) \cdot_* y_2(x) +_* \lambda(x) \cdot_* y_2^*(x) \end{array} \right)$$

$$= \lambda(x) \cdot_* y_1(x) \cdot_* \left(\lambda^*(x) \cdot_* y_2(x) \right.$$

$$\left. +_* \lambda(x) \cdot_* y_2^*(x) \right)$$

$$-_* \lambda(x) \cdot_* y_2(x) \cdot_* \left(\lambda^*(x) \cdot_* y_1(x) \right.$$

$$\left. +_* \lambda(x) \cdot_* y_1(x) \right)$$

$$= \lambda(x) \cdot_* \lambda^*(x) \cdot_* y_1(x) \cdot_* y_2(x)$$

$$+_* (\lambda(x))^{2_*} \cdot_* y_1(x) \cdot_* y_2^*(x)$$

$$-_* \lambda(x) \cdot_* \lambda^*(x) \cdot_* y_1(x) \cdot_* y_2(x)$$

$$-_* (\lambda(x))^{2*} \cdot_* y_2(x) \cdot_* y_1^*(x)$$

$$= (\lambda(x))^{2*} \cdot_* (y_1(x) \cdot_* y_2^*(x) -_* y_1^*(x) \cdot_* y_2(x))$$

$$= (\lambda(x))^{2*} \cdot_* W(y_1, y_2)(x), \quad x \in J.$$

3. We have

$$
\begin{aligned}
W_*(y_1, \lambda \cdot_* y_1) &= \det_* \begin{pmatrix} y_1(x) & \lambda(x) \cdot_* y_1(x) \\ y_1^*(x) & \lambda^*(x) \cdot_* y_1(x) +_* \lambda(x) \cdot_* y_1^*(x) \end{pmatrix} \\
&= y_1(x) \cdot_* \left(\lambda^*(x) \cdot_* y_1(x) +_* \lambda(x) \cdot_* y_1^*(x) \right) \\
&\quad -_* \lambda(x) \cdot_* y_1(x) \cdot_* y_1^*(x) \\
&= \lambda(x) \cdot_* (y_1(x))^{2*} +_* \lambda(x) \cdot_* y_1(x) \cdot_* y_1^*(x) \\
&\quad -_* \lambda(x) \cdot_* y_1(x) \cdot_* y_1^*(x) \\
&= \lambda^*(x) \cdot_* (y_1(x))^{2*}, \quad x \in J.
\end{aligned}
$$

This completes the proof.

Theorem 4.2.14 *Let $y_1, \ldots, y_n : J \to \mathbb{R}_*$ be $(n-1)$-times multiplicative differentiable on J. If*

$$W_*(y_1, \ldots, y_n)(x_0) \neq 0_* \tag{4.11}$$

for some $x_0 \in J$, then $y_1, \ldots, y_n$ are multiplicative linearly independent on J.

Proof 4.2.15 *Assume that $y_1, \ldots, y_n$ are multiplicative linearly dependent on J. Then there exist $c_1, \ldots, c_n \in J$ such that*

$$(c_1, \ldots, c_n) \neq (0_*, \ldots, 0_*) \tag{4.12}$$

and

$$c_1 \cdot_* y_1(x) +_* \cdots +_* c_n \cdot_* y_n(x) = 0_*, \quad x \in J. \tag{4.13}$$

Multiplicative differentiating, we get the system

$$c_1 \cdot_* y_1(x) +_* \cdots +_* c_n \cdot_* y_n(x) \qquad = 0_*$$

$$c_1 \cdot_* y_1^*(x) +_* \cdots +_* c_n \cdot_* y_n^*(x) \qquad = 0_*$$

$$\tag{4.14}$$

$$\vdots$$

$$c_1 \cdot_* y_1^{*(n-1)}(x) +_* \cdots +_* c_n \cdot_* y_n^{*(n-1)}(x) \quad = 0_*, \quad x \in J.$$

The last system we consider with respect to $c_1, \ldots, c_n$. Its multiplicative determinant is $W_*(y_1, \ldots, y_n)(x)$, $x \in J$. Since we have that (4.12) holds, we conclude that

$$W_*(y_1, \ldots, y_n)(x) = 0_*, \quad x \in J,$$

which is a contradiction because we have that (4.11) holds. Therefore $y_1, \ldots, y_n$ are multiplicative linearly independent on J. This completes the proof.

Theorem 4.2.16 *Let $y_1, \ldots, y_n : J \to \mathbb{R}_*$ be $(n-1)$ times multiplicative differentiable and multiplicative linearly dependent on J. Then*

$$W_*(y_1, \ldots, y_n)(x) = 0_*, \quad x \in J. \tag{4.15}$$

Proof 4.2.17 *Since $y_1, \ldots, y_n$ are multiplicative linearly dependent on J, there exist $c_1, \ldots, c_n \in \mathbb{R}_*$ such that (4.12) and (4.13) hold. Multiplicative differentiating, we get the system (4.14), which has a non-trivial solution $c_1, \ldots, c_n$. Therefore, (4.15) holds. This completes the proof.*

Example 4.2.18 *Consider the functions*

$$y_1(x) = x,$$

$$y_2(x) = e^x,$$

$$y_3(x) = x \cdot_* e^x, \quad x \in J.$$

Then

$$y_1^*(x) = 1_*,$$

$$y_2^*(x) = e^x,$$

$$y_3^*(x) = e^x +_* x \cdot_* e^x$$

$$= (x +_* 1_*) \cdot_* e^x,$$

$$y_1^{**}(x) = 0_*,$$

$$y_2^{**}(x) = e^x,$$

$$y_3^{**}(x) = e^x +_* (x +_* 1_*) \cdot_* e^x$$

$$= (x +_* 2_*) \cdot_* e^x, \quad x \in \mathbb{R}_*.$$

Then

$$W_*(y_1, y_2, y_3)(x) = \det_* \begin{pmatrix} y_1(x) & y_2(x) & y_3(x) \\ y_1^*(x) & y_2^*(x) & y_3^*(x) \\ y_1^{**}(x) & y_2^{**}(x) & y_3^{**}(x) \end{pmatrix}$$

$$= \det_* \begin{pmatrix} x & e^x & x \cdot_* e^x \\ 1_* & e^x & (x +_* 1_*) \cdot_* e^x \\ 0_* & e^x & (x +_* 1_*) \cdot_* e^x \end{pmatrix}$$

$$= x \cdot_* e^x \cdot_* (x +_* 2_*) \cdot_* e^x +_* x \cdot_* e^x \cdot_* e^x$$

$$-_* x \cdot_* (x +_* 1_*) \cdot_* e^x \cdot_* e^x$$

$$-_* e^x \cdot_* (x +_* 2_*) \cdot_* e^x$$

$$= -_*x^{2*} \cdot_* e^{x^2}$$
$$\neq 0_* \quad for \quad x \neq 0_*.$$

Hence and Theorem 4.2.14, it follows that the functions y_1, y_2, y_3 are multiplicative linearly independent.

Exercise 4.2.19 *Using Theorem 4.2.14, prove that the functions*

$$y_1(x) = e^x,$$

$$y_2(x) = e^{e^{2} \cdot_* x},$$

$$y_3(x) = e^{e^{3} \cdot_* x}, \quad x \in \mathbb{R}_*,$$

are multiplicative linearly independent.

Theorem 4.2.20 *Two solutions y_1 and y_2 of the equation (4.1) are multiplicative linearly independent if and only if*

$$W_*(y_1, y_2)(x_0) \neq 0_* \tag{4.16}$$

for some $x_0 \in J$.

Proof 4.2.21   *1. Let y_1 and y_2 be two solutions of the equation (4.1) that are multiplicative linearly independent. Then the system*

$$c_1 \cdot_* y_1(x) +_* c_2 \cdot_* y_2(x) = 0_*$$
$$c_1 \cdot_* y_1^*(x) +_* c_2 \cdot_* y_2^*(x) = 0, \quad x \in J, \tag{4.17}$$

has only the multiplicative zero solution. Therefore, there exists $x_0 \in J$ for which (4.16) holds.

   2. Suppose that (4.16) holds for some $x_0 \in J$ and for two solutions y_1 and y_2 of the equation (4.1). Let $c_1, c_2 \in \mathbb{R}_*$ be such that

$$c_1 \cdot_* y_1(x) +_* c_2 \cdot_* y_2(x) = 0_*, \quad x \in J.$$

Hence, we get the system (4.17). In particular, we obtain the system

$$c_1 \cdot_* y_1(x_0) +_* c_2 \cdot_* y_2(x_0) = 0_*$$

$$c_1 \cdot_* y_1^*(x_0) +_* c_2 \cdot_* y_2^*(x_0) = 0_*,$$

Then, applying (4.16), we conclude that

$$c_1 = 0_*$$

$$c_2 = 0_*.$$

Therefore, y_1 and y_2 are multiplicative linearly independent. This completes the proof.

4.3 The Multiplicative Abel Theorem

Theorem 4.3.1 *Let y_1 and y_2 be two solutions of the equation (4.1). Then, for the multiplicative Wronskian, the following multiplicative Abel identity holds.*

$$W_*(y_1, y_2)(x) = W_*(y_1, y_2)(x_0) \cdots_* e^{-* \int_{x_0*}^x (a_1(s)/_* a_0(s)) \cdots_* d_* s}, \quad x, x_0 \in J.$$
$$(4.18)$$

Proof 4.3.2 *Take $x_0 \in J$ arbitrarily. Since y_1 is a solution to the equation (4.1), we have*

$$y_1^{**}(x) +_* (a_1(x)/_* a_0(x)) \cdot_* y_1^*(x) +_* (a_2(x)/_* a_0(x)) \cdot_* y_1(x) = 0_*, \quad x \in J,$$

whereupon

$$y_1^{**}(x) = -_*(a_1(x)/_* a_0(x)) \cdot_* y_1^*(x) -_* (a_2(x)/_* a_0(x)) \cdot_* y_1(x), \quad x \in J.$$

As above,

$$y_2^{**}(x) = -_*(a_1(x)/_* a_0(x)) \cdot_* y_2^*(x) -_* (a_2(x)/_* a_0(x)) \cdot_* y_2(x), \quad x \in J.$$

Also, we have

$$W_*(y_1, y_2)(x) = \det\begin{pmatrix} y_1(x) & y_2(x) \\ y_1^*(x) & y_2^*(x) \end{pmatrix}, \quad x \in \mathbb{R}_*.$$

Then

$$W_*^*(y_1, y_2)(x) = \det_*\begin{pmatrix} y_1^*(x) & y_2^*(x) \\ y_1^*(x) & y_2^*(x) \end{pmatrix} +_* \det_*\begin{pmatrix} y_1(x) & y_2(x) \\ y_1^{**}(x) & y_2^{**}(x) \end{pmatrix}$$

$$= \det_*\begin{pmatrix} y_1(x) & y_2(x) \\ y_1^{**}(x) & y_2^{**}(x) \end{pmatrix}$$

$$= \det_*\begin{pmatrix} y_1(x) \\ -_*(a_1(x)/_*a_0(x)) \cdot_* y_1^*(x) -_* (a_2(x)/_*a_0(x)) \cdot_* y_1(x) \\[1em] y_2(x) \\ -_*(a_1(x)/_*a_0(x)) \cdot_* y_2^*(x) -_* (a_2(x)/_*a_0(x)) \cdot_* y_2(x) \end{pmatrix}$$

$$= \det_*\begin{pmatrix} y_1(x) & y_2(x) \\ -_*(a_1(x)/_*a_0(x)) \cdot_* y_1^*(x) & -_*(a_1(x)/_*a_0(x)) \cdot_* y_2^*(x) \end{pmatrix}$$
$$+_* \det_*\begin{pmatrix} y_1(x) & y_2(x) \\ -_*(a_2(x)/_*a_0(x)) \cdot_* y_1(x) & -_*(a_2(x)/_*a_0(x)) \cdot_* y_2(x) \end{pmatrix}$$

$$= \det_*\begin{pmatrix} y_1(x) & y_2(x) \\ -_*(a_1(x)/_*a_0(x)) \cdot_* y_1^*(x) & -_*(a_1(x)/_*a_0(x)) \cdot_* y_2^*(x) \end{pmatrix}$$

$$= -_*(a_1(x)/_*a_0(x)) \cdot_* y_1(x) \cdot_* y_2^*(x) +_* (a_1(x)/_*a_0(x)) \cdot_* y_1^*(x) \cdot_* y_2(x)$$

$$= -_*(a_1(x)/_*a_0(x)) \cdot_* (y_1(x) \cdot_* y_2^*(x) -_* y_1^*(x) \cdot_* y_2(x))$$

$$= -_*(a_1(x)/_*a_0(x)) \cdot_* W_*(y_1, y_2)(x)$$

$$= -_* e^{\frac{\log(a_1(x))}{\log(a_0(x))} \log(W_*(y_1, y_2)(x))}$$

$$= e^{-\frac{\log(a_1(x))}{\log(a_0(x))} \log(W_*(y_1, y_2)(x))}, \quad x \in \mathbb{R}_*,$$

whereupon

$$e^{x(\log(W_*(y_1,y_2)))'(x)} = e^{-\frac{\log(a_1(x))}{\log(a_0(x))}\log(W_*(y_1,y_2)(x))},$$

$x \in J$. *Hence, we get the equation*

$$x\left(\log(W_*(y_1,y_2))\right)'(x) = -\frac{\log(a_1(x))}{\log(a_0(x))}\log\left(W_*(y_1,y_2)(x)\right),$$

$x \in J$. *Then*

$$\log(W_*(y_1,y_2))(x) = \log(W_*(y_1,y_2)(x_0))e^{-\int_{x_0}^x \frac{\log(a_1(s))}{s\log(a_0(s))}ds}$$

$$= \log(W_*(y_1,y_2)(x_0))e^{-\int_{x_0}^x \frac{1}{s}\log e^{\frac{\log(a_1(s))}{\log(a_0(s))}}ds}$$

$$= \log(W_*(y_1,y_2)(x_0))e^{-\int_{x_0}^x \frac{1}{s}\log(a_1(s)a_0(s))ds}$$

$$= \frac{\log(W_*(y_1,y_2)(x_0))}{e^{-\int_{x_0}^x \frac{1}{s}\log(a_1(s)a_0(s))ds}}, \quad x \in J,$$

and

$$W_*(y_1,y_2)(x) = e^{e^{\frac{\log(W_*(y_1,y_2)(x_0))}{\int_{x_0}^x \frac{1}{s}\log(a_1(s)a_0(s))ds}}}$$

$$= e^{e^{\frac{\log e^{\log(W_*(y_1,y_2)(x_0))}}{\int_{x_0}^x \frac{1}{s}\log(a_1(s)a_0(s))ds}}}$$

$$= (W(y_1,y_2)(x_0))/_* \left(e^{\int_{x_0}^x \frac{1}{s}\log(a_1(s)a_0(s))ds}\right)$$

$$= (W(y_1,y_2)(x_0)) \cdot_* e^{-\int_{x_0}^x \frac{1}{s}\log(a_1(s)a_0(s))ds}, \quad x \in J.$$

This completes the proof.

If we know a particular solution y_1 of the equation (4.1), employing the multiplicative Abel identity, we can find the other solution y_2 of the equation (4.1). Indeed, we have

$$
W_*(y_1,y_2)(x) = \det_* \begin{pmatrix} y_1(x) & y_2(x) \\ y_1^*(x) & y_2^*(x) \end{pmatrix}
$$

$$
= y_1(x) \cdot_{(} {*} y_2^*(x) -_* y_1^*(x) \cdot_* y_2(x)
$$

$$
= y_1(x) \cdot_* e^{x(\log y_2)'(x)} -_* e^{x(\log y_1)'(x)} \cdot_* y_2(x)
$$

$$
= e^{x(\log(y_1(x)))(\log y_2)'(x)} -_* e^{x(\log y_1)'(x)(\log(y_2(x)))}
$$

$$
= e^{x((\log(y_1(x)))(\log y_2)'(x)-(\log y_1)'(x)(\log(y_2(x))))}
$$

$$
= e^{x(\log(y_1(x)))^2 \frac{(\log(y_1(x)))(\log y_2)'(x)-(\log y_1)'(x)(\log(y_2(x)))}{(\log(y_1(x)))^2}}
$$

$$
= e^{x(\log(y_1(x)))^2 \left(\frac{\log y_2}{\log y_1}\right)'(x)}, \quad x \in J,
$$

whereupon

$$
\log(W_*(y_1,y_2)(x)) = x(\log(y_1(x)))^2 \left(\frac{\log y_2}{\log y_1}\right)'(x), \quad x \in J.
$$

Now, we apply the multiplicative Abel identity to find

$$
x(\log(y_1(x)))^2 \left(\frac{\log y_2}{\log y_1}\right)'(x) = ce^{-\int_{x_0}^x \frac{\log(a_1(s))}{s\log(a_0(s))}ds}, \quad x \in J,
$$

and

$$
\left(\frac{\log y_2}{\log y_1}\right)'(x) = c\frac{e^{-\int_{x_0}^x \frac{\log(a_1(s))}{s\log(a_0(s))}ds}}{x(\log(y_1(x)))^2}, \quad x \in J.
$$

Hence,

$$
\frac{\log(y_2(x))}{\log(y_1(x))} = c\int_{x_0}^x \frac{e^{-\int_{x_0}^y \frac{\log(a_1(s))}{s\log(a_0(s))}ds}}{y(\log(y_1(y)))^2}dy + c_1, \quad x \in J,
$$

and

$$\log(y_2(x)) = c\log(y_1(x))\int_{x_0}^x \frac{e^{-\int_{x_0}^y \frac{\log(a_1(s))}{s\log(a_0(s))}ds}}{y(\log(y_1(y)))^2}dy$$

$$+_*c_1\log(y_1(x)), \quad x \in J.$$

Therefore,

$$y_2(x) = \exp\left(c\log(y_1(x))\int_{x_0}^x \frac{e^{-\int_{x_0}^y \frac{\log(a_1(s))}{s\log(a_0(s))}ds}}{y(\log(y_1(y)))^2}dy\right.$$

$$\left.+_*c_1\log(y_1(x))\right), \quad x \in J.$$

Here $c, c_1 \in \mathbb{R}_*$. The last representation of y_2 we can write in the form

$$y_2(x) = \exp\left(c\log(y_1(x))\int \frac{e^{-\int \frac{\log(a_1(x))}{x\log(a_0(x))}dx}}{x(\log(y_1(x)))^2}dx\right.$$

$$\left.+_*c_1\log(y_1(x))\right), \quad x \in J. \tag{4.19}$$

Now, we will represent (4.19) in a multiplicative form. We have

$$y_2(x) = e^{c\log(y_1(x))\int_{x_0}^x \frac{1}{y(\log(y_1(y)))^2 e^{\int_{x_0}^y \frac{\log(a_1(s))}{s\log(a_0(s))}ds}}dy}$$

$$\times e^{c_1\log(y_1(x))}$$

$$= e^{c\log(y_1(x))\int_{x_0}^x \frac{1}{y(\log(y_1(y)))^2 \int_{x_0*}^y (a_1(s)/*a_0(s))\cdot*d*s}dx}$$

$$+_*e^{c_1}\cdot_* y_1(x)$$

$$= e^c\cdot_* y_1(x)\cdot_* e^{\int_{x_0}^x \frac{1}{y}\frac{-*\int_{x_0*}^y (a_1(s)/*a_0(s))\cdot*d*s}{(\log(y_1(y)))^2}dy}$$

$$+_* e^{c_1} \cdot_* y_1(x)$$

$$= e^c \cdot_* y_1(x) \cdot_* e^{\int_{x_0}^x \frac{1}{y} \frac{\log_e \frac{-*\int_{x_0}^y *(a_1(s)/*a_0(s))\cdot*d*s}{\log_e^{(\log(y_1(y)))^2}}}{} dy}$$

$$+_* e^{c_1} \cdot_* y_1(x)$$

$$= e^c \cdot_* y_1(x) \cdot_* e^{\int_{x_0}^x \frac{1}{y} \frac{\log_e \frac{-*\int_{x_0}^y *(a_1(s)/*a_0(s))\cdot*d*s}{(\log(y_1(y))^{2*})}}{} dy}$$

$$+_* e^{c_1} \cdot_* y_1(x)$$

$$= e^c \cdot_* y_1(x) \cdot_* e^{\int_{x_0}^x \frac{1}{y} \log_e \frac{\log_e \frac{-*\int_{x_0}^y *(a_1(s)/*a_0(s))\cdot*d*s}{(\log(y_1(y))^{2*})}}{} dy}$$

$$+_* e^{c_1} \cdot_* y_1(x)$$

$$= e^c \cdot_* y_1(x) \cdot_* e^{\int_{x_0}^x \frac{1}{x} \log\left(e^{-*\int_{x_0}^y *(a_1(s)/*a_0(s))\cdot*d*s} /_* (y_1(y))^{2*}\right) dy}$$

$$+_* e^{c_1} \cdot_* y_1(x)$$

$$= e^c \cdot_* y_1(x) \cdot_* \int_{x_0*}^x \left(e^{-*\int_{x_0*}^y (a_1(s)/*a_0(s))\cdot*d*s} /_* (y_1(y))^{2*}\right) \cdot_* d_8 y$$

$$+_* e^{c_1} \cdot_* y_1(x), \quad x \in J,$$

i.e.,

$$y_2(x) = e^c \cdot_* y_1(x) \cdot_* \int_{x_0*}^x \left(e^{-*\int_{x_0*}^y (a_1(s)/*a_0(s))\cdot*d*s} /_* (y_1(y))^{2*}\right) \cdot_* d_8 y$$

$$+_* e^{c_1} \cdot_* y_1(x), \quad x \in J.$$

Example 4.3.3 *Let* $y_1(x) = x$, $x \in \mathbb{R}_*$, *be a solution to the equation*

$$\left(x^{2*} +_* e\right) \cdot_* y^{**} -_* e^2 \cdot_* x \cdot_* y^* +_* e^2 \cdot_* y = 0_*, \quad x \in \mathbb{R}_*.$$

Here

$$
\begin{aligned}
a_0(x) &= x^{2*} +_* e \\
&= e^{(\log x)^2} +_* e \\
&= e^{(\log x)^2 + 1},
\end{aligned}
$$

$$
\begin{aligned}
a_1(x) &= -_* e^2 \cdot_* x \\
&= -_* e^{2\log x} \\
&= e^{-2\log x},
\end{aligned}
$$

$$
a_2(x) = e^2, \quad x \in \mathbb{R}_*.
$$

Then

$$
\begin{aligned}
\int \frac{\log(a_1(x))}{x \log(a_0(x))} dx &= \int \frac{\log\left(e^{-2\log x}\right)}{x \log e^{(\log x)^2 + 1}} dx \\
&= -\int \frac{2\log x}{x((\log x)^2 + 1)} dx \\
&= -\int \frac{2\log x}{(\log x)^2 + 1} d\log x \\
&= -\int \frac{d((\log x)^2 + 1)}{(\log x)^2 + 1} \\
&= -\log((\log x)^2 + 1), \quad x \in \mathbb{R}_*,
\end{aligned}
$$

and

$$
\int \frac{e^{-\int \frac{\log(a_1(x))}{x \log(a_0(x))} dx}}{x(\log(y_1(x)))^2} dx = -\int \frac{e^{\log((\log x)^2 + 1)}}{x(\log x)^2} dx
$$

$$= \int \frac{(\log x)^2 + 1}{x(\log x)^2} dx$$

$$= \int \frac{(\log x)^2 + 1}{(\log x)^2} d(\log x)$$

$$= \int d(\log x) + \int \frac{1}{(\log x)^2} d(\log x)$$

$$= \log x - \frac{1}{\log x}, \quad x \in J.$$

Now, using (4.19), we get

$$
y_2(x) = \exp\left(c \log(y_1(x)) \int \frac{e^{-\int \frac{\log(a_1(x))}{x \log(a_0(x))} dx}}{x(\log(y_1(x)))^2} dx \right.
$$

$$
\left. +_* c_1 \log(y_1(x)) \right)
$$

$$
= e^{c \log x \left(\log x - \frac{1}{\log x} \right) + c_1 \log x}
$$

$$
= e^{c(\log x)^2 - c + c_1 \log x}
$$

$$
= e^{c((\log x)^2 - 1) + c_1 \log x}
$$

$$
= e^c \cdot_* e^{(\log x)^2 - 1} +_* e^{c_1 \log x}
$$

$$
= e^c \cdot_* \left(\frac{e^{(\log x)^2}}{e} \right) +_* e^{c_1} \cdot_* x
$$

$$
= e^c \cdot_* (x^{2*} -_* e) +_* e^{c_1} \cdot_* x, \quad x \in J.
$$

Here $c, c_1 \in \mathbb{R}_*$.

Example 4.3.4 *Let $y_1(x) = x^{2*}$, $x \in \mathbb{R}_*$, be a solution to the equation*

$$
x^{2*} \cdot_* y^{**} -_* e^2 \cdot_* x \cdot_* y^* +_* e^2 \cdot_* y = 0_*, \quad x \in \mathbb{R}_*.
$$

We will find the second solution y_2. *Here*

$$y_1(x) = x^{2*}$$

$$= e^{(\log x)^2},$$

$$a_0(x) = x^{2*}$$

$$= e^{(\log x)^2},$$

$$a_1(x) = -_* e^2 \cdot_* x$$

$$= -_* e^{2\log x}$$

$$= e^{-2\log x},$$

$$a_2(x) = e^2, \quad x \in \mathbb{R}_*.$$

Then

$$\int \frac{\log(a_1(x))}{x\log(a_0(x))} dx = \int \frac{\log e^{-2\log x}}{x\log\left(e^{(\log x)^2}\right)} dx$$

$$= -\int \frac{2\log x}{x(\log x)^2} dx$$

$$= -2\int \frac{d(\log x)}{\log x}$$

$$= -2\log|\log x|$$

$$= -\log(\log x)^2, \quad x \in \mathbb{R}_*,$$

and

$$\int \frac{e^{-\int \frac{\log(a_1(x))}{x\log(a_0(x))} dx}}{x(\log(y_1(x)))^2} dx = \int \frac{e^{\log(\log x)^2}}{x\left(\log\left(e^{(\log x)^2}\right)\right)^2} dx$$

$$= \int \frac{(\log x)^2}{x(\log x)^4} dx$$

$$= \int \frac{d(\log x)}{(\log x)^2}$$

$$= -\frac{1}{\log x}, \quad x \in \mathbb{R}_*.$$

Now, employing (4.19), we get

$$y_2(x) = \exp\left(c\log(y_1(x)) \int \frac{e^{-\int \frac{\log(a_1(x))}{x\log(a_0(x))}dx}}{x(\log(y_1(x)))^2} dx\right.$$

$$\left. +_* c_1 \log(y_1(x))\right)$$

$$= e^{c\log\left(e^{(\log x)^2}\right)\left(-\frac{1}{\log x}\right) + c_1 \log\left(e^{(\log x)^2}\right)}$$

$$= e^{c(\log x)^2\left(-\frac{1}{\log x}\right) + c_1(\log x)^2}$$

$$= e^{-c\log x + c_1(\log x)^2}$$

$$e^{-c\log x} +_* e^{c_1(\log x)^2}$$

$$= e^{-c} \cdot_* x +_* e^{c_1} \cdot_* x^{2*}, \quad x \in \mathbb{R}_*.$$

Here $c, c_1 \in \mathbb{R}_*$.

Example 4.3.5 *Let $J = (2, \infty)$ and*

$$y_1(x) = e/_*(x -_* e), \quad x \in J,$$

be a solution to the equation

$$(x^{2*} -_* x) \cdot_* y^{**} +_* (e^3 \cdot_* x -_* e) \cdot_* y^* +_* y = 0_*, \quad x \in J.$$

We will find the second solution y_2. Here

$$y_2(x) \;=\; e/_* \left(\frac{x}{e}\right)$$

$$=\; e^{\frac{\log e}{\log \frac{x}{e}}}$$

$$=\; e^{\frac{1}{\log x - 1}},$$

$$a_0(x) \;=\; x^{2_*} -_* x$$

$$=\; e^{(\log x)^2} -_* x$$

$$=\; \frac{e^{(\log x)^2}}{x},$$

$$a_1(x) \;=\; e^3 \cdot_* x -_* e$$

$$=\; e^{3\log x} -_* e$$

$$=\; e^{3\log x - 1},$$

$$a_2(x) \;=\; 1_*$$

$$=\; e, \quad x \in J.$$

Then

$$\int \frac{\log(a_0(x))}{x \log(a_1(x))} dx \;=\; \int \frac{\log e^{3\log x - 1}}{x \log\left(\frac{e^{(\log x)^2}}{x}\right)} dx$$

$$=\; \int \frac{3\log x - 1}{x\left(\log e^{(\log x)^2} - \log x\right)} dx$$

$$=\; \int \frac{3\log x - 1}{x \log x (\log x - 1)} dx$$

$$= \int \frac{3\log x - 1}{\log x(\log x - 1)} d(\log x)$$

$$= \int \frac{2\log x + \log x - 1}{\log x(\log x - 1)} d(\log x)$$

$$= 2\int \frac{d(\log x - 1)}{\log x - 1} + \int \frac{d(\log x)}{\log x}$$

$$= 2\log|\log x - 1| + \log|\log x|$$

$$= \log(\log x - 1)^2 + \log|\log x|$$

$$= \log\left((\log x - 1)^2 |\log x|\right), \quad x \in J,$$

and

$$\int \frac{e^{-\int \frac{\log(a_1(x))}{x\log(a_0(x))} dx}}{x(\log(y_1(x)))^2} dx = \int \frac{e^{-\log\left((\log x - 1)^2 |\log x|\right)}}{x\left(\log e^{\frac{1}{\log x - 1}}\right)^2} dx$$

$$= \int \frac{(\log x - 1)^2}{x(\log x - 1)^2 \log x} dx$$

$$= \int \frac{dx}{x\log x}$$

$$= \int \frac{d(\log x)}{\log x}$$

$$= \log|\log x|, \quad x \in J.$$

Now, applying (4.19), we find

$$y_2(x) = \exp\left(c\log(y_1(x)) \int \frac{e^{-\int \frac{\log(a_1(x))}{x\log(a_0(x))} dx}}{x(\log(y_1(x)))^2} dx \right.$$

$$\left. +_*c_1 \log(y_1(x))\right)$$

$$= e^{c}\log e^{\frac{1}{\log x-1}}\log|\log x|+c_1\log e^{\frac{1}{\log x-1}}$$

$$= e^{c\frac{1}{\log x-1}\log|\log x|+\frac{c_1}{\log x-1}}$$

$$= e^{c}\cdot_* e^{\frac{\log|\log x|}{\log x-1}} +_* e^{c_1}\cdot_* e^{\frac{1}{\log x-1}}$$

$$= e^{c}\cdot_* (\log_* x/_*(x-_*c))+_* e^{c_1}/_*(x-_*e), \quad x\in J.$$

Exercise 4.3.6 *Given the solution* $y_1(x)=e+_*e/_*x$, $x\in\mathbb{R}_*$, *find the second solution of the following equation:*

$$x^{2*}\cdot_*(x+_*e)\cdot_* y^{**}-_*e^2\cdot_* y=0_*, \quad x\in\mathbb{R}_*.$$

Answer

$$y_2(x)=x/_*e^2+_*e-_*((x+_*e)/_*x)\cdot_*\log_*|x+_*e|_*, \quad x\in\mathbb{R}_*.$$

Theorem 4.3.7 *Let* y_1 *and* y_2 *be two linearly independent solutions of the equation* (4.1) *and* $y_1\neq 0_*$ *on* J. *Then* $y=y_2/_*y_1$ *satisfies the equation*

$$y_1(x)\cdot_* y^{**}+_*\left(e^2\cdot_* y_1^*(x)+_*(a_1(x)/_*a_0(x))\cdot_* y_1(x)\right)\cdot_* y^*=0_*,$$

$x\in\mathbb{R}_*.$

Proof 4.3.8 *Since* y_1 *and* y_2 *are solutions of the equation* (4.1), *we have the following:*

$$y_1^{**}(x) = -_*(a_1(x)/_*a_0(x))\cdot_* y_1^*(x)-_*(a_2(x)/_*a_0(x))\cdot_* y_1(x),$$

$$y_2^{**}(x) = -_*(a_1(x)/_*a_0(x))\cdot_* y_2^*(x)-_*(a_2(x)/_*a_0(x))\cdot_* y_2(x),$$

$x\in J.$ *Hence,*

$$y^*(x)=(y_2^*(x)\cdot_* y_1(x)-_* y_2(x)\cdot_* y_1^*(x))/_*\left((y_1(x))^{2*}\right), \quad x\in J,$$

and

$$
y^{**}(x) = \left((y_1(x))^{2*} \cdot_* \left(y_2^{**}(x) \cdot_* y_1(x) +_* y_1^*(x) \cdot_* y_2^*(x) \right. \right.
$$

$$
\left. -_* y_1^*(x) \cdot_* y_2^*(x) -_* y_2(x) \cdot_* y_1^*(x) \right)
$$

$$
\left. -_* e^2 \cdot_* y_1(x) \cdot_* y_1^*(x) \cdot_* \left(y_2^*(x) \cdot_* y_1(x) -_* y_2(x) \cdot_* y_1^*(x) \right) \right)
$$

$$
/_* \left(y_1(x) \right)^{4*}
$$

$$
= \left((y_1(x)) \cdot_* \left(y_2^{**}(x) \cdot_* y_1(x) +_* y_1^*(x) \cdot_* y_2^*(x) \right. \right.
$$

$$
\left. -_* y_1^*(x) \cdot_* y_2^*(x) -_* y_2(x) \cdot_* y_1^*(x) \right)
$$

$$
\left. -_* e^2 \cdot_* y_1^*(x) \cdot_* \left(y_2^*(x) \cdot_* y_1(x) -_* y_2(x) \cdot_* y_1^*(x) \right) \right)
$$

$$
/_* \left(y_1(x) \right)^{3*}, \quad x \in J,
$$

whereupon

$$
y_1(x) \cdot_* y^{**}(x) = \left((y_1(x)) \cdot_* \left(y_2^{**}(x) \cdot_* y_1(x) +_* y_1^*(x) \cdot_* y_2^*(x) \right. \right.
$$

$$
\left. -_* y_1^*(x) \cdot_* y_2^*(x) -_* y_2(x) \cdot_* y_1^*(x) \right)
$$

$$
\left. -_* e^2 \cdot_* y_1^*(x) \cdot_* \left(y_2^*(x) \cdot_* y_1(x) -_* y_2(x) \cdot_* y_1^*(x) \right) \right)
$$

$$
/_* \left(y_1(x) \right)^{2*}
$$

$$
= y_2^{**}(x) -_* (y_2(x)/_* y_1(x)) \cdot_* y_1^{**}(x)
$$

$$-_*e^2 \cdot_* y_1^*(x) \cdot_* \Big((y_2^*(x) \cdot_* y_1(x)$$

$$-_*y_2(x) \cdot_* y_1^*(x))/_*(y_1(x))^{2*} \Big)$$

$$= \quad y_2^{**}(x) -_* y_1^{**}(x) \cdot_* y(x)$$

$$-_*e^2 \cdot_* y_1^*(x) \cdot_* (y_2/_*y_1)^*(x)$$

$$= \quad -_*(a_1(x)/_*a_0(x)) \cdot_* y_2^*(x) -_* (a_2(x)/_*a_0(x)) \cdot_* y_2(x)$$

$$+_* \Big((a_1(x)/_*a_0(x)) \cdot_* y_1^*(x)$$

$$+_*(a_2(x)/_*a_1(x)) \cdot_* y_1(x) \Big) \cdot_* y(x)$$

$$-_*e^2 \cdot_* y_1^*(x) \cdot_* y^*(x)$$

$$= \quad -_*(a_1(x)/_*a_0(x)) \cdot_* y_2^*(x)$$

$$+_*(a_1(x)/_*a_0(x)) \cdot_* y_1^*(x) \cdot_* (y_2(x)/_*y_1(x))$$

$$-_*e^2 \cdot_* y_1^*(x) \cdot_* y^*(x)$$

$$= \quad -_*(a_1(x)/_*a_0(x)) \cdot_* \Big(y_2^*(x) \cdot_* y_1(x)$$

$$-_*y_1^*(x) \cdot_* y_2(x) \Big)/_*y_1(x)$$

$$-_*e^2 \cdot_* y_1^*(x) \cdot_* y^*(x)$$

$$= \quad -_*(a_1(x)/_*a_0(x)) \cdot_* y_1(x) \cdot_* y^*(x)$$

$$-_* e^2 \cdot_* y_1^*(x) \cdot_* y^*(x)$$

$$= -_*\left((a_1(x)/_* a_0(x)) \cdot_* y_1(x) +_* e^2 \cdot_* y_1^*(x) \right) \cdot_* y^*(x),$$

$x \in J$. *This completes the proof.*

Suppose that a_0 is multiplicative differentiable on J. Note that the equation (4.1) can be transformed into a first-order nonlinear MDE. Take $x_0 \in J$ arbitrarily. Set

$$y(x) = e^{\int_{x_0 *}^x a_0(t) \cdot_* z(t) \cdot_* d_* t}, \quad x \in J.$$

Then

$$y(x) = e^{\int_{x_0 *}^x e^{\log(a_0(t)) \log(z(t))} \cdot_* d_* t}$$

$$= e^{e^{\int_{x_0}^x \frac{1}{t} \log(a_0(t)) \log(z(t)) dt}}, \quad x \in J.$$

Hence,

$$y'(x) = \frac{1}{x} \log(a_0(x)) \log(z(x)) e^{\int_{x_0}^x \frac{1}{t} \log(a_0(t)) \log(z(t)) dt}$$

$$\times e^{\int_{x_0}^x \frac{1}{t} \log(a_0(t)) \log(z(t)) dt},$$

$$y^*(x) = \exp\left(\log(a_0(x)) \log(z(x)) \right.$$

$$\left. \times \log e^{e^{\int_{x_0}^x \frac{1}{t} \log(a_0(t)) \log(z(t)) dt}} \right)$$

$$= e^{\log(a_0(x)) \log(z(x)) \log(y_0(x))}$$

$$= a_0(x) \cdot_* z(x) \cdot_* y(x),$$

$$y^{**}(x) = a_0^*(x) \cdot_* z(x) \cdot_* y(x)$$

$$+ _* a_0(x) \cdot _* z^*(x) \cdot _* y(x)$$

$$+ _* a_0(x) \cdot _* z(x) \cdot _* y^*(x)$$

$$= a_0^*(x) \cdot _* z(x) \cdot _* y(x)$$

$$+ _* a_0(x) \cdot _* z^*(x) \cdot _* y(x)$$

$$+ _* (a_0(x))^{2*} \cdot _* (z(x))^{2*} \cdot _* y(x)$$

$$= \left(a_0^*(x) \cdot _* z(x) + _* a_0(x) \cdot _* z^*(x) \right.$$

$$\left. + _* (a_0(x))^{2*} \cdot _* (z(x))^{2*} \right) \cdot _* y(x), \quad x \in J.$$

Then

$$0_* = a_0(x) \cdot _* y^{**}(x) + _* a_1(x) \cdot _* y^*(x) + _* a_2(x) \cdot _* y(x)$$

$$= a_0(x) \cdot _* \left(a_0^*(x) \cdot _* z(x) + _* a_0(x) \cdot _* z^*(x) \right.$$

$$\left. + _* (a_0(x))^{2*} \cdot _* (z(x))^{2*} \right) \cdot _* y(x)$$

$$+ _* a_1(x) \cdot _* a_0(x) \cdot _* z(x) \cdot _* y(x)$$

$$+ _* a_2(x) \cdot _* y(x), \quad x \in \mathbb{R}_*,$$

whereupon

$$0_* = (a_0^*(x) /_* a_0(x)) \cdot _* z(x) + _* z^*(x) + _* a_0(x) \cdot _* (z(x))^{2*}$$

$$+ _* (a_0^*(x) /_* a_0(x)) \cdot _* z(x) + _* \left(a_2(x) /_* (a_0(x))^{2*} \right)$$

$$= z^*(x) +_* ((a_0^*(x)/_*a_0(x)) +_* (a_1(x)/_*a_0(x))) \cdot_* z(x)$$

$$+_* a_0(x) \cdot_* (z(x))^{2*} +_* (a_2(x)/_*(a_0(x))^{2*}), \quad x \in J.$$

Thus, we get the equation

$$z^* +_* a_0(x) \cdot_* z^{2*} +_* ((a_0^*(x)/_*a_0(x)) +_* (a_1(x)/_*a_0(x))) \cdot_* z$$

$$+_* (a_2(x)/_*(a_0(x))^{2*}) = 0_*, \quad x \in J.$$

4.4 A Particular Case

In this section, we will investigate the equation

$$y^{**} +_* a_1(x) \cdot_* y +_* a_2(x) \cdot_* y = 0_*, \tag{4.20}$$

where $a_1, a_2 : J \to \mathbb{R}_*$, $a_1, a_2 \in \mathscr{C}(\mathbb{R}_*)$.

Theorem 4.4.1 *Let y_1 and y_2 be solutions of the equation (4.20). If y_1 and y_2 multiplicative vanish at the same point, then y_1 is a constant multiplicative multiple of y_2.*

Proof 4.4.2 *Let $x_0 \in J$ be such that*

$$y_1(x_0) = 0_*,$$

$$y_2(x_0) = 0_*.$$

Then

$$W_*(y_1, y_2)(x_0) = 0_*.$$

Hence, applying the multiplicative Abel identity, we obtain

$$W_*(y_1, y_2)(x) = 0_*, \quad x \in J.$$

Thus, y_1 is a constant multiplicative multiple of y_2. This completes the proof.

Theorem 4.4.3 *Let y_1 and y_2 be solutions to the equation (4.20) that have maxima or minima at the same point in J. Then y_1 and y_2 are multiplicative linearly dependent.*

Proof 4.4.4 *Let $x_0 \in J$ be such that*

$$y_1^*(x_0) \;=\; 0_*,$$

$$y_2^*(x_0) \;=\; 0_*.$$

Then
$$W_*(y_1, y_2)(x_0) = 0_*. \tag{4.21}$$

Let also, $c_1, c_2 \in \mathbb{R}_*$ be such that

$$c_1 \cdot_* y_1(x) +_* c_2 \cdot_* y_2(x) = 0_*, \quad x \in J.$$

Then

$$c_1 \cdot_* y_1^*(x) +_* c_2 \cdot_* y_2^*(x) = 0_*, \quad x \in J.$$

In particular, we get

$$c_1 \cdot_* y_1(x_0) +_* c_2 \cdot_* y_2(x_0) \;=\; 0_*$$

$$c_1 \cdot_* y_1^*(x_0) +_* c_2 \cdot_* y_2^*(x_0) \;=\; 0_*.$$

The last system we consider with respect to c_1 and c_2. Since (4.21) holds, the last system has a solution:

$$(c_1, c_2) \neq (0_*, 0_*).$$

Therefore, y_1 and y_2 are multiplicative linearly dependent. This completes the proof.

Theorem 4.4.5 *Let y_1 and y_2 be multiplicative linearly independent solutions of (4.20). If $W_*(y_1, y_2)$ is a constant on J, then*

$$a_1(x) = 0_*, \quad x \in J. \tag{4.22}$$

Proof 4.4.6 *Since y_1 and y_2 are multiplicative linearly independent on J, there is a $x_0 \in J$ such that*

$$W_*(y_1, y_2)(x_0) \neq 0_*. \tag{4.23}$$

By the multiplicative Abel identity, we have

$$W_*(y_1, y_2)(x) = W_*(y_1, y_2)(x_0) \cdot_* e^{-* \int_{x_0*}^x a_1(t) \cdot_* d_* t}, \quad x \in J. \tag{4.24}$$

Because $W_*(y_1, y_2)$ is a constant on J, we have that

$$(W_*(y_1, y_2))^*(x) = 0_*, \quad x \in J.$$

Now, multiplicative differentiating the equation (4.24), we find

$$\begin{aligned} 0_* &= (W_*(y_1, y_2))^*(x) \\ \\ &= -_* a_1(x) \cdot_* W_*(y_1, y_2)(x_0) \cdot_* e^{-* \int_{x_0*}^x a_1(t) \cdot_* d_* t}, \quad x \in J. \end{aligned}$$

Hence, using (4.23), we obtain (4.22). This completes the proof.

Theorem 4.4.7 *Let y_1 and y_2 be multiplicative linearly independent solutions to the equation (4.20). Then y_1 and y_2 cannot have a common point of inflexion in J unless a_1 and a_2 multiplicative vanish simultaneously there.*

Proof 4.4.8 *Assume that there is $x_0 \in J$ such that*

$$y_1^{**}(x_0) = 0_*,$$

$$y_2^{**}(x_0) = 0_*.$$

Because y_1 and y_2 are solutions to the equation (4.20), we have

$$\begin{aligned} 0_* &= y_1^{**}(x_0) \\ \\ &= -_* a_1(x_0) \cdot_* y_1^*(x_0) -_* a_2(x_0) \cdot_* y_1(x_0) \end{aligned}$$

and

$$0_* \;=\; y_2^{**}(x_0)$$

$$=\; -_*a_1(x_0)\cdot_* y_2^*(x_0) -_* a_2(x_0)\cdot_* y_2(x_0).$$

Thus, we get the system

$$a_1(x_0)\cdot_* y_1^*(x_0) +_* a_2(x_0)\cdot_* y_1(x_0) \;=\; 0_*$$

$$a_1(x_0)\cdot_* y_2^*(x_0) +_* a_2(x_0)\cdot_* y_2(x_0) \;=\; 0_*.$$

The last system we consider with respect to $a_{(}x_0)$ and $a_2(x_0)$. Its multiplicative determinant is

$$W_*(y_1,y_2)(x_0) \neq 0_*,$$

because y_1 and y_2 are multiplicative linearly independent solutions of the equation (4.20). Therefore,

$$a_1(x_0) \;=\; 0_*,$$

$$a_2(x_0) \;=\; 0_*.$$

This completes the proof.

Theorem 4.4.9 *Let y_1 and y_2 be solutions to the equation (4.20) such that*

$$W_*(y_1,y_2)(x_0) = y_1(x_0) = 0_*$$

for some $x_0 \in J$. Then either

$$y_1(x) = 0_*, \quad x \in J,$$

or

$$y_2(x) = (y_2^*(x_0)/_* y_1(x_0))\cdot_* y_1(x), \quad x \in J.$$

Proof 4.4.10 *We have that*

$$W_*(y_1,y_2)(x_0) \;=\; \det_* \begin{pmatrix} y_1(x_0) & y_2(x_0) \\ y_1^*(x_0) & y_2^*(x_0) \end{pmatrix}$$

$$= y_1(x_0) \cdot_* y_2^*(x_0) -_* y_2(x_0) \cdot_* y_1^*(x_0)$$

$$= -_* y_2(x_0) \cdot_* y_1^*(x_0)$$

$$= 0_*.$$

Hence,

$$y_1^*(x_0) = 0_* \tag{4.25}$$

or

$$y_2(x_0) = 0_*.$$

If (4.25) holds, then

$$y_1(x) = 0, \quad x \in J.$$

Let now,

$$y_1^*(x_0) \neq 0_*.$$

By the multiplicative Abel identity, we find

$$W_*(y_1, y_2)(x) = W_*(y_1, y_2)(x_0) \cdot_* e^{-_* \int_{x_0*}^x a_1(t) \cdot_* d_* t}$$

$$= 0_*, \quad x \in J.$$

Hence,

$$0_* = W_*(y_1, y_2)(x)$$

$$= \det_* \begin{pmatrix} y_1(x) & y_2(x) \\ y_1^*(x) & y_2^*(x) \end{pmatrix}$$

$$= y_1(x) \cdot_* y_2^*(x) -_* y_1^*(x) \cdot_* y_2(x), \quad x \in J,$$

whereupon

$$y_2(x) \cdot_* y_1^*(x) = y_1(x) \cdot_* y_2^*(x), \quad x \in J. \tag{4.26}$$

Consider the function

$$z(x) = y_2(x) /_* y_1(x), \quad x \in J.$$

Then, using (4.26), we get

$$z^*(x) = (y_2^*(x) \cdot_* y_1(x) -_* y_2(x) \cdot_* y_1^*(x))/_*(y_1(x))^{2*}$$

$$= 0_*, \quad x \in J.$$

Thus, z is a multiplicative constant on J. Therefore

$$y_2(x)/_*y_1(x) = y_2(x_0)/_*y_1(x_0), \quad x \in J. \tag{4.27}$$

Hence and (4.26), we arrive at

$$y_2(x)/_*y_1(x) = y_2^*(x)/_*y_1^*(x), \quad x \in J.$$

Now, applying (4.27), we find

$$y_2^*(x)/_*y_1^*(x) = y_2^*(x_0)/_*y_1^*(x_0), \quad x \in J.$$

Because we have (4.26), we have

$$y_2(x) = (y_2^*(x)/_*y_1^*(x)) \cdot_* y_1(x)$$

$$= (y_2^*(x_0)/_*y_1^*(x_0)) \cdot_* y_1(x), \quad x \in J.$$

This completes the proof.

Theorem 4.4.11 *Let y_1 and y_2 be linearly independent solutions of the equation (4.20) and W_* be their multiplicative Wronskian. Then*

$$y^{**} +_* a_1 \cdot_* y^* +_* a_2 \cdot_* y = (W_*/_*y_1) \cdot_* \left(\left(y_1^{2*}/_*W_*\right) \cdot_* (y/_*y_1)^* \right)^*.$$

Proof 4.4.12 *By the multiplicative Abel identity, we have*

$$W_*^*/_*W_* = -_*a_1.$$

Then

$$(W_*/_*y_1) \cdot_* \left(\left(y_1^{2*}/_*W_*\right) \cdot_* (y/_*y_1)^* \right)^*$$

$$= \ (W_* /_* y_1) \cdot_* \left(\left(y_1^{2*} /_* W_* \right) \right.$$

$$\cdot_* \left(\left(y^* \cdot_* y_1 -_* y \cdot_* y_1^* \right) /_* y_1^{2*} \right)^*$$

$$= \ (W_* /_* y_1) \cdot_* \left(\left(y^* \cdot_* y_1 -_* y \cdot_* y_1^* \right) /_* W_* \right)^*$$

$$= \ (W_* /_* y_1) \cdot_* \left(\left(y^* \cdot_* y_1 -_* y \cdot_* y_1^* \right)^* \cdot_* W_* \right.$$

$$-_* \left(y^* \cdot_* y_1 -_* y \cdot_* y_1^* \right) \cdot_* W_*^* \right) /_* W_*^{2*}$$

$$= \ \left(\left(y^* \cdot_* y_1 -_* y \cdot_* y_1^* \right)^* \cdot_* W_*^{2*} \right.$$

$$-_* \left(y^* \cdot_* y_1 -_* y \cdot_* y_1^* \right) \cdot_* W_* \cdot_* W_*^* \right) /_* \left(y_1 \cdot_* W_*^{2*} \right)$$

$$= \ \left(\left(y^* \cdot_* y_1 -_* y \cdot_* y_1^* \right)^* \right.$$

$$-_* \left(y^* \cdot_* y_1 -_* y \cdot_* y_1^* \right) \cdot_* \left(W_*^* /_* W_* \right) \right) /_* y_1$$

$$= \ \left(\left(y^* \cdot_* y_1 -_* y \cdot_* y_1^* \right)^* \right.$$

$$+_* a_1 \cdot_* \left(y^* \cdot_* y_1 -_* y \cdot_* y_1^* \right) \right) /_* y_1$$

$$= \ \left(y^{**} \cdot_* y_1 +_* y^* \cdot_* y_1^* -_* y^* \cdot_* y_1^* -_* y \cdot_* y_1^{**} \right.$$

$$+_* a_1 \cdot_* \left(y^* \cdot_* y_1 -_* y \cdot_* y_1^* \right) \right) /_* y_1$$

$$= \ \left(y^{**} \cdot_* y_1 -_* y \cdot_* y_1^{**} \right.$$

$$+_* a_1 \cdot_* \left(y^* \cdot_* y_1 -_* y \cdot_* y_1^* \right) \Big/_* y_1$$

$$= \ y^{**} -_* (y \cdot_* y_1^{**}) /_* y_1$$

$$+_* a_1 \cdot_* y^* -_* a_1 \cdot_* y \cdot_* \cdot_* (y_1^* /_* y_1)$$

$$= \ y^{**} +_* a_1 \cdot_* y^*$$

$$-_* \left(y \cdot_* \left(-_* a_1 \cdot_* y_1^* -_* a_2 \cdot_* y_1 \right) \right) \Big/_* y_1$$

$$-_* a_1 \cdot_* y \cdot_* (y_1^* /_* y_1)$$

$$= \ y^{**} +_* a_1 \cdot_* y^* +_* a_1 \cdot_* y \cdot_* (y_1^* /_* y_1)$$

$$+_* a_2 \cdot_* y -_* a_1 \cdot_* y \cdot_* (y_1^* /_* y_1)$$

$$= \ y^{**} +_* a_1 \cdot_* y^* +_* a_2 \cdot_* y.$$

This completes the proof.

Theorem 4.4.13 *Let y_1 and y_2 be two multiplicative linearly independent solutions of the equation (4.20). Then*

$$a_1(x) \ = \ \left(y_1^{**}(x) \cdot_* y_2(x) -_* y_2^{**}(x) \cdot_* y_1(x) \right) \Big/_* W_*(y_1, y_2)(x),$$

$$a_2(x) \ = \ \left(y_2^{**}(x) \cdot_* y_1^*(x) -_* y_1^{**}(x) \cdot_* y_2^*(x) \right) \Big/_* W_*(y_1, y_2)(x),$$

$x \in J.$

Proof 4.4.14 *Since y_1 and y_2 are multiplicative linearly independent, we have that*

$$W_*(y_1, y_2)(x) \neq 0_*, \quad x \in J.$$

Because y_1 and y_2 are solutions to the equation (4.20), we have

$$y_1^{**}(x) +_* a_1(x) \cdot_* y_1^*(x) +_* a_2(x) \cdot_* y_1(x) \;=\; 0_*$$

$$y_2^{**}(x) +_* a_1(x) \cdot_* y_2^*(x) +_* a_2(x) \cdot_* y_2(x) \;=\; 0_*, \quad x \in J.$$

Hence, we get the system

$$a_1(x) \cdot_* y_1^*(x) +_* a_2(x) \cdot_* y_1(x) \;=\; -_* y_1^{**}(x)$$

$$a_1(x) \cdot_* y_2^*(x) +_* a_2(x) \cdot_* y_2(x) \;=\; -_* y_2^{**}(x), \quad x \in J,$$

from where we get the desired result. This completes the proof.

Example 4.4.15 Let

$$y_1(x) \;=\; e,$$

$$y_2(x) \;=\; \cos_* x, \quad x \in \mathbb{R}_*.$$

Then

$$y_1^*(x) \;=\; 0_*,$$

$$y_2^*(x) \;=\; -_* \sin_* x, \quad x \in \mathbb{R}_*.$$

Hence,

$$W_*(y_1, y_2)(x) \;=\; \det_*\begin{pmatrix} y_1(x) & y_2(x) \\ y_1^*(x) & y_2^*(x) \end{pmatrix}$$

$$= \det_*\begin{pmatrix} e & \cos_* x \\ 0_* & -_* \sin_* x \end{pmatrix}$$

$$= e \cdot_* (-_* \sin_* x) -_* 0_* \cdot_* \cos_* x$$

$$= -_* \sin_* x, \quad x \in \mathbb{R}_*.$$

Next,

$$y_1^{**}(x) \;=\; 0_*,$$

$$y_2^{**}(x) \;=\; -_* \cos_* x, \quad x \in \mathbb{R}_*.$$

Therefore,

$$a_1(x) \;=\; (0_* \cdot_* \cos_* x +_* \cos_* x \cdot_* e)/_*(-_* \sin_* x)$$

$$\;=\; -_* \cot_* x,$$

$$a_2(x) \;=\; (-_* \cos_* x \cdot_* 0_* -_* 0_* \cdot_* (-_* \sin_* x))/_*(-_* \sin_* x)$$

$$\;=\; 0_*, \quad x \in \mathbb{R}_*.$$

Then we get the equation

$$y^{**} -_* \cot_* x \cdot_* y^* = 0_*, \quad x \in \mathbb{R}_*.$$

Exercise 4.4.16 *Find the equation (4.20) if the following functions*

$$y_1(x) \;=\; x,$$

$$y_2(x) \;=\; e^x, \quad x \in \mathbb{R}_*,$$

are its solutions.

Answer

$$(x -_* e) \cdot_* y^{**} -_* x \cdot_* y^* +_* y = 0_*, \quad x \in \mathbb{R}_*.$$

4.5   The Multiplicative Constant Case

In this section, we will investigate the equation

$$y^{**} +_* e^a \cdot_* y^* +_* e^b \cdot_* y = 0_*, \tag{4.28}$$

where $a, b \in \mathbb{R}$. We will search for a solution of the equation (4.28) in the form

$$y(x) = e^{e^r \cdot_8 x}, \quad x \in J,$$

where $r \in \mathbb{R}$ will be determined below. We have

$$y(x) = e^{e^{\log e^r} \log x}$$

$$= e^{e^{r \log x}}, \quad x \in J.$$

Then

$$y'(x) = \frac{r}{x} e^{r \log x} e^{e^{r \log x}},$$

$$y^*(x) = e^{r e^{r \log x}}$$

$$= e^r \cdot_* e^{e^{r \log x}}$$

$$= e^r \cdot_* y(x),$$

$$(y^*)'(x) = \frac{r^2}{x} e^{r \log x} e^{e^{r \log x}},$$

$$y^{**}(x) = e^{r^2 e^{r \log x}}$$

$$= e^{r^2} \cdot_* e^{e^{r \log x}}$$

$$= e^{r^2} \cdot_* y(x), \quad x \in J.$$

Then, by the equation (4.28), we get

$$0_* = e^{r^2} \cdot_* y(x) +_* e \cdot_* e^r \cdot_* y(x) +_* e^b \cdot_* y(x)$$

$$= \left(e^{r^2} +_* e \cdot_* e^r +_* e^b \right) \cdot_* y(x)$$

$$= \left(e^{r^2} +_* e^{ar} +_* e^b \right) \cdot_* y(x)$$

$$= e^{r^2 + ar + b} \cdot_* y(x), \quad x \in J.$$

Hence, we get the equation

$$r^2 + ar + b = 0. \tag{4.29}$$

We have the following cases.

1. Let
$$a^2 - 4b > 0.$$

   Then
   $$r_{1,2} = \frac{-a \pm \sqrt{a^2 - 4b}}{2}$$

   and the equation (4.28) has two distinct solutions

   $$y_1(x) = e^{e^{r_1} \cdot_* x},$$

   $$y_2(x) = e^{e^{r_2} \cdot_* x}, \quad x \in J,$$

   and the general solution of the equation (4.28) is given by

   $$y(x) = c_1 \cdot_* e^{e^{r_1} \cdot_* x} +_* c_2 \cdot_* e^{e^{r_2} \cdot_* x}, \quad x \in J,$$

   where $c_1, c_2 \in \mathbb{R}_*$.

2. Let
   $$a^2 - 4b = 0.$$

   Then the equation (4.29) has a unique root

   $$r_1 = -\frac{a}{2}$$

   and
   $$y_1(x) = e^{e^{r_1} \cdot_* x}, \quad x \in J,$$

   is a particular solution of the equation (4.28). We will search the general solution of the equation (4.28) in the form

   $$y(x) = c(x) \cdot_* e^{e^{r_1} \cdot_* x}, \quad x \in J,$$

   where $c : J \to \mathbb{R}_*$ and $c \in \mathscr{C}_*^2(J)$. We have

   $$y^*(x) = c^*(x) \cdot_* e^{e^{r_1} \cdot_* x} +_* c(x) \cdot_* e^{r_1} \cdot_* e^{e^{r_1} \cdot_* x}$$

$$= \left(c^*(x) +_* e^{r_1 q} \cdot_* c(x) \right) \cdot_* e^{e^{r_1} \cdot_* x},$$

$$y^{**}(x) = \left(c^{**}(x) +_* e^{r_1} \cdot_* c^*(x) \right) \cdot_* e^{e^{r_1} \cdot_* x}$$

$$+_* e^{r_1} \cdot_* \left(c^*(x) +_* e^{r_1} \cdot_* c(x) \right) \cdot_* e^{e^{r_1} \cdot_* x}$$

$$= \left(c^{**}(x) +_* e^{2r_1} \cdot_* c^*(x) +_* e^{r_1^2} \cdot_* c(x) \right) \cdot_* e^{e^{r_1} \cdot_* x},$$

$x \in J$. Hence, using the equation (4.28), we get

$$0_* = y^{**}(x) +_* e \cdot_* y^*(x) +_* e^b \cdot_* y(x)$$

$$= \left(c^{**}(x) +_* e^{2r_1} \cdot_* c^*(x) +_* e^{r_1^2} \cdot_* c(x) \right) \cdot_* e^{e^{r_1} \cdot_* x}$$

$$+_* e^a \cdot_* \left(c^*(x) +_* e^{r_1} \cdot_* c(x) \right) \cdot_* e^{e^{r_1} \cdot_* x}$$

$$+_* e^b \cdot_* c(x) \cdot_* e^{e^{r_1} \cdot_* x}$$

$$= \left(c^{**}(x) +_* \left(e^{2r_1} +_* e^a \right) \cdot_* c^*(x) \right.$$

$$+_* \left(e^{r_1^2} +_* e^a \cdot_* e^{r_1} +_* e^b \right) \cdot_* c(x) \right) \cdot_* e^{e^{r_1} \cdot_* x}$$

$$= \left(c^{**}(x) +_* e^{2r_1 + a} \cdot_* c^*(x) \right.$$

$$+_* e^{r_1^2 + a r_1 + b} \cdot_* c(x) \right) \cdot_* e^{e^{r_1} \cdot_* x}$$

$$= c^{**}(x) \cdot_* e^{e^{r_1} \cdot_* x}, \quad x \in J.$$

Therefore,

$$c^{**}(x) = 0, \quad x \in J.$$

Hence,

$$c(x) = c_1 \cdot_* x +_* c_2, \quad x \in J,$$

where $c_1, c_2 \in \mathbb{R}_*$. In this case, the general solution of the equation (4.28) is given by

$$y(x) = (c_1 \cdot_* x +_* c_2) \cdot_* e^{e^{r_1} \cdot_* x}, \quad x \in \mathbb{R}_*.$$

3. Let

$$a^2 - 4b < 0.$$

Then the solutions of the equation (4.29) can be represented in the form

$$r_{1,2} = \mu \pm i\nu,$$

where

$$\mu = -\frac{a}{2},$$

$$\nu = \frac{\sqrt{4b - a^2}}{2}.$$

We have

$$\mu^2 - \nu^2 + a\mu + b = \frac{a^2}{4} - b + \frac{a^2}{4} - \frac{a^2}{2} + b$$

$$= 0,$$

$$a + 2\mu = a - a$$

$$= 0.$$

We will show that

$$y_1(x) = e^{e^{\mu} \cdot_* x} \cdot_* \cos_* \left(e^{\nu} \cdot_* x \right),$$

$$y_2(x) = e^{e^{\mu} \cdot_* x} \cdot_* \sin_* \left(e^{\nu} \cdot_* x \right), \quad x \in J,$$

are particular solutions of (4.28). We have

$$y_1(x) = e^{e^{\mu \log x}} \cdot_* \cos_* \left(e^{v \log x} \right)$$

$$= e^{e^{\mu \log x}} \cdot_* e^{\cos(v \log x)}$$

$$= e^{e^{\mu \log x} \cos(v \log x)}, \quad x \in J,$$

and

$$y_2(x) = e^{e^{\mu \log x}} \cdot_* \sin_* \left(e^{v \log x} \right)$$

$$= e^{e^{\mu \log x}} \cdot_* e^{\sin(v \log x)}$$

$$= e^{e^{\mu \log x} \sin(v \log x)}, \quad x \in J.$$

We have

$$y_1'(x) = \frac{d}{dx} \left(e^{\mu \log x} \cos(v \log x) \right) e^{e^{\mu \log x} \cos(v \log x)}$$

$$= \left(\frac{\mu}{x} e^{\mu \log x} \cos(v \log x) - \frac{v}{x} e^{\mu \log x} \sin(v \log x) \right)$$

$$\times e^{e^{\mu \log x} \cos(v \log x)}$$

$$= \frac{1}{x} (\mu \cos(v \log x) - v \sin(v \log x)) e^{\mu \log x}$$

$$\times e^{e^{\mu \log x} \cos(v \log x)},$$

$$y_1^*(x) = e^{(\mu \cos(v \log x) - v \sin(v \log x)) e^{\mu \log x}},$$

$$\left(y_1^* \right)'(x) = \frac{d}{dx} \left((\mu \cos(v \log x) - v \sin(v \log x)) e^{\mu \log x} \right)$$

$$\times e^{(\mu\cos(v\log x)-v\sin(v\log x))e^{\mu\log x}}$$

$$= \left(-\frac{\mu v}{x}\sin(v\log x) - \frac{v^2}{x}\cos(v\log x)\right.$$

$$\left. +\frac{\mu^2}{x}\cos(v\log x) - \frac{\mu v}{x}\sin(v\log x)\right)$$

$$\times e^{(\mu\cos(v\log x)-v\sin(v\log x))e^{\mu\log x}}$$

$$= \left(\frac{\mu^2-v^2}{x}\cos(v\log x) - \frac{2\mu v}{x}\sin(v\log x)\right)e^{\mu\log x}$$

$$\times e^{(\mu\cos(v\log x)-v\sin(v\log x))e^{\mu\log x}},$$

$$y_1^{**}(x) = e^{\left(\frac{\mu^2-v^2}{x}\cos(v\log x)-\frac{2\mu v}{x}\sin(v\log x)\right)e^{\mu\log x}}, \quad x \in J.$$

Hence,

$$y_1^{**}(x) +_* e^a \cdot_* y_1^*(x) +_* e^b \cdot_* y_1(x)$$

$$= \exp\left(\left((\mu^2-v^2)\cos(v\log x) - 2\mu v\sin(v\log x)\right.\right.$$

$$+a\mu\cos(v\log x) - av\sin(v\log x)$$

$$\left.\left. +b\cos(v\log x)\right)e^{\mu\log x}\right)$$

$$= \exp\left(\left((\mu^2-v^2+a\mu+b)\cos(v\log x)\right.\right.$$

$$\left.\left. -(a+2\mu)v\sin(v\log x)\right)e^{\mu\log x}\right)$$

$$= e^0$$

$$= 0_*, \quad x \in J,$$

i.e., y_1 is a solution of the equation (4.28). Next,

$$y_2'(x) = \frac{d}{dx}\left(e^{\mu\log x}\sin(v\log x)\right)e^{e^{\mu\log x}\sin(v\log x)}$$

$$= \left(\frac{\mu}{x}e^{\mu\log x}\sin(v\log x)+\frac{v}{x}e^{\mu\log x}\cos(v\log x)\right)$$

$$\times e^{e^{\mu\log x}\sin(v\log x)}$$

$$= \frac{1}{x}(\mu\sin(v\log x)+v\cos(v\log x))e^{\mu\log x}$$

$$\times e^{e^{\mu\log x}\sin(v\log x)},$$

$$y_1^*(x) = e^{(\mu\sin(v\log x)+v\cos(v\log x))e^{\mu\log x}},$$

$$\left(y_1^*\right)'(x) = \frac{d}{dx}\left((\mu\sin(v\log x)+v\cos(v\log x))e^{\mu\log x}\right)$$

$$\times e^{(\mu\sin(v\log x)+v\cos(v\log x))e^{\mu\log x}}$$

$$= \left(+\frac{\mu v}{x}\cos(v\log x)-\frac{v^2}{x}\sin(v\log x)\right.$$

$$\left.+\frac{\mu^2}{x}\sin(v\log x)+\frac{\mu v}{x}\cos(v\log x)\right)$$

$$\times e^{(\mu\sin(v\log x)+v\cos(v\log x))e^{\mu\log x}}$$

$$= \left(\frac{\mu^2-v^2}{x}\sin(v\log x)+\frac{2\mu v}{x}\cos(v\log x)\right)e^{\mu\log x}$$

$$\times e^{(\mu\sin(v\log x)+v\cos(v\log x))e^{\mu\log x}},$$

$$y_1^{**}(x) = e^{\left(\frac{\mu^2-v^2}{x}\sin(v\log x)+\frac{2\mu v}{x}\cos(v\log x)\right)e^{\mu\log x}}, \quad x\in J.$$

Hence,

$$y_1^{**}(x) +_* e^a \cdot_* y_1^*(x) +_* e^b \cdot_* y(x)$$

$$= \exp\left(\left((\mu^2 - v^2)\sin(v\log x) + 2\mu v\cos(v\log x)\right.\right.$$

$$+ a\mu\sin(v\log x) + av\cos(v\log x)$$

$$\left.\left. + b\sin(v\log x)\right)e^{\mu\log x}\right)$$

$$= \exp\left(\left((\mu^2 - v^2 + a\mu + b)\sin(v\log x)\right.\right.$$

$$\left.\left. + (a + 2\mu)v\cos(v\log x)\right)e^{\mu\log x}\right)$$

$$= e^0$$

$$= 0_*, \quad x \in J,$$

i.e., y_2 is a solution of the equation (4.28). Therefore, the general solution of the equation (4.28) is given by

$$y(x) \quad = \quad e^{e^\mu \cdot_* x} \cdot_* \left(c_1 \cdot_* \cos_* (e^v \cdot_* x)\right.$$

$$\left. +_* c_2 \cdot_* \sin_* \left(e^v \cdot_8 x\right)\right), \quad x \in J.$$

Definition 4.5.1 *The equation (4.29) is said to be the characteristic equation of the equation (4.28).*

Example 4.5.2 *Consider the equation*

$$y^{**} +_* y^* -_* e^2 \cdot_* y = 0_* \quad on \quad \mathbb{R}_*.$$

The characteristic equation is

$$r^2 + r - 2 = 0.$$

Then

$$r_1 = -2,$$

$$r_2 = 1.$$

Therefore, the general solution of the considered equation is

$$y(x) = c_1 \cdot_* e^{e^{-2} \cdot_* x} +_* c_2 \cdot_* e^x, \quad x \in \mathbb{R}_*.$$

Here c_1 and c_2 are multiplicative constants.

Example 4.5.3 *Consider the equation*

$$y^{**} +_* e^2 \cdot_* y^* +_* e^{10} \cdot_* y = 0_* \quad on \quad \mathbb{R}_*.$$

The characteristic equation is

$$r^2 + 2r + 10 = 0.$$

Then

$$r_{1,2} = -1 \pm 3i.$$

The general solution of the considered equation is

$$y(x) = e^{e^{-1} \cdot_* x} \cdot_* \left(c_1 \cdot_* \cos_* \left(e^3 \cdot_* x \right) +_* c_2 \cdot_* \sin_* \left(e^3 \cdot_* x \right) \right),$$

$x \in \mathbb{R}_*$. *Here c_1 and c_2 are multiplicative constants.*

Example 4.5.4 *Consider the equation*

$$y^{**} -_* e^2 \cdot_* y^* +_* y = 0_* \quad on \quad \mathbb{R}_*.$$

The characteristic equation is

$$r^2 - 2r + 1 = 0.$$

We have

$$r = 1$$

and then the general solution of the considered equation is

$$y(x) = (c_1 \cdot_* x +_* c_2) \cdot_* e^x, \quad x \in \mathbb{R}_*.$$

Here c_1 and c_2 are multiplicative constants.

Exercise 4.5.5 *Find the general solution of the following equation:*

$$y^{**} +_* e^4 \cdot_* y^* +_* e^3 \cdot_* y = 0_* \quad on \quad \mathbb{R}_*.$$

Answer

$$y(x) = c_1 \cdot_* e^{e^{-1} \cdot_* x} +_* c_2 \cdot_* e^{e^{-3} \cdot_* x}, \quad x \in \mathbb{R}_*.$$

Here c_1 and c_2 are multiplicative constants.

4.6   The Method of Variation of Parameters

Consider the equation (4.2) and suppose that y_1 and y_2 are two multiplicative linearly independent solutions to the equation (4.1). We will search for a particular solution y_p of the equation (4.2) in the form

$$y_p(x) = c_1(x) \cdot_* y_1(x) +_* c_2(x) \cdot_* y_2(x), \quad x \in J.$$

We have

$$y_p^*(x) \quad = \quad c_1^*(x) \cdot_* y_1(x) +_* c_1(x) \cdot_* y_1^{**}(x)$$

$$+_* c_2^*(x) \cdot_* y_2(x) +_* c_2(x) \cdot_* y_2^*(x), \quad x \in J.$$

As a first condition, assume that

$$c_1^*(x) \cdot_* y_1(x) +_* c_2^*(x) \cdot_* y_2(x) = 0_*, \quad x \in J. \tag{4.30}$$

Then

$$y_p^*(x) = c_1(x) \cdot_* y_1^*(x) +_* c_2(x) \cdot_* y_2^*(x), \quad x \in J,$$

and

$$y_p^{**}(x) \quad = \quad c_1^*(x) \cdot_* y_1^*(x) +_* c_1(x) \cdot_* y_1^{**}(x)$$

$$+_* c_2^*(x) \cdot_* y_2^*(x) +_* c_2(x) \cdot_* y_2^{**}(x), \quad x \in J.$$

Substituting into (4.2) and using that

$$a_0(x) \cdot_* y_1^{**}(x) +_* a_1(x) \cdot_* y_1^{*}(x) +_* a_2(x) \cdot_* y_1(x) \;=\; 0_*,$$

$$a_0(x) \cdot_* y_2^{**}(x) +_* a_1(x) \cdot_* y_2^{*}(x) +_* a_2(x) \cdot_* y_2(x) \;=\; 0_*, \quad x \in J,$$

we get

$$f(x) \;=\; a_0(x) \cdot_* y_p^{**}(x) +_* a_1(x) \cdot_* y_p^{*}(x) +_* a_2(x) \cdot_* y_p(x)$$

$$=\; a_0(x) \cdot_* \left(c_1^{*}(x) \cdot_* y_1^{*}(x) +_* c_1(x) \cdot_* y_1^{**}(x) \right.$$

$$\left. +_* c_2^{*}(x) \cdot_* y_2^{*}(x) +_* c_2(x) \cdot_* y_2^{**}(x) \right)$$

$$+_* a_1(x) \cdot_* \left(c_1(x) \cdot_* y_1^{*}(x) +_* c_2(x) \cdot_* y_2^{*}(x) \right)$$

$$+_* a_2(x) \cdot_* \left(c_1(x) \cdot_* y_1(x) +_* c_2(x) \cdot_* y_2(x) \right)$$

$$=\; c_1(x) \cdot_* \left(a_0(x) \cdot_* y_1^{**}(x) +_* a_1(x) \cdot_* y_1^{*}(x) \right.$$

$$\left. +_* a_2(x) \cdot_* y_1(x) \right)$$

$$+_* c_2(x) \left(a_0(x) \cdot_* y_2^{**}(x) +_* a_1(x) \cdot_* y_2^{*}(x) \right.$$

$$\left. +_* a_2(x) \cdot_* y_2(x) \right)$$

$$+_* a_0(x) \cdot_* \left(c_1^{*}(x) \cdot_* y_1^{*}(x) +_* c_2^{*}(x) \cdot_* y_2^{*}(x) \right)$$

$$=\; a_0(x) \cdot_* \left(c_1^{*}(x) \cdot_* y_1^{*}(x) +_* c_2^{*}(x) \cdot_* y_2^{*}(x) \right),$$

$x \in J$, whereupon

$$c_1^{*}(x) \cdot_* y_1^{*}(x) +_* c_2^{*}(x) \cdot_* y_2^{*}(x) = f(x) /_* a_0(x), \quad x \in J.$$

By the last equation and (4.30), we arrive at the system

$$c_1^*(x) \cdot_* y_1^*(x) +_* c_2^*(x) \cdot_* y_2^*(x) \;=\; f(x)/_* a_0(x)$$

$$_1^*(x) \cdot_* y_1(x) +_* c_2^*(x) \cdot_* y_2(x) \;=\; 0, \quad x \in J,$$

whereupon

$$c_1^*(x) \;=\; -_*\left((y_2(x) \cdot_* f(x))/_* a_0(x) \right)/_* W_*(y_1, y_2)(x),$$

$$c_2^*(x) \;=\; \left((y_1(x) \cdot_* f(x))/_* a_0(x) \right)/_* W_*(y_1, y_2)(x), \quad x \in J.$$

Take $x_0 \in J$ arbitrarily. Then

$$c_1(x) \;=\; -_* \int_{x_0*}^{x} \left((y_2(t) \cdot_* f(t))/_* a_0(t) \right)/_* W_*(y_1, y_2)(t) \cdot_* d_*t,$$

$$c_2^*(x) \;=\; \int_{x_0*}^{x} \left((y_1(t) \cdot_* f(t))/_* a_0(t) \right)/_* W_*(y_1, y_2)(t) \cdot_* d_*t, \quad x \in J.$$

Consequently,

$$y_p(x) = -_* y_1(x) \cdot_* \int_{x_0*}^{x} \left((y_2(t) \cdot_* f(t))/_* a_0(t) \right)/_* W_*(y_1, y_2)(t) \cdot_* d_*t,$$

$$+_* y_2(x) \cdot_* \int_{x_0*}^{x} \left((y_1(t) \cdot_* f(t))/_* a_0(t) \right)/_* W_*(y_1, y_2)(t) \cdot_* d_*t, \quad x \in J,$$

and the general solution of the equation (4.2) is given by

$$y(x) = b_1 \cdot_* y_1(x) +_* b_2 \cdot_* y_2(x) +_* y_p(x)$$

$$= b_1 \cdot_* y_1(x) +_* b_2 \cdot_* y_2(x)$$

$$-_* y_1(x) \cdot_* \int_{x_0*}^{x} \left((y_2(t) \cdot_* f(t))/_* a_0(t) \right)/_* W_*(y_1, y_2)(t) \cdot_* d_*t,$$

$$+_* y_2(x) \cdot_* \int_{x_0*}^{x} \left((y_1(t) \cdot_* f(t))/_* a_0(t) \right)/_* W_*(y_1, y_2)(t) \cdot_* d_*t, \quad x \in J.$$

By the construction of the functions c_1 and c_2, we have that

$$c_1(x_0) \quad = \quad 0_*,$$

$$c_2(x_0) \quad = \quad 0_*.$$

Hence,

$$y_p(x_0) \quad = \quad c_1(x_0) \cdot_* y_1(x_0) +_* c_2(x_0) \cdot_* y_2(x_0)$$

$$= \quad 0_*$$

and

$$y_p^*(x_0) \quad = \quad c_1(x_0) \cdot_* y_1^*(x_0) +_* c_2(x_0) \cdot_* y_2^*(x_0)$$

$$= \quad 0_*.$$

Example 4.6.1 *Consider the equation*

$$y^{**} +_* y = e^x \cdot_* \sin_* x, \quad x \in \mathbb{R}_*.$$

The corresponding homogeneous equation is

$$y^{**} +_* y = 0_*$$

and its characteristic equation is as follows:

$$r^2 + 1 = 0.$$

Then

$$r_{1,2} = \pm i.$$

Then

$$y_1(x) \quad = \quad \sin_* x,$$

$$y_2(x) \quad = \quad \cos_* x, \quad x \in \mathbb{R}_*,$$

are particular solutions of the homogeneous equation. We have

$$y_1^*(x) \;=\; \cos_* x,$$

$$y_2^*(x) \;=\; -_* \sin_* x, \quad x \in \mathbb{R}_*,$$

and

$$W_*(y_1, y_2)(x) \;=\; \det_* \begin{pmatrix} y_1(x) & y_2(x) \\ y_1^*(x) & y_2^*(x) \end{pmatrix}$$

$$=\; \det_* \begin{pmatrix} \sin_* x & \cos_* x \\ \cos_* x & -_* \sin_* x \end{pmatrix}$$

$$=\; -_* \left((\sin_* x)^{2*} +_* (\cos_* x)^{2*} \right)$$

$$=\; -_* e$$

$$\neq\; 0_*, \quad x \in \mathbb{R}_*.$$

Therefore, y_1 and y_2 are multiplicative linearly independent on J. Here

$$f(x) \;=\; e^4 \cdot_8 \sin_* x,$$

$$a_0(x) \;=\; e, \quad x \in \mathbb{R}_*.$$

Then

$$c_1(x) \;=\; -_* \int_{*0_*}^x \left(\cos_* t \cdot_* e^4 \cdot_* \sin_* t \right) /_* (-_* e) \cdot_* d_* t$$

$$=\; e^3 \cdot_* \int_{*0_*}^x \sin_* t \cdot_* \cos_* t \cdot_* d_* t$$

$$=\; e^3 \cdot_* \int_{*0_*}^x \sin_* t \cdot_* d_* \sin_* t$$

$$=\; e^3 \cdot_* \left((\sin_* t)^{2*} /_* e^2 \right) \Big|_{t=0_*}^{t=x}$$

$$=\; e^{\frac{3}{2}} \cdot_* (\sin_* x)^{2*}, \quad x \in \mathbb{R}_*,$$

and

$$c_2(x) = \int_{*0_*}^{*x} \left(\sin_* t \cdot_* e^4 \cdot_* \sin_* t \right) /_* (-_* e) \cdot_* d_* t$$

$$= -_* e^3 \cdot_* \int_{*0_*}^{*x} ('\sin_* t)^{2*} \cdot_* d_* t$$

$$= -_* e^3 \cdot_* \int_{*0_*}^{*x} \left(e -_* \cos_* \left(e^2 \cdot_* t \right) \right) /_* e^2 \cdot_* d_* t$$

$$= -_* e^{\frac{3}{2}} \cdot_* \int_{*0_*}^{*x} \left(e -_* \cos_* \left(e^2 \cdot_* t \right) \right) \cdot_* d_* t$$

$$= -_* e^{\frac{3}{2}} \cdot_* \left(\int_{*0_*}^{*x} d_* t -_* \int_{*0_*}^{*x} \cos_* \left(e^2 \cdot_* t \right) \cdot_* d_* t \right)$$

$$= -_* e^{\frac{3}{2}} \cdot_* \left(x -_* (e /_* e^2) \cdot_* \sin_* \left(e^2 \cdot_* t \right) \Big|_{t=0_*}^{|t=x} \right)$$

$$= -_* e^{\frac{3}{2}} \cdot_* \left(x -_* e^{\frac{1}{2}} \cdot_* \sin_* \left(e^2 \cdot_* x \right) \right)$$

$$= -_* e^{\frac{3}{2}} \cdot_* x +_* e^{\frac{3}{4}} \cdot_* \sin_* \left(e^2 \cdot_* x \right), \quad x \in J.$$

Consequently, the general solution of the considered equation is

$$y(x) = b_1 \cdot_* y_1(x) +_* b_2 \cdot_* y_2(x)$$

$$+_* c_1(x) \cdot_* y_1(x) +_* c_2(x) \cdot_* y_2(x)$$

$$= (b_1 +_* c_1(x)) \cdot_* y_1(x) +_* (b_2 +_* c_2(x)) \cdot_* y_2(x)$$

$$= \left(b_1 +_* e^{\frac{3}{2}} \cdot_* (\sin_* x)^{2*} \right) \cdot_* \sin_* x$$

$$+_* \left(b_2 -_* e^{\frac{3}{2}} \cdot_* x +_* e^{\frac{3}{4}} \cdot_* \sin_* \left(e^2 \cdot_* x \right) \right) \cdot_* \cos_* x,$$

$x \in \mathbb{R}_*.$ *Here b_1 and b_2 are multiplicative constants.*

Example 4.6.2 *Consider the equation*

$$y^{**} +_* y = e/_* \sin_* x, \quad x \in \mathbb{R}_*.$$

The corresponding homogeneous equation is

$$y^{**} +_* y = 0_*.$$

As in the previous example, we have that the functions

$$y_1(x) \quad = \quad \sin_* x,$$

$$y_2(x) \quad = \quad \cos_* x, \quad x \in \mathbb{R}_*,$$

are multiplicative linearly independent solutions of the homogeneous equation and

$$W_*(y_1, y_2)(x) = -_* e, \quad x \in \mathbb{R}_*.$$

Here

$$f(x) \quad = \quad e/_* \sin_* x,$$

$$a_0(x) \quad = \quad e, \quad x \in \mathbb{R}_*.$$

Then

$$
\begin{aligned}
c_1(x) \quad &= \quad -_* -_* \int_{*0_*}^{x} \left(\cos_* t \cdot_* (e/_* \sin_* t) \right) /_* (-_* e) \cdot_* t_* \\[2mm]
&= \quad \int_{*0_*}^{x} \cos_* t \cdot_* \sin_* t \cdot_* d_* t \\[2mm]
&= \quad \int_{*0_*}^{x} \sin_* t \cdot_* (\sin_* t) \\[2mm]
&= \quad \left((\sin_* t)^{2*} /_* e^2 \right) \Big|_{t=0_*}^{t=x} \\[2mm]
&= \quad e^{\frac{1}{2}} \cdot_* (\sin_* x)^{2*}, \quad x \in \mathbb{R}_*,
\end{aligned}
$$

and

$$c_2(x) \quad = \quad \int_{*0_*}^{x} \left(\sin_* t \cdot_* (e/_* \sin_* t) \right) /_* (-_* e) \cdot_* t$$

$$= -_* \int_{*0_*}^x d_*t$$

$$= -_*x, \quad x \in \mathbb{R}_*.$$

Consequently, the general solution of the considered equation is

$$y(x) = b_1 \cdot_* y_1(x) +_* b_2 \cdot_* y_2(x)$$

$$+_* c_1(x) \cdot_* y_1(x) +_* c_2(x) \cdot_* y_2(x)$$

$$= (b_1 +_* c_1(x)) \cdot_* y_1(x) +_* (b_2 +_* c_2(x)) \cdot_* y_2(x)$$

$$= \left(b_1 +_* e^{\frac{1}{2}} \cdot_* (\sin_* x)^{2*} \right) \cdot_* \sin_* x$$

$$+_* (b_2 -_* x) \cdot_* \cos_* x, \quad x \in \mathbb{R}_*,$$

where b_1 and b_2 are multiplicative constants.

Example 4.6.3 *Consider the MIVP*

$$y^{**} +_* y = e^4 \cdot_* e^x, \quad x \in \mathbb{R}_*,$$

$$y(0_*) = e^4, \quad y^*(0_*) = e^{-3}.$$

Here

$$f(x) = e^4 \cdot_* e^4,$$

$$a_0(x) = e, \quad x \in \mathbb{R}_*.$$

Let

$$I = e^4 \cdot_* \int_{*0_*}^x e^t \cdot_* \sin_* t \cdot_* d_*t.$$

Then

$$I = e^4 \cdot_* \int_{*0_*}^x \sin_* t \cdot_* d_*(e^t)$$

$$= \quad e^4 \cdot_* \sin_* t \cdot_* e^t \Big|_{t=0_*}^{t=x}$$

$$-_* e^4 \cdot_* \int_{*0_*}^{x} \cos_* t \cdot_* e^t \cdot_* d_* t$$

$$= \quad e^4 \cdot_* \sin_* x \cdot_* e^x$$

$$-_* e^4 \cdot_* \int_{*0_*}^{x} \cos_* t \cdot_* d_*)(e^t)$$

$$= \quad e^4 \cdot_* \sin_* x \cdot_* e^x$$

$$-_* e^4 \cdot_* \cos_* t \cdot_* e^t \Big|_{t=0_*}^{t=x}$$

$$-_* e^4 \cdot_* \int_{*0_*}^{x} \sin_* t \cdot_* e^t \cdot_* d_* t$$

$$= \quad e^4 \cdot_* (\sin_* x -_* \cos_* x) \cdot_* e^x$$

$$+_* e^4 -_* I, \quad x \in \mathbb{R}_*,$$

whereupon

$$e^2 \cdot_* I = e^4 \cdot_* (\sin_* x -_* \cos_* x) \cdot_* e^x +_* e^4,$$

$x \in \mathbb{R}_*$, *and*

$$I = e^2 \cdot_* (\sin_* x -_* \cos_* x) \cdot_* e^x +_* e^2,$$

$x \in \mathbb{R}_*$. *Moreover,*

$$e^4 \cdot_* \int_{*0_*}^{x} \cos_* t \cdot_* d_* t = e^4 \cdot_* \sin_* x \cdot_* e^x$$

$$-_* e^2 \cdot_* (\sin_* x -_* \cos_* x) \cdot_* e^x -_* e^2$$

$$= \quad e^2 \cdot_* (\sin_* x +_* \cos_* x) -_* e^2,$$

$x \in \mathbb{R}_*$. *Using the computations in the previous examples, we find*

$$
\begin{aligned}
c_1(x) &= -_* \int_{*0_*}^x \left(\cos_* t \cdot_* \left(e^4 \cdot_* e^t \right) \right) /_* (-_* e) \cdot_* d_* t \\
&= e^4 \cdot_* \int_{*0_*}^x \cdot_* \cos_* t \cdot_* e^t \cdot_* d_* t \\
&= e^2 \cdot_* \left((\sin_* x +_* \cos_* x) \cdot_* e^x -_* e \right), \quad x \in \mathbb{R}_*,
\end{aligned}
$$

and

$$
\begin{aligned}
c_2(x) &= \int_{*0_*}^x \left(\sin_* t \cdot_* \left(e^4 \cdot_* e^t \right) \right) /_* (-_* e) \cdot_* d_* t \\
&= -_* e^4 \cdot_* \int_{*0_*}^x \cdot_* \sin_* t \cdot_* e^t \cdot_* d_* t \\
&= -_* e^2 \cdot_* \left((\sin_* x -_* \cos_* x) \cdot_* e^x -_* e \right), \quad x \in \mathbb{R}_*.
\end{aligned}
$$

Then for the general solution of the considered equation, we have

$$
\begin{aligned}
y(x) &= \left(b_1 +_* e^2 \cdot_* \left((\sin_* x +_* \cos_* x) \cdot_* e^x -_* e \right) \right) \cdot_* \sin_* x \\
&\quad +_* \left(b_2 -_* *e^2 \cdot_* \left((\sin_* x -_* \cos_* x) \cdot_* e^x -_* e \right) \right) \cdot_* \cos_* x,
\end{aligned}
$$

$x \in \mathbb{R}_*$. *Hence,*

$$
\begin{aligned}
y(0_*) &= b_2 \\
&= e^4.
\end{aligned}
$$

Next,

$$
\begin{aligned}
y^*(x) &= e^2 \cdot_* (\cos_* x -_* \sin_* x) \cdot_* e^x \cdot_* \sin_* x \\
&\quad +_* \left(b_1 +_* e^2 \left((\sin_* x +_* \cos_* x) \cdot_* e^x \cdot_* e \right) \cdot_* \cos_* x
\end{aligned}
$$

$$+_*e^2\cdot_*\left(\left(\sin_* x+_*\cos_* x\right)\cdot_* e^x\right)\cdot_*\sin_* x$$

$$-_*e^2\cdot_*\left(\cos_* x+_*\sin_* x\right)\cdot_* e^x\cdot_*\cos_* x$$

$$-_*e^2\left(\sin_* x-_*\cos_* x\right)\cdot_* e^x\cdot_*\cos_* x$$

$$-_*\left(e^4-_*e^2\cdot_*\left(\left(\sin_* x-_*\cos_* x\right)\cdot_* e^x+_* e\right)\right)\cdot_*\sin_* x,$$

$x\in\mathbb{R}_*$, and

$$y^*(0_*)\;=\;b_2$$

$$=\;e^{-3}.$$

Thus,

$$y(x)\;=\;\left(e^{-3}+_*e^2\cdot_*\left(\left(\sin_* x+_*\cos_* x\right)\cdot_* e^x-_* e\right)\right)\cdot_*\sin_* x$$

$$+_*\left(e^4-_**e^2\cdot_*\left(\left(\sin_* x-_*\cos_* x\right)\cdot_* e^x-_* e\right)\right)\cdot_*\cos_* x$$

$$=\;e^{-5}\cdot_*\sin_* x+_*e^2\cdot_*\left(\left(\sin_* x\right)^{2*}+_*\sin_* x\cdot_*\cos_* x\right)\cdot_* e^x$$

$$+_*e^2\cdot_*\cos_* x-_*e^2\cdot_*\left(\sin_* x\cdot_*\cos_* x-_*\left(\cos_* x\right)^{2*}\right)\cdot_* e^x$$

$$=\;e^{-5}\cdot_*\sin_* x+_*e^2\cdot_*\cos_* x+_*e^2\cdot_* e^x,$$

$x\in\mathbb{R}_*$.

Exercise 4.6.4 *Find the general solution of the equation*

$$y^{**}-_* e^2\cdot_* y^*-_* e^3\cdot_* y=e^{e^4\cdot_* x},\quad x\in\mathbb{R}_*.$$

Answer

$$y(x)=b_1\cdot_* e^{-_* x}+_* b_2\cdot_* e^{e^3\cdot_* x}+_* e^{\frac{1}{5}}\cdot_* e^{e^4\cdot_* x},\quad x\in\mathbb{R}_*,$$

where b_1 and b_2 are multiplicative constants.

Exercise 4.6.5 *Find the solution of the following MIVP*

$$y^{**} -_* e^2 \cdot_* y^* = e^2 \cdot_8 e^x, \quad x \in \mathbb{R}_*,$$

$$y(e) = e^{-1}, \quad y^*(e) = 1.$$

Answer

$$y(x) = e^{e^2 \cdot_* x -_* e} -_* e^2 \cdot_* e^x +_* e^e -_* e, \quad x \in \mathbb{R}_*.$$

4.7 The Multiplicative Cauchy-Euler Equation

In this section, we will investigate the equation

$$\left(e^c \cdot_* x +_* e^d\right)^{2_*} \cdot_* y^{**} +_* e^a \cdot_* \left(e^c \cdot_* x +_* e^d\right) \cdot_* y^* \tag{4.31}$$

$$+_* e^b \cdot_* y = f(x), \quad x \in J_1,$$

where $a, b, c, d \in \mathbb{R}$, $c \neq 0$, and

$$J_1 = \{x \in \mathbb{R}_* : e^c \cdot_* x +_* e^d > 0_*\}.$$

Definition 4.7.1 *The equation* (4.31) *is said to be the multiplicative Cauchy-Euler equation.*

Set

$$e^z = e^c \cdot_* x +_* e^d, \quad x \in J_1.$$

Then

$$x = \left(e^z -_* e^d\right) /_* e^c$$

and

$$z = \log_* \left(e^c \cdot_* x +_* e^d\right), \quad x \in J_1.$$

Denote

$$g(z) = y(x), \quad x \in J_1.$$

We have that $z = z(x)$ and

$$z^*(x) \;=\; e^c/_*(e^c\cdot_* x +_* e^d)$$

$$=\; e^c\cdot_* e^{-*z(x)},$$

$$z^{**}(x) \;=\; -_*e^c\cdot_* e^{-*z(x)}\cdot_* z^*(x)$$

$$=\; -_*e^c\cdot_* e^{-*z(x)}\cdot_* e^c\cdot_* e^{-*z(x)}$$

$$=\; -_*e^{c^2}\cdot_* e^{-*(z(x))^2}, \quad x \in J_1,$$

and

$$\left(e^c\cdot_* x +_* e^d\right)^{2*} \;=\; \left(e^{z(x)}\right)^{2*}$$

$$=\; e^{\left(\log e^{z(x)}\right)^2}$$

$$=\; e^{(z(x))^2}, \quad x \in J_1.$$

Next,

$$y^*(x) \;=\; g^*(z)\cdot_* z^*(x)$$

$$=\; e^c\cdot_* e^{-*z(x)}\cdot_* g^*(z),$$

$$y^{**}(x) \;=\; \left(e^c\cdot_* e^{-*z(x)}\cdot_* g^*(z)\right)^*_z\cdot_* z^*(x)$$

$$=\; e^c\cdot_* e^{-*z(x)}\cdot_* e^c$$

$$\cdot_* \left(\left(e^{-*z(x)}\right)^*_z\cdot_* g^*(z) +_* e^{-*z(x)}\cdot_* g^{**}(z)\right)$$

$$=\; e^{c^2}\cdot_* e^{-*(z(x))^2}\cdot_* \left(g^{**}(z) -_* g^*(z)\right), \quad x \in J_1.$$

Then, the equation (4.31) takes the form

$$f\left(\left(e^z -_* e^d\right)/_* e^c\right) = f(x)$$

$$= e^{(z(x))^2} \cdot_* e^{c^2} \cdot_* e^{-*(z(x))^2} \cdot_* \left(g^{**}(z) -_* g^*(z)\right)$$

$$+_* e^a \cdot_* e^{z(x)} \cdot_* e^c \cdot_* e^{-*z(x)} \cdot_* g^*(z)$$

$$+_* e^b \cdot_* g(z)$$

$$= e^{c^2} \cdot_* \left(g^{**}(z) -_* g^*(z)\right) +_* e^{ac} \cdot_* g^*(z)$$

$$+_* e^b \cdot_* g(z)$$

$$= e^{c^2} \cdot_* g^{**}(z) +_* \left(e^{ac} -_* e^{c^2}\right) \cdot_* g^*(z)$$

$$+_* e^b \cdot_* g(z)$$

$$= e^{c^2} \cdot_* g^{**}(z) +_* e^{ac-c^2} \cdot_* g^*(z)$$

$$+_* e^b \cdot_* g(z),$$

i.e., we get the equation

$$e^{c^2} \cdot_* g^{**}(z) +_* e^{ac-c^2} \cdot_* g^*(z)$$

$$+_* e^b \cdot_* g(z) = f\left(\left(e^z -_* e^d\right)/_* e^c\right).$$

Example 4.7.2 *Consider the equation*

$$x^{2*} \cdot_* y^{**} +_* x \cdot_* y^* -_* y = 0_*, \quad x \in \mathbb{R}_*.$$

Here

$$a = 1,$$

$$b = -1,$$

$$c = 1,$$

$$d = 0.$$

Then, we get the equation

$$e \cdot_* g^{**}(z) +_* e^0 \cdot_* g^*(z) -_* g(z) = 0_*$$

or

$$g^{**}(z) -_* g(z) = 0_*.$$

The characteristic equation of the last equation is

$$r^2 - 1 = 0,$$

whereupon

$$r_{1,2} = \pm 1.$$

Therefore,

$$g(z) = b_1 \cdot_* e^{e \cdot_* z} +_* b_2 \cdot_* e^{e^{-1} \cdot_* z}$$

$$= b_1 \cdot_* e^z +_* b_2 \cdot_* e^{z^{-1}}$$

and

$$y(x) = b_1 \cdot_* x +_* b_2 \cdot_* e^{(\log x)^{-1}}$$

$$= b_1 \cdot_* x +_* b_2 \cdot_* x^{-1_*}, \quad x \in \mathbb{R}_*.$$

Here b_1 and b_2 are multiplicative constants.

Example 4.7.3 *Consider the equation*

$$\left(x -_* e^2\right)^{2*} \cdot_* y^{**} -_* e^3 \cdot_* \left(x -_* e^2\right) \cdot_* y^* +_* e^4 \cdot_* y = 0_*,$$

$x \in \mathbb{R}_*$. *Here*

$$a = -3,$$

$$b = 4,$$

$$c = 1,$$

$$d = 2.$$

Then, we get the equation

$$g^{**} +_* e^{-4} \cdot_* g^* +_* e^4 \cdot g = 0_*.$$

Its characteristic equation is

$$r^2 - 4r + 4 = 0,$$

whereupon
$$r_{1,2} = 2.$$

Therefore,

$$g(z) = e^{e^2 \cdot_* z} \cdot_* (b_1 \cdot_* z +_* b_2)$$

$$= (x -_* e^2)^{2*} \cdot_* \left(b_1 \cdot_* \log_* |x -_* e^2|_* +_* b_2\right), \quad x \in \mathbb{R}_*,$$

and

$$y(x) = (x -_* e^2)^{2*} \cdot_* \left(b_1 \cdot_* \log_* |x -_* e^2|_* +_* b_2\right), \quad x \in \mathbb{R}_*.$$

Here b_1 and b_2 are multiplicative constants.

Exercise 4.7.4 *Find the general solution of the following equation:*

$$x^{2*} \cdot_* y^{**} -_* e^4 \cdot_* x \cdot_* y^* +_* e^6 \cdot_* y = 0_*.$$

Answer
$$y(x) = b_1 \cdot_* x^{2*} +_* b_2 \cdot_* x^{3*},$$

where b_1 and b_2 are multiplicative constants.

4.8   Advanced Practical Problems

Problem 4.8.1 *Prove that the function*

$$y(x) = x^{2*} +_* (1_*/_*x) \cdot_* \left(e^{-\frac{2}{3}} \cdot_* \log_* x -_* (\log_* x)^{2*} \right), \quad x \in \mathbb{R}_*,$$

is a solution to the equation

$$x^{3*} \cdot_* y^{**} -_* e^2 \cdot_* x \cdot_* y = e^6 \cdot_* \log_* x, \quad x \in \mathbb{R}_*.$$

Problem 4.8.2 *Prove that the function*

$$y(x) = x^{2*} \cdot_* \left(\cos_*(\log_* |x|_*) +_* \sin_*(\log_* |x|_*) +_* e^3 \right), \quad x \in \mathbb{R}_*,$$

is a solution to the equation

$$x^{2*} \cdot_* y^{**} -_* e^3 \cdot_* x \cdot_* y^* +_* e^5 \cdot_* y = e^3 \cdot_* x^{2*}, \quad x \in \mathbb{R}_*.$$

Problem 4.8.3 *Prove that the function*

$$y(x) = x^{3*} \cdot_* \log_* |x|_* -_* e^2 \cdot_* x^{2*}, \quad x \in \mathbb{R}_*,$$

is a solution to the equation

$$x^{2*} \cdot_* y^{**} -_* e^6 \cdot_* y = e^5 \cdot_* x^{3*} +_* e^8 \cdot_* x^{2*}, \quad x \in \mathbb{R}_*.$$

Problem 4.8.4 *Prove that the function*

$$y(x) = x^{2*} +_* (1_*/_*x) +_* e^{\frac{1}{10}} \cdot_* \cos_*(\log_* x), \quad x \in \mathbb{R}_*,$$

is a solution to the equation

$$x^{2*} \cdot_* y^{**} -_* e^2 \cdot_* y = \sin_*(\log_* x), \quad x \in \mathbb{R}_*.$$

Problem 4.8.5 *Prove that the function*

$$y(x) = (x -_* e^2)^{2*} +_* x -_* e, \quad x \in \mathbb{R}_*,$$

is a solution to the equation

$$(x -_* e^2)^{2*} \cdot_* y^{**} -_* e^3 \cdot_* (x -_* e^2) \cdot_* y^* +_* e^4 \cdot_* y = x, \quad x \in \mathbb{R}_*.$$

Problem 4.8.6 *Using the definition, investigate for multiplicative linear dependence and multiplicative linear independence in the following systems of functions.*

    1.

$$u^1(x) = e^6 \cdot_* x +_* e^9,$$

$$u^2(x) = e^8 \cdot_* x +_* e^{12}, \quad x \in \mathbb{R}_*.$$

    2.

$$u^1(x) = \sin_* x,$$

$$u^2(x) = \cos_* x, \quad x \in \mathbb{R}_*.$$

    3.

$$u^1(x) = e,$$

$$u^2(x) = x,$$

$$u^3(x) = x^{2_*}, \quad x \in \mathbb{R}_*.$$

    4.

$$u^1(x) = e^4 -_* x,$$

$$u^2(x) = e^2 \cdot_* x +_* e^3,$$

$$u^3(x) = e^6 \cdot_* x +_* e^8, \quad x \in \mathbb{R}_*.$$

    5.

$$u^1(x) = x^{2_*} +_* e^2,$$

$$u^2(x) = e^3 \cdot_* x^{2_*} -_* e,$$

$$u^3(x) = x +_* e^4, \quad x \in \mathbb{R}_*.$$

Answer

1. *Multiplicative linearly dependent.*
2. *Multiplicative linearly independent.*
3. *Multiplicative linearly independent.*
4. *Multiplicative linearly dependent.*
5. *Multiplicative linearly independent.*

Problem 4.8.7 *Using Theorem 4.2.14, prove that the functions*

1.

$$y_1(x) = \sin_* x,$$

$$y_2(x) = \cos_* x,$$

$$y_3(x) = \sin_* \left(e^2 \cdot_* x \right), \quad x \in \mathbb{R}_*.$$

2.

$$y_1(x) = x^{\frac{1}{2}_*},$$

$$y_2(x) = (x +_* e)^{\frac{1}{2}_*},$$

$$y_3(x) = (x +_* e^2)^{\frac{1}{2}_*}, \quad x \in \mathbb{R}_*.$$

3.

$$y_1(x) = x^{4_*},$$

$$y_2(x) = x^{5_*},$$

$$y_3(x) = x^{6*},$$

$$y_4(x) = x^{7*}, \quad x \in \mathbb{R}_*.$$

4.

$$y_1(x) = e +_* x,$$

$$y_2(x) = e -_* x,$$

$$y_3(x) = x^{2*},$$

$$y_4(x) = x^{4*}, \quad x \in \mathbb{R}_*.$$

5.

$$y_1(x) = x^{2*} -_* e,$$

$$y_2(x) = x^{2*} +_* e,$$

$$y_3(x) = x,$$

$$y_4(x) = x^{2*} +_* x +_* e, \quad x \in \mathbb{R}_*.$$

are multiplicative linearly independent.

Problem 4.8.8 *Given the solution* y_1, *find the second solution of the following equations.*

1. $y_1(x) = e^x /_* x$, $x \in \mathbb{R}_*$,

$$x \cdot_* y^{**} +_* e^2 \cdot_* y -_* x \cdot_* y = 0_*, \quad x \in \mathbb{R}_*.$$

2. $y_1(x) = \tan_* x$, $x \in \mathbb{R}_*$,

$$y^{**} -_* e^2 \cdot_* \left(e +_* (\tan_* x)^{2*} \right) \cdot_* y = 0_*, \quad x \in \mathbb{R}_*.$$

3. $y_1(x) = e^x -_* e,\ x \in \mathbb{R}_*,$

$$(e^x +_* e) \cdot_* y^{**} -_* e^2 \cdot_* y^* -_* e^x \cdot_* y = 0_*, \quad x \in \mathbb{R}_*.$$

4. $y_1(x) = \sin_* x,\ x \in \mathbb{R}_*,$

$$y^{**} -_* \tan_* x \cdot_* y^* +_* e^2 \cdot_* y = 0_*, \quad x \in \mathbb{R}_*.$$

5. $y_1(x) = e^{x^{2*}},\ x \in \mathbb{R}_*,$

$$y^{**} +_* e^4 \cdot_* x \cdot_* y^* +_* \left(e^4 \cdot_* x^{2*} +_* e^2\right) \cdot_* y = 0_*, \quad x \in \mathbb{R}_*.$$

Answer

1. $x \cdot_* y_2(x) = e^{-_* x},\ x \in \mathbb{R}_*.$

2. $y_2(x) = e +_* x \cdot_* \tan_* x,\ x \in \mathbb{R}_*.$

3. $y_2(x) = e/_* (e^x +_* e),\ x \in \mathbb{R}_*.$

4. $y_2(x) = e^2 -_* \sin_* x \cdot_* \log\left((e +_* \sin_* x)/_* (e -_* \sin_* x)\right),$
$x \in \mathbb{R}_*.$

5. $y_2(x) = x \cdot_* e^{-_* x^{2*}},\ x \in \mathbb{R}_*.$

Problem 4.8.9 *Find the general solution of the following equations.*

1.
$$y^{**} -_* e^2 \cdot_* y^* = 0_* \quad on \quad \mathbb{R}_*.$$

2.
$$e^2 \cdot_* y^{**} -_* e^5 \cdot_* y^* +_* e^2 \cdot_* y = 0_* \quad on \quad \mathbb{R}_*.$$

3.
$$y^{**} -_* e^4 \cdot_* y^* +_* e^5 \cdot_* y = 0_* \quad on \quad \mathbb{R}_*.$$

4.
$$y^{**} +_* e^4 \cdot_* y = 0_* \quad on \quad \mathbb{R}_*.$$

5.
$$e^4 \cdot_* y^{**} +_* e^4 \cdot_* y^* +_* y = 0_* \quad on \quad \mathbb{R}_*.$$

Answer

1.
$$y(x) = c_1 +_* c_2 \cdot_* e^{e^2 \cdot_* x}, \quad x \in \mathbb{R}_*.$$

2.
$$y(x) = c_1 \cdot_* e^{e^2 \cdot_* x} +_* c_2 \cdot_* e^{e^{\frac{1}{2}} \cdot_* x}, \quad x \in \mathbb{R}_*.$$

3.
$$y(x) = e^{e^2 \cdot_* x} \cdot_* \left(c_1 \cdot_* \cos_* x +_* c_2 \cdot_* \sin_* x \right), \quad x \in \mathbb{R}_*.$$

4.
$$y(x) = c_1 \cdot_* \cos_* \left(e^2 \cdot_* x \right) +_* c_2 \cdot_* \sin_* \left(e^2 \cdot_* x \right), \quad x \in \mathbb{R}_*.$$

5.
$$y(x) = e^{e^{-\frac{1}{2}} \cdot_* x} \cdot_* (c_1 +_* c_2 \cdot_* x), \quad x \in \mathbb{R}_*.$$

Here c_1 and c_2 are multiplicative constants.

Problem 4.8.10 *Find the general solutions of the following equations.*

1.
$$y^{**} +_* y = e^4 \cdot_* x \cdot_* e^x, \quad x \in \mathbb{R}_*.$$

2.
$$y^{**} -_* y = e^2 \cdot_* e^x -_* x^{2*}, \quad x \in \mathbb{R}_*.$$

3.
$$y^{**} +_* y^* -_* e^2 \cdot_* y = e^3 \cdot_* x \cdot_* e^x, \quad x \in \mathbb{R}_*.$$

4.
$$y^{**} -_* e^3 \cdot_* y^* +_* e^2 \cdot_* y = \sin_* x, \quad x \in \mathbb{R}_*.$$

5.
$$y^{**} -_* e^5 \cdot_* y^* +_* e^4 \cdot_* y = e^4 \cdot_* x^{2*} \cdot_* e^{e^2 \cdot_* x}, \quad x \in \mathbb{R}_*.$$

6.
$$y^{**} -_* e^3 \cdot_* y^* +_* e^2 \cdot_* y = x \cdot_* \cos_* x, \quad x \in \mathbb{R}_*.$$

7.
$$y^{**} +_* e^3 \cdot_* y^* -_* e^4 \cdot_* y = e^{e^{-4} \cdot_* x} +_* x \cdot_* e^{-_* x}, \quad x \in \mathbb{R}_*.$$

8.
$$y^{**} +_* e^2 \cdot_* y^* -_* e^3 \cdot_* y = x^{2*} \cdot_* e^x, \quad x \in \mathbb{R}_*.$$

9.
$$y^{**} -_* e^4 \cdot_* y^* +_* e^8 \cdot_* y = e^{e^2 \cdot_* x} +_* \sin_* \left(e^2 \cdot_* x \right) \quad x \in \mathbb{R}_*.$$

10.
$$y^{**} -_* e^9 \cdot_* y = e^{e^3 \cdot_* x} \cdot_* \cos_* x, \quad x \in \mathbb{R}_*.$$

11.
$$y^{**} -_* e^2 \cdot_* y^* +_* y = e^6 \cdot_* x \cdot_* e^x, \quad x \in \mathbb{R}_*.$$

12.
$$y^{**} +_* y = x \cdot_* \sin_* x, \quad x \in \mathbb{R}_*.$$

13.
$$y^{**} +_* e^4 \cdot_* y^* +_* e^4 \cdot_* y = x \cdot_* e^{e^2 \cdot_* x}, \quad x \in \mathbb{R}_*.$$

14.
$$y^{**} -_* e^5 \cdot_* y^* = e^3 \cdot_* x^{2*} +_* \sin_* \left(e^5 \cdot_* x \right), \quad x \in \mathbb{R}_*.$$

15.
$$y^{**} -_* e^2 \cdot_* y^* +_* y = e^x /_* x, \quad x \in \mathbb{R}_*.$$

16.
$$y^{**} +_* e^3 \cdot_* y^* +_* e^2 \cdot_* y = e /_* (e^x +_* e), \quad x \in \mathbb{R}_*.$$

17.
$$y^{**} +_* e^4 \cdot_* y = e^2 \cdot_* \tan_* x, \quad x \in \mathbb{R}_*.$$

18.

$$y^{**} +_* e^2 \cdot_* y^* +_* y = e^3 \cdot_* e^{-*x} \cdot_* (x +_* e)^{\frac{1}{2}*}, \quad x \in \mathbb{R}_*.$$

19.

$$y^{**} +_* e^2 \cdot_* y^* +_* y = x \cdot_* e^x +_* e/_* (x \cdot_* e`x), \quad x \in \mathbb{R}_*.$$

20.

$$y^{**} +_* e^2 \cdot_* y^* +_* e^5 \cdot_* y$$

$$= e^{-*x} \cdot_* \left((\cos_* x)^{2*} +_* \tan_* x \right), \quad x \in \mathbb{R}_*.$$

Answer

1.

$$y(x) = b_1 -_* \cos_* x +_* b_2 \cdot_* \sin_* x +_* \left(e^2 \cdot_* x -_* e^2 \right) \cdot_* e^x, \quad x \in \mathbb{R}_*.$$

2.

$$y(x) = b_1 \cdot_* e^x +_* b_2 \cdot_* e^{-*x} +_* x \cdot_* e^x +_* x^{2*} +_* e^2, \quad x \in \mathbb{R}_*.$$

3.

$$y(x) = \left(x^{2*} /_* e^2 -_* x /_* e^3 \right) \cdot_* e^x +_* b_1 \cdot_* e^{e^{-2} \cdot_* x}$$

$$+_* b_2 \cdot_* e^x, \quad x \in \mathbb{R}_*.$$

4.

$$y(x) = b_1 \cdot_* e^x +_* b_2 \cdot_* e^{e^2 \cdot_* x} +_* e^{\frac{1}{10}} \cdot_* \sin_* x$$

$$+_* e^{\frac{3}{10}} \cdot_* \cos_* x, \quad x \in \mathbb{R}_*.$$

5.

$$y(x) = b_1 \cdot_* \cos_* x +_* b_2 \cdot_* \sin_* x -_* e^2 \cdot_* x \cdot_* \cos_* x, \quad x \in \mathbb{R}_*.$$

6.

$$y(x) \;=\; b_1 \cdot_* e^x +_* b_2 \cdot_* e^{e^{4}\cdot_* x}$$

$$-_* \left(e^2 \cdot_* x^{2_*} -_* e^2 \cdot_* x +_* e^3\right) \cdot_* e^{e^{2}\cdot_* x} \quad x \in \mathbb{R}_*.$$

7.

$$y(x) \;=\; b_1 \cdot_* e^x +_* b_2 \cdot_* e^{e^{2}\cdot_* x}$$

$$+_* \left(e^{\frac{1}{10}} \cdot_* x -_* e^{\frac{3}{25}}\right) \cdot_* \cos_* x$$

$$-_* \left(e^{\frac{3}{10}} \cdot_* x +_* e^{\frac{17}{P50}}\right) \cdot_* \sin_* x, \quad x \in \mathbb{R}_*.$$

8.

$$y(x) \;=\; b_1 \cdot_* e^x +_* b_2 \cdot_* e^{e^{-4}\cdot_* x}$$

$$-_* \left(x /_* e^5\right) \cdot_* e^{e^{-4}\cdot_* x}$$

$$-_* \left(x /_* e^6 +_* e^{\frac{1}{30}}\right) \cdot_* e^{-_* x}, \quad x \in \mathbb{R}_*.$$

9.

$$y(x) \;=\; \left(x^{3_*} /_* e^{10} -_* x^{2_*} /_* e^{16} +_* x /_* e^{32}\right) \cdot_* e^x$$

$$+_* b_1 \cdot_* e^x +_* b_2 \cdot_* e^{e^{-3}\cdot_* x}, \quad x \in \mathbb{R}_*.$$

10.

$$y(x) \;=\; e^{e^{2}\cdot_* x} \cdot_* \left(b_1 \cdot_* \cos_* \left(e^2 \cdot_* x\right)\right.$$

$$\left. +_* b_2 \cdot_* \sin_* \left(e^2 \cdot_* x\right)\right)$$

$$=_* e^{\frac{1}{4}} \cdot_* e^{e^2 \cdot_* x} +_* e^{\frac{1}{10}} \cdot_* \cos_* \left(e^2 \cdot_8 x \right)$$

$$+_* e^{\frac{1}{20}} \cdot_* \sin_* \left(e^2 \cdot_* x \right), \quad x \in \mathbb{R}_*.$$

11.

$$y(x) \;=\; b_1 \cdot_* e^{e^3 \cdot_* x} +_* b_2 \cdot_* e^{e^{-3} \cdot_* x}$$

$$+_* e^{e^3 \cdot_* x} \left(e^{\frac{6}{37}} \cdot_* \sin_* x -_* e^{\frac{1}{37}} \cdot_* \cos_* x \right), \quad x \in \mathbb{R}_*.$$

12.

$$y = \left(b_1 +_* b_2 \cdot_* x +_* x^{3*} \right) \cdot_* e^x, \quad x \in \mathbb{R}_*.$$

13.

$$y(x) \;=\; \left(b_1 -_* x^{2*}/_* e^4 \right) \cdot_* \cos_* x$$

$$+_* \left(b_2 +_* x/_* e^4 \right) \cdot_* \sin_* x, \quad x \in \mathbb{R}_*.$$

14.

$$y(x) \;=\; \left(b_1 +_* b_2 \cdot_* x \right) \cdot_* e^{e^{-2} \cdot_* x}$$

$$+_* \left(x/_* e^{\frac{1}{16}} -_* e^{\frac{1}{32}} \right) \cdot_* e^{e^2 \cdot_* x}, \quad x \in \mathbb{R}_*.$$

15.

$$y(x) = e^x \cdot_* \left(x \cdot_* \log_* |x|_* +_* b_1 \cdot_* x +_* b_2 \right), \quad x \in \mathbb{R}_*.$$

16.

$$y(x) \;=\; \left(e^{-*x} +_* e^{e^{-2} \cdot_* x} \right) \cdot_* \log_* \left(e^x +_* e \right)$$

$$+_* b_1 \cdot_* e^{-*x} +_* b_2 \cdot_* e^{e^{-2} \cdot_* x}, \quad x \in \mathbb{R}_*.$$

17.

$$y(x) \ = \ \sin_*\left(e^2 \cdot_* x\right) \cdot_* \log_* |\sin_* x|_*$$

$$-_* x \cdot_* \cos_*(e^2 \cdot_* x)$$

$$+_* b_1 \cdot_* \sin_*(e^2 \cdot_* x) +_* b_2 \cdot_* \cos_*(e^2 \cdot_* x), \quad x \in \mathbb{R}_*.$$

18.

$$y(x) \ = \ e^{-_* x} \cdot_* \left(e^{\frac{4}{5}} \cdot_* (x +_* e)^{\frac{5}{2}} {}_* \right.$$

$$\left. +_* b_1 +_* b_2 \cdot_* x \right), \quad x \in \mathbb{R}_*.$$

19.

$$y(x) \ = \ \left(b_1 +_* b_2 \cdot_* x +_* x \cdot_* \log_* |x|_* \right) \cdot_* e^{-_* x}$$

$$+_* \left((x -_* e)/_* e^4 \right) \cdot_* e^x, \quad x \in \mathbb{R}_*.$$

20.

$$y(x) \ = \ \left(e^{\frac{1}{8}} +_* (b_1 -_* x/_* e^2) \cdot_* \cos_*(e^2 \cdot_* x) \right.$$

$$+_* \left(b_2 +_* x/_* e^8 +_* e^{\frac{1}{2}} \cdot_* \log_* |\cos_* x|_* \right)$$

$$\left. \cdot_* \sin_*(e^2 \cdot_* x) \right) \cdot_* e^{-_* x}, \quad x \in \mathbb{R}_*.$$

Problem 4.8.11 *Find the solutions of the following MIVPs.*

1.

$$y^{**} -_* e^2 \cdot_* y^* +_* y \ = \ 0_*, \quad x \in \mathbb{R}_*,$$

$$y(e^2) = e, \quad y^*(e^2) \ = \ e^{-2}.$$

2.

$$y^{**} +_* e^2 \cdot_* y +_* e^2 \cdot_* y \ = \ x \cdot_* e^{-_* x}, \quad x \in \mathbb{R}_*,$$

$$y(1) = y^*(1) \ = \ 1.$$

Answer

1.

$$y(x) = (e^7 -_* e^3 \cdot_* x) \cdot_* e^{x -_* e^2}, \quad x \in \mathbb{R}_*.$$

2.

$$y(x) = e^{-_* x} \cdot_* (x -_* \sin_* x), \quad x \in \mathbb{R}_*.$$

Problem 4.8.12 *Find the general solutions of the following equations.*

1.
$$x^{2_*} \cdot_* y^{**} -_* x \cdot_* y^* -_* e^3 \cdot_* y = 0_*, \quad x \in \mathbb{R}_*.$$

2.
$$x^{2_*} \cdot_* y^{**} -_* x \cdot_* y^* +_* y = e^8 \cdot_* x^{3_*}, \quad x \in \mathbb{R}_*.$$

3.
$$x^{2_*} \cdot_* y^{**} +_* x \cdot_* y^* +_* e^4 \cdot_* y = e^{10} \cdot_* x, \quad x \in \mathbb{R}_*.$$

4.
$$x^{3_*} \cdot_* y^{**} -_* e^2 \cdot_* x \cdot_* y = e^6 \cdot_* \log_* x, \quad x \in \mathbb{R}_*.$$

5.
$$x^{2_*} \cdot_* y^{**} -_* e^3 \cdot_* x \cdot_* y^* +_* e^5 \cdot_* y = e^3 \cdot_* x62_*, \quad x \in \mathbb{R}_*.$$

6.
$$x^{2_*} \cdot_* y^{**} -_* e^6 \cdot_* y = e^5 \cdot_* x^{3_*} +_* e^8 \cdot_* x^{2_*}, \quad x \in \mathbb{R}_*.$$

7.
$$x^{2_*} \cdot_* y^{**} -_* e^2 \cdot_* y = \sin_* \log_* x, \quad x \in \mathbb{R}_*.$$

Answer

1.
$$y(x) = b_1 \cdot_* x^{3*} +_* b_2 \cdot_* x^{-1*}, \quad x \in \mathbb{R}_*.$$

2.
$$y(x) = x \cdot_* (b_1 +_* b_2 \cdot_* \log_* |x|_*) +_* e^2 \cdot_* x^{3*}, \quad x \in \mathbb{R}_*.$$

3.
$$y(x) \;=\; b_1 \cdot_* \cos_* \left(e^2 \cdot_* \log_* |x|_* \right)$$

$$+_* b_2 \cdot_* \sin_* \left(e^2 \cdot_* \log_* |x|_* \right), \quad x \in \mathbb{R}_*.$$

4.
$$y(x) = b_1 \cdot_* x^{2*}$$

$$+_* (e/_* x) \cdot_* \left(b_2 -_* e^{\frac{2}{3}} \cdot_* \log_* x -_* (\log_* x)^{2*} \right), \quad x \in \mathbb{R}_*.$$

5.
$$y(x) \;=\; x^{2*} \cdot_* \left(b_1 \cdot_* \cos_* (\log_* |x|_*) \right.$$

$$\left. +_* b_2 \cdot_* \sin_* (\log_* |x|_*) +_* es \right), \quad x \in \mathbb{R}_*.$$

6.
$$y(x) \;=\; b_1 \cdot_* x^{3*} +_* b_2 \cdot_* x^{-2*}$$

$$+_* x^{3*} \cdot_* \log_* |x|_* -_* e^2 \cdot_* x^{2*}, \quad x \in \mathbb{R}_*.$$

7.
$$y(x) \;=\; b_1 \cdot_* x^{2*} +_* b_2 \cdot_* x^{-1*} +_* e^{\frac{1}{10}} \cdot_* \cos_* (\log_* x)$$

$$-_* e^{\frac{3}{10}} \cdot_* \sin_* (\log_* x), \quad x \in \mathbb{R}_*.$$

Here b_1 and b_2 are multiplicative constants.

5

Existence and Uniqueness of Solutions

5.1 Introduction

Let $x_0, y_0 \in \mathbb{R}_*$, $J \subset \mathbb{R}_*$ be an interval containing the point x_0. In this chapter we will investigate the initial value problem

$$y^* = f(x,y), \quad x \in J, \tag{5.1}$$

$$y(x_0) = y_0, \tag{5.2}$$

where $f(x,y)$ is assumed to be continuous function in a domain $D \subset \mathbb{R}_*^2$ containing the point (x_0, y_0).

Definition 5.1.1 *We will say that the function y, defined in J, is a solution to the initial value problem (5.1), (5.2) if*

1. $y(x_0) = y_0$,
2. $(x, y(x)) \in D$ for any $x \in J$,
3. $y^*(x)$ exists for any $x \in J$,
4. $y(x) = f(x, y(x))$ for any $x \in J$.

Theorem 5.1.2 *Let f be a continuous function in D. Then y is a solution to the problem (5.1), (5.2) in J if and only if*

$$y(x) = y_0 +_* \int_{*x_0}^x f(t, y(t)) \cdot_* d_* t, \quad x \in J. \tag{5.3}$$

Proof 5.1.3   *1. Let y be a solution of the problem (5.1), (5.2)*

DOI: 10.1201/9781003393344-5

in J. Let also, $x \in J$ be arbitrarily chosen. We multiplicative integrate the equation (5.1) on $[x_0, x]$ and we get

$$\int_{*x_0}^{x} y^*(t) \cdot_* d_* t = \int_{*x_0}^{x} f(t, y(t)) \cdot_* d_* t, \quad x \in J,$$

whereupon

$$y(x) -_* y(x_0) = \int_{*x_0}^{x} f(t, y(t)) \cdot_* d_* t, \quad x \in J,$$

or

$$y(x) = y_0 +_* \int_{*x_0}^{x} f(t, y(t)) \cdot_* d_* t, \quad x \in J,$$

i.e., y satisfies (5.3) in J.

2. Let y be a solution of (5.3) in J. Then

$$\begin{aligned} y(x_0) &= y_0 +_* \int_{*x_0}^{x_0} f(t, y(t)) \cdot_* d_* t \\ &= y_0, \end{aligned}$$

i.e., y satisfies (5.2). Also, since f is continuous in D we have that $y^*(x)$ exists for any $x \in J$ and

$$y^*(x) = f(x, y(x)) \qquad \text{for} \qquad \forall x \in J.$$

We note that $(x, y(x)) \in D$ for every $x \in J$. Therefore y is a solution of the problem (5.1), (5.2).

Example 5.1.4 *Let us consider the initial value problem*

$$y^*(x) = \sin_* x +_* e, \qquad y(1) = 1.$$

Here

$$x_0 = 1,$$

$$y_0 = 1,$$

$$f(x, y(x)) = \sin x +_* e, \quad x \in J.$$

We have that

$$y(x) = -_* \cos_* x +_* x +_* e$$

is its solution in $\mathbb{R}_*$. *Let* $x \in \mathbb{R}_*$ *be arbitrarily chosen. Then*

$$y_0 +_* \int_{*x_0}^x f(t, y(t)) \cdot_* d_* t = \int_{*1}^x (\sin t +_* e) \cdot_* d_* t$$

$$= \int_{*1}^x \sin_* t \cdot_* d_* t +_* \int_{*1}^x \cdot_* d_* t$$

$$= -_* \cos_* t \Big|_{t=1}^{t=x} +_* x$$

$$= -_* \cos_* x +_* e +_* x$$

$$= y(x). \quad x \in \mathbb{R}_*.$$

Exercise 5.1.5 *Prove that*

$$\int_{*x_0}^x \int_{*x_0}^{x_1} \cdots \int_{*x_0}^{x_n} f(t, y(t)) \cdot_* d_* t \cdot_* dx_n \cdot_* \cdots_* dx_1$$

$$= \int_{*x_0}^x ((x -_* t)^{n_*} /_* n!_*) \cdot_* f(t, y(t)) \cdot_* d_* t,$$

where f *is a continuous function in a domain* $D \subset \mathbb{R}_*^2$.

Exercise 5.1.6 *Prove that the initial value problem*

$$y^{**} = f(x, y), \quad x \in \mathbb{R}_*,$$

$$y(x_0) = y_0,$$

$$y^*(x_0) = y_1,$$

where $f(x, y)$ *is a continuous function in a domain* D *containing the point* (x_0, y_0) *is equivalent to the integral equation*

$$y(x) = y_0 +_* (x -_* x_0) \cdot_* y_1 +_* \int_{*x_0}^x (x -_* t) \cdot_* f(t, y(t)) \cdot_* d_* t,$$

$x \in J.$

Definition 5.1.7 *The function $f(x,y)$ is said to satisfy a multiplicative uniform Lipschitz condition if*

$$|f(x,y_1) -_* f(x,y_2)|_* \leq L \cdot_* |y_1 -_* y_2|_* \qquad (5.4)$$

for all (x,y_1), $(x,y_2) \in D$ and for some $L \in \mathbb{R}_*$, $L \geq 1$. The multiplicative constant L is called a multiplicative Lipshitz constant.

Example 5.1.8 *Let $D = \mathbb{R}^2_*$. We consider the function*

$$f(x,y) = x +_* y, \quad (x,y) \in D.$$

Then for any (x,y_1), $(x,y_2) \in D$, we have

$$|f(x,y_1) -_* f(x,y_2)|_* = |x +_* y_1 -_* x -_* y_2|_*$$

$$= |y_1 -_* y_2|_*,$$

i.e., the function $f(x,y)$ satisfies the Lipschitz condition in D with multiplicative Lipschitz constant $L = e$.

Example 5.1.9 *Let $D = \mathbb{R}^2_*$,*

$$f(x,y) = e^2 \cdot_* (y /_* (e +_* x^2)), \quad (x,y) \in D.$$

Then for any (x,y_1), $(x,y_2) \in D$, we have

$$|f(x,y_1) -_* f(x,y_2)|_* = e^2 \cdot_* \left| (y_1 /_* (e +_* x^2)) -_* (y_2 /_* (e +_* x^2)) \right|_*$$

$$= e(e^2 /_* (1e +_* x^2)) \cdot_* |y_1 -_* y_2|_*$$

$$\leq e^2 \cdot_* |y_1 -_* y_2|_*,$$

i.e., the function f satisfies the multiplicative Lipschitz condition in D with Lipschitz constant $L = e^2$.

Example 5.1.10 *Let*

$$D = \{(x,y) \in \mathbb{R}^2_* : |x|_* \leq e, \quad |y|_* \leq e^2\},$$

and

$$f(x,y) = (x/_*(e +_* y^2)), \quad (x,y) \in D.$$

Then

$$|f(x,y_1) -_* f(x,y_2)|_* \; = \; |(x/_*(e +_* y_1^2)) -_* (x/_*(e +_* y_2^2))|_*$$

$$= \; |x|_* \cdot_* (|y_1^{2*} -_* y_2^{2*}|_* /_* ((e +_* y_1^{2*}) \cdot_* (e +_* y_2^{2*})))$$

$$\leq \; |x|_* |y_1 -_* y_2|_* (|y_1|_* +_* |y_2|_*)$$

$$\leq \; e^4 \cdot_* |y_1 -_* y_2|_*,$$

i.e., the function f satisfies the multiplicative Lipschitz condition in D with multiplicative Lipschitz constant $L = e^4$.

Exercise 5.1.11 *Let*

$$D = \{(x,y) \in \mathbb{R}_*^2 : |x|_* \leq e^2, \quad |y|_* \leq e\}.$$

Prove that the function

$$f(x,y) = x^{2*} \cdot_* \cos_* y -_* y \cdot_* \sin_* x, \quad (x,y) \in D,$$

satisfies the multiplicative Lipschitz condition in D. Find the multiplicative Lipschitz constant L.

Answer $L = 5$.

Exercise 5.1.12 *Let*

$$D = \{(x,y) \in \mathbb{R}_*^2 : |x -_* 1|_* \leq e^3, \quad |y|_* \leq e^2\}.$$

Prove that the function

$$f(x,y) = x^{2*} \cdot_* y^{2*} +_* x \cdot_* y +_* e, \quad (x,y) \in D,$$

satisfies the multiplicative Lipschitz condition. Find the multiplicative Lipschitz constant L.

Answer $L = e^{68}$.

Exercise 5.1.13 *Let*

$$D = \{(x,y) \in \mathbb{R}^2_* : |x|_* \le e, \quad |y|_* \le e\}.$$

Prove that the function

$$f(x,y) = |x \cdot_* y|_*, \quad (x,y) \in D,$$

satisfies the multiplicative Lipschitz condition. Find the multiplicative Lipschitz constant L.

Answer $L = e$.

Theorem 5.1.14 *Let the domain D be multiplicative convex and the function $f(x,y)$ be multiplicative differentiable with respect to y in D. Then for the multiplicative Lipschitz condition (5.4) to be satisfied, it is necessary and sufficient that*

$$\sup_D \left| (\partial_* f /_* \partial_* y)(x,y) \right|_* \le L. \tag{5.5}$$

Proof 5.1.15   *1. Let (5.5) holds. Then for any (x,y_1), $(x,y_2) \in D$ we have*

$$|f(x,y_1) -_* f(x,y_2)|_* = \left| (\partial_* f /_* \partial_* y)(x,y^1) \right|_* \cdot_* |y_1 -_* y_2|$$

$$\le L \cdot_* |y_1 -_* y_2|_*,$$

where y^1 is between y_1 and y_2. Since the domain D is multiplicative convex, we have that $(x,y^1) \in D$. Consequently, the function f satisfies the multiplicative Lipschitz condition (5.4).

2. Let the function f satisfy the Lipschitz condition (5.4). Then for any (x,y_1), $(x,y_2) \in D$, we have

$$\lim_{y_2 \to y_1} \left| ((f(x,y_2) -_* f(x,y_1)) /_* (y_2 -_* y_1)) \right|_* \le L$$

or

$$\left|(\partial_* f/{_*}\partial_* y)(x,y_1)\right|_* \leq L,$$

whereupon

$$\sup_D \left|(\partial_* f/{_*}\partial_* y)(x,y_1)\right|_* \leq L.$$

This completes the proof.

Example 5.1.16 *Let*

$$D = \{(x,y) \in \mathbb{R}^2_* : |x|_* \leq a, \quad |y|_* \leq b\},$$

$a, b \geq 1$,

$$f(x,y) = x \cdot_* \sin_* y +_* y \cdot_* \cos_* x, \qquad (x,y) \in D.$$

We have that f is multiplicative differentiable in D with respect to y and

$$(\partial_* f/{_*}\partial_* y)(x,y) = x \cdot_* \cos_* y +_* \cos_* x, \quad (x,y) \in D.$$

Hence,

$$\left|(\partial_* f/{_*}\partial_* y)(x,y)\right|_* = |x \cdot_* \cos_* y +_* \cos_* x|_*$$

$$\leq |x|_* \cdot_* |\cos_* y|_* +_* |\cos_* x|_*$$

$$\leq a +_* e, \quad (x,y) \in D.$$

Consequently, the function f satisfies the Lipschitz condition in D with a constant $L = a +_* e$.

Example 5.1.17 *Let*

$$D = \{(x,y) \in \mathbb{R}^2_* : |x|_* \leq e^2, \quad |y|_* \leq e^3\},$$

and

$$f(x,y) = y^{2_*} \cdot_* e^{x +_* y}, \quad (x,y) \in D.$$

Then f is multiplicative differentiable with respect to y and

$$(\partial_* f/_* \partial_* y)(x,y) = e^2 \cdot_* y \cdot_* e^{x+_* y} +_* y^{2_*} \cdot_* e^{x+_* y},$$

$(x,y) \in D$. *Hence,*

$$\left| (\partial_* f/_* \partial_* y)(x,y) \right|_* = \left| e^2 \cdot_* y \cdot_* e^{x+_* y} +_* y^{2_*} e^{x+_* y} \right|_*$$

$$\leq e^2 \cdot_* |y|_* \cdot_* e^{x+_* y} +_* y^{2_*} \cdot_* e^{x+_* y}$$

$$\leq e^6 \cdot_* e^{e^5} +_* e^9 \cdot_* e^{e^5}$$

$$= e^{15} \cdot_* e^{e^5}.$$

Therefore, the function f satisfies the multiplicative Lipschitz condition in D with multiplicative Lipschitz constant $L = e^{15} \cdot_* e^{e^5}$.

Example 5.1.18 *Let*

$$D = \{(x,y) \in \mathbb{R}_*^2 : |x|_* \leq e, \quad |y|_* < \infty\},$$

and

$$f(x,y) = p(x) \cdot_* y +_* q(x), \quad (x,y) \in D,$$

where p and q are continuous functions in the interval $|x|_* \leq e$. Then there exists a constant $M > 1$ such that

$$\max_{|x|_* \leq e} |p(x)|_* = M.$$

We note that f is multiplicative differentiable with respect to y and

$$(\partial_* f/_* \partial_* y)(x,y) = p(x), \quad (x,y) \in D.$$

Hence, for every $(x,y) \in D$, we have

$$|(\partial_* f/_* \partial_* y)(x,y)| = |p(x)|$$

$$\leq M, \quad (x,y) \in D.$$

Therefore, the function f satisfies the multiplicative Lipschitz condition in D with multiplicative Lipschitz constant $L = M$.

Exercise 5.1.19 *Let*

$$D = \{(x,y) \in \mathbb{R}^2_* : |x|_* \leq a, \quad |y|_* \leq b\},$$

where $a, b > 1$, *and*

$$f(x,y) = y^{3*} \cdot_* e^{-*x \cdot_* y}, \quad (x,y) \in D.$$

Prove that the function f *satisfies the multiplicative Lipschitz condition in* D. *Find the multiplicative Lipschitz constant* L.

Answer

$$L = (e^3 \cdot_* b^{2*} +_* a \cdot_* b^{3*}) \cdot_* e^{a \cdot_* b}.$$

Example 5.1.20 *Let*

$$D = \{(x,y) \in \mathbb{R}^2_* : |x|_* \leq e, \quad |y|_* \leq e^2\},$$

and

$$f(x,y) = \begin{cases} ((x \cdot_* y)/_* (x^{2*} +_* y^{2*})), & (x,y) \neq (1,1), \quad (x,y) \in D, \\ 1, & (x,y) = (1,1). \end{cases}$$

We assume that f *satisfies the multiplicative Lipschitz condition with multiplicative Lipschitz constant* L. *Then for every* (x,y_1), $(x,y_2) \in D$ *we have*

$$|f(x,y_1) -_* f(x,y_2)|_* \leq L \cdot_* |y_1 -_* y_2|_*.$$

In particular, for every $(x,y) \in D$, *we have*

$$|f(x,y)|_* \leq L \cdot_* |y|_*$$

or

$$\left|((x \cdot_* y)/_* (x^{2*} +_* y^{2*}))\right|_* \leq L \cdot_* |y|_*,$$

or

$$\left|(x/_* (x^{2*} +_* y^{2*}))\right|_* \leq L.$$

Hence, for $x = y$, *we have*

$$(e/_* |x| \leq L \qquad \text{for} \qquad \text{any} \qquad |x|_* \leq e,$$

which is a contradiction. Therefore, the function f *doesn't satisfy the multiplicative Lipschitz condition in* D.

Exercise 5.1.21 *Let*

$$D = \{(x,y) \in \mathbb{R}_*^2 : |x|_* \le e, \quad |y|_* < \infty\},$$

and

$$f(x,y) = \begin{cases} \sin_* y /_* x, & x \ne 1, \quad (x,y) \in D, \\ \\ 1, & x = 1, \quad (1,y) \in D. \end{cases}$$

Prove that the function f doesn't satisfy the Lipschitz condition in D.

5.2   The Multiplicative Gronwall Type Integral Inequalities

Theorem 5.2.1 *(The Multiplicative Gronwall Type Integral Inequality) Let y, a, b be multiplicative nonnegative continuous functions in the interval* $|x -_* x_0|_* \le c$ *and*

$$y(x) \le a(x) +_* \left| \int_{*x_0}^x b(t) \cdot_* y(t) \cdot_* d_* t \right|_* \qquad \text{for} \qquad |x -_* x_0|_* \le c. \tag{5.6}$$

Then the following multiplicative integral inequality holds

$$y(x) \le a(x) +_* \left| \int_{*x_0}^x a(t) \cdot_* b(t) e^{|_* \int_{*t}^x b(s) \cdot_* d_* s|_*} \cdot_* d_* t \right|_* \tag{5.7}$$

for $|x -_* x_0| \le c$. *Here* $c > 1$ *is a multiplicative constant.*

Proof 5.2.2 *We will prove the result for* $x \in [x_0, x_0 +_* c]$. *The proof in the case when* $x \in [x_0 -_* c, x_0]$ *is similar and we leave to the reader as an exercise. We suppose that* $x \in [x_0, x_0 +_* a]$. *Then the inequality (5.6) takes the form*

$$y(x) \le a(x) +_* \int_{*x_0}^x b(t) \cdot_* y(t) \cdot_* d_* t. \tag{5.8}$$

Let

$$z(x) = \int_{*x_0}^x b(t) \cdot_* y(t) \cdot_* d_* t.$$

Then, using (5.8), we have

$$y(x) \leq a(x) +_* z(x).$$ (5.9)

Also, $z(x_0) = 0$ and

$$z^*(x) = b(x) \cdot_* y(x)$$

$$\leq b(x) \cdot_* \left(a(x) +_* \int_{*x_0}^x b(t) \cdot_* y(t) \cdot_* d_* t \right)$$

$$= b(x) \cdot_* a(x) +_* b(x) \cdot_* z(x),$$

which we multiplicative multiply by $e^{-* \int_{*x_0}^x b(t) \cdot_* d_* t}$ and we find

$$z^*(x) \cdot_* e^{-* \int_{*x_0}^x b(t) \cdot_* d_* t} \leq b(x) \cdot_* a(x) \cdot_* e^{-* \int_{*x_0}^x b(t) \cdot_* d_* t}$$

$$+_* b(x) \cdot_* z(x) \cdot_* e^{-* \int_{*x_0}^x b(t) \cdot_* d_* t}$$

or

$$\left(z(x) \cdot_* e^{-* \int_{*x_0}^x b(t) \cdot_* d_* t} \right)^* \leq b(x) \cdot_* a(x) \cdot_* e^{-* \int_{*x_0}^x b(t) \cdot_* d_* t}.$$

The last inequality we multiplicative integrate on $[x_0, x_0 +_* a]$ and we become

$$\int_{*x_0}^x \left(z(t) \cdot_* e^{-* \int_{*x_0}^t b(s) \cdot_* d_* s} \right)^* \cdot_* d_* t$$

$$\leq \int_{*x_0}^x b(t) \cdot_* a(t) \cdot_* e^{-* \int_{*x_0}^t b(s) \cdot_* d_* s} \cdot_* d_* t$$

or

$$z(x) \cdot_* e^{-* \int_{*x_0}^x b(t) \cdot_* d_* t} \leq \int_{*x_0}^x b(t) \cdot_* a(t) \cdot_* e^{-* \int_{*x_0}^t b(s) \cdot_* d_* s} \cdot_* d_* t,$$

whereupon

$$z(x) \leq e^{\int_{*x_0}^x b(t) \cdot_* d_* t} \cdot_* \int_{*x_0}^x a(t) \cdot_* b(t) \cdot_* e^{-* \int_{*x_0}^t b(s) \cdot_* d_* s} \cdot_* d_* t$$

$$= \int_{*x_0}^x a(t) \cdot_* b(t) \cdot_* e^{\int_t^x b(s) \cdot_* d_* s} \cdot_* d_* t.$$

From the last inequality and from (5.9), we get the inequality (5.7). This completes the proof.

Example 5.2.3 *We suppose that y is continuous multiplicative non-negative function in* $[1,e]$ *and*

$$y(x) \le e^x +_* \int_{*1}^x y(t) \cdot_* d_* t, \quad x \in [1,e].$$

Here

$$a(x) = e^x,$$

$$b(x) = 1, \quad x \in [1,e].$$

Then, using the inequality (5.7), we obtain

$$
\begin{aligned}
y(x) & \le e^x +_* e^{\int_{*1}^x d_* t} \cdot_* \int_{*1}^x e^t \cdot_* t \cdot_* e^{-*\int_{*1}^t \cdot_* d_* s} \cdot_* d_* t \\[2mm]
& = e^x +_* e^x \cdot_* \int_0^x t \cdot_* e^t \cdot_* e^{-*t} \cdot_* d_* t \\[2mm]
& = e^x +_* e^x \cdot_* \int_{*1}^x t \cdot_* d_* t \\[2mm]
& = \left(e +_* (x^{2*}/_* e^2)\right) \cdot_* e^x, \quad x \in [1,e].
\end{aligned}
$$

Example 5.2.4 *Let y be multiplicative nonnegative continuous function in* $[1,e^{\frac{\pi}{2}}]$ *and*

$$y(x) \le \sin_* x +_* \int_{*1}^x \cos_* t \cdot_* y(t) \cdot_* d_* t, \quad x \in [1,e^{\frac{\pi}{2}}].$$

We will apply the multiplicative Gronwall type integral inequality (5.7). Then

$$
\begin{aligned}
y(x) & \le \sin_* x +_* e^{\int_0^x \cos_* t \cdot_* d_* t} \int_0^x \sin_* t \cdot_* \cos_* t \cdot_* e^{-*\int_0^t \cos_* s \cdot_* d_* s} \cdot_* d_* t \\[2mm]
& = \sin_* x e^{\sin_* x} \cdot_* \int_{*1}^x \sin_* t \cdot_* \cos_* t \cdot_* e^{-*\sin_* t} \cdot_* d_* t \\[2mm]
& = \sin_* x +_* e^{\sin_* x} \cdot_* \int_0^x \sin_* t \cdot_* e^{-*\sin_* t} \cdot_* d_* (\sin_* t)
\end{aligned}
$$

$$= \quad \sin_* x +_* e^{\sin_* x} \cdot_* \left(-_* \sin_* t \cdot_* e^{-_* \sin_* t} \Big|_{*t=0}^{t=x} \right.$$

$$\left. +_* \int_{*1}^x \cos_* t \cdot_* e^{-_* \sin_* t} \cdot_* d_* t \right)$$

$$= \quad \sin_* x +_* e^{\sin_* x} \cdot_* \left(-_* \sin_* x \cdot_* e^{-_* \sin_* x} \right.$$

$$\left. -_* e^{-_* \sin_* t} \Big|_{*t=0}^{t=x} \right)$$

$$= \quad \sin_* x +_* e^{\sin_* x} \left(-_* \sin_* x \cdot_* e^{-_* \sin_* x} -_* e^{-_* \sin_* x} +_* e \right)$$

$$= \quad \sin_* x -_* \sin_* x -_* e +_* e^{\sin_* x}$$

$$= \quad -_* e +_* e^{\sin_* x}, \quad x \in [1, e^{\frac{\pi}{2}}].$$

Example 5.2.5 *Let y be multiplicative nonnegative continuous function in $[1, e^2]$ and*

$$y(x) \leq x^{2*} +_* \int_{*1}^x t \cdot_* y(t) \cdot_* d_* t, \quad x \in [1, e^2].$$

Here

$$a(x) = x^{2*},$$

$$b(x) = x, \quad x \in [1, e^2].$$

We will apply the multiplicative Gronwall type integral inequality (5.7). We have

$$y(x) \leq x^{2*}$$

$$+_* e^{\int_0^x t \cdot_* d_* t} \cdot_* \int_0^x t^{3*} \cdot_* e^{-_* \int_0^t s \cdot_* d_* s} \cdot_* d_* t$$

$$= x^{2*} +_* e^{(x^{2*}/_* e^2)} \cdot_* \int_{*1}^x t^{3*} \cdot_* e^{-_*(t^{2*}/_* e^2)} \cdot_* d_* t$$

$$= x^{2*} +_* e^{(x^{2*}/_* e^2)} \cdot_* \int_{*1}^x t^{2*} \cdot_* e^{-*(t^{2*}/_* e^2)} \cdot_* d_*(t^{2*}/_* e^2)$$

$$= x^{2*} -_* e^{(x^{2*}/_* e^2)} \cdot_* t^{2*} \cdot_* e^{-*(t^{2*}/_* e^2)} \Big|_{*t=1}^{t=x}$$

$$+_* e^2 \cdot_* e^{(x^{2*}/_* e^2)} \cdot_* \int_0^x t \cdot_* e^{-*(t^{2*}/_* e^2)} \cdot_* d_* t$$

$$= x^{2*} -_* x^{2*} -_* e^2 \cdot_* e^{(x^{2*}/_* e^2)} \cdot_* \int_{*1}^x e^{-*(t^{2*}/_* e^2)} \cdot_* d_* \left(-_*(t^{2*}/_* e^2)\right)$$

$$= -_* e^2 \cdot_* e^{(x^{2*}/_* e^2)} \cdot_* e^{-*(t^{2*}/_* e^2)} \Big|_{*t=1}^{t=x}$$

$$= -_* e^2 +_* e^2 \cdot_* e^{(x^{2*}/_* e^2)}, \qquad x \in [1, e^2].$$

Exercise 5.2.6 *Let* $y(x)$ *be multiplicative nonnegative continuous function in* $[1, e^{\frac{\pi}{2}}]$ *and*

$$y(x) \leq \cos_* x +_* \int_{*1}^x \sin_* t \cdot_* y(t) \cdot_* d_* t, \quad x \in [1, e^{\frac{\pi}{2}}].$$

Prove that

$$y(x) \leq e, \qquad x \in [1, e^{\frac{\pi}{2}}].$$

Theorem 5.2.7 *Let* y, a *and* b *be multiplicative nonnegative continuous functions in* $|x -_* x_0|_* \leq c$ *and the inequality* (5.6) *holds. Let also, a be multiplicative nondecreasing in* $[x_0, x_0 +_* c]$ *and multiplicative nonincreasing in* $[x_0 -_* c, x_0]$. *Then*

$$y(x) \leq a(x) \cdot_* e^{\left| \int_{*x_0}^x b(t) \cdot_* d_* t \right|} \quad \text{for} \quad |x -_* x_0| \leq c. \qquad (5.10)$$

Here $c > 1$ *is a multiplicative constant.*

Proof 5.2.8 *We will prove the inequality* (5.10) *in the case when* $x \in [x_0, x_0 +_* c]$. *The case when* $x \in [x_0 -_* c, x_0]$ *is similar, and we leave it to* *the reader for an exercise. Let $x \in [x_0, x_0 +_* c]$. Since a is multiplicative nondecreasing in $[x_0, x_0 +_* c]$ we have that*

$$a(t) \leq a(x)$$

for every $t \leq x$, $t, x \in [x_0, x_0 +_* a]$. From here and from the multiplicative Gronwall type integral inequality (5.7), we get

$$y(x) \leq a(x) +_* \int_{*x_0}^x a(t) \cdot_* b(t) \cdot_* e^{\int_{*t}^{x_0} b(s) \cdot_* d_* s} \cdot_* d_* t$$

$$\leq a(x) \cdot_* \left(1 +_* \int_{*x_0}^x b(t) \cdot_* e^{\int_{*t}^x b(s) \cdot_* d_* s} \cdot_* d_* t\right)$$

$$= a(x) \cdot_* \left(e -_* e^{\int_{*t}^x b(s) \cdot_* d_* s} \Big|_{*t=x_0}^{t=x}\right)$$

$$= a(x) \cdot_* \left(e -_* e +_* e^{\int_{*x_0}^x b(s) \cdot_* d_* s}\right)$$

$$= a(x) \cdot_* e^{\int_{*x_0}^x b(s) \cdot_* d_* s}.$$

This completes the proof.

Theorem 5.2.9 *Let y be multiplicative nonnegative continuous function in $|x -_* x_0|_* \leq c$ and*

$$y(x) \leq c_0 +_* c_1 \cdot_* |x -_* x_0|_* +_* c_2 \cdot_* \left|\int_{*x_0}^x y(t) \cdot_* d_* t\right|_*, \qquad (5.11)$$

in $|x -_* x_0|_* \leq c$, where c, c_0, c_1 and c_2 are multiplicative nonnegative constants. Then

$$y(x) \leq (c_0 +_* (c_1/_* c_2)) \cdot_* e^{c_2 \cdot_* |x -_* x_0|_*} -_* (c_1/_* c_2) \qquad (5.12)$$

in $|x -_* x_0|_* \leq c$.

Proof 5.2.10 *We will prove the inequality (5.12) for $x \in [x_0, x_0 +_* c]$. The case when $x \in [x_0 -_* c, x_0]$ is similar, and we leave it to the reader for an exercise. Let $x \in [x_0, x_0 +_* c]$. Then the inequality (5.11) takes the form*

$$y(x) \leq c_0 +_* c_1 \cdot_* (x -_* x_0) +_* c_2 \cdot_* \int_{*x_0}^x y(t) \cdot_* d_* t.$$

Now we will apply the multiplicative Gronwall type integral inequality (5.7). Here

$$a(x) = c_0 +_* c_1 \cdot_* (x -_* x_0),$$

$$b(x) = c_2.$$

Then

$$y(x) \leq c_0 +_* c_1 \cdot_* (x -_* x_0) +_* c_2 \cdot_* e^{c_2 \cdot_* \int_{*x_0}^x \cdot_* d_* t}$$

$$\cdot_* \int_{*x_0}^x (c_0 +_* c_1 \cdot_* (t -_* x_0)) \cdot_* e^{-_* c_2 \cdot_* \int_{*x_0}^t d_* s} \cdot_* d_* t$$

$$= c_0 +_* c_1 \cdot_* (x -_* x_0) +_* c_2 \cdot_* e^{c_2 \cdot_* x}$$

$$\cdot_* \int_{*x_0}^x (c_0 +_* c_1 \cdot_* (t -_* x_0)) \cdot_* e^{-_* c_2 \cdot_* t} \cdot_* d_* t$$

$$= c_0 +_* c_1 \cdot_* (x -_* x_0) -_* e^{c_2 \cdot_* x} \cdot_* \left((c_0 +_* c_1 \cdot_* (t -_* x_0)) \right.$$

$$\cdot_* e^{-_* c_2 \cdot_* t} \Big|_{*t=x_0}^{t=x}$$

$$\left. -_* c_1 \cdot_* e^{c_2 \cdot_* x} \cdot_* \int_{*x_0}^x e^{-_* c_2 \cdot_* t} \cdot_* d_* t \right)$$

$$= c_0 +_* c_1 \cdot_* (x -_* x_0) -_* e^{c_2 \cdot_* x} \cdot_* (c_0 +_* c_1 \cdot_* (x -_* x_0))$$

$$\cdot_* e^{-_* c_2 \cdot_* x} +_* e^{c_2 \cdot_* x} c_0 \cdot_* e^{-_* c_2 \cdot_* x_0}$$

$$-_* e^{c_2 \cdot_* x} (c_1 /_* c_2) \cdot_* e^{-_* c_2 \cdot_* t} \Big|_{*t=x_0}^{t=x}$$

$$= c_0 \cdot_* e^{c_2 \cdot_* (x -_* x_0)} -_* (c_1 /_* c_2) \cdot_* e^{c_2 \cdot_* x}$$

$$\cdot_* \left(e^{-_* c_2 \cdot_* x} -_* e^{-_* c_2 \cdot_* x_0} \right)$$

$$= (c_0 +_* (c_1 /_* c_2)) \cdot_* e^{c_2 \cdot_* (x -_* x_0)} -_* (c_1 /_* c_2).$$

This completes the proof.

Theorem 5.2.11 *Let y be multiplicative nonnegative continuous func-tion in the interval* $|x -_* x_0|_* \leq d$, *and* $c \geq 1$ *be a given multiplicative constant. Let also,*

$$y(x) \leq c \cdot_* \left| \int_{*x_0}^x (y(t))^{\alpha_*} \cdot_* d_*t \right|_*, \quad 1 < e^\alpha < e, \quad |x -_* x_0|_* \leq d.$$

$$(5.13)$$

Then

$$y(x) \leq (c \cdot_* (e -_* e^\alpha) \cdot_* |x -_* x_0|_*)^{\frac{1}{1-\alpha}_*} \quad (5.14)$$

in $|x -_* x_0|_* \leq d$. *Here d is a multiplicative nonnegative constant.*

Proof 5.2.12 *We will prove the inequality* (5.14) *for* $x \in [x_0, x_0 +_* d]$. *The case when* $x \in [x_0 -_* d, x_0]$ *is similar, and we leave it to the reader for an exercise. Let* $x \in [x_0, x_0 +_* d]$. *Then the inequality* (5.14) *takes the form*

$$y(x) \leq c \cdot_* \int_{*x_0}^x (y(t))^{\alpha_*} \cdot_* d_*t. \quad (5.15)$$

Let

$$z(x) = c \cdot_* \int_{*x_0}^x (y(t))^{\alpha_*} \cdot_* d_*t.$$

Then, using (5.15), *we have*

$$y(x) \leq z(x). \quad (5.16)$$

Also, using (5.16),

$$z(x_0) = 1,$$

$$z^*(x) = c \cdot_* (y(x))^{\alpha_*}$$

$$\leq c \cdot_* (z(x))^{\alpha_*}$$

or

$$(z^*(x) /_* (z(x))^{\alpha_*} \leq c,$$

which we multiplicative integrate on $[x_0, x]$ *and we find*

$$\int_{*x_0}^x (z^*(t)(z(t))^\alpha) \cdot_* d_*t \leq c \cdot_* \int_{*x_0}^x \cdot_* d_*t$$

or

$$\int_{*x_0}^x ((d_* z(t))/_*(z(t))^{\alpha_*}) \le c \cdot_* (x -_* x_0),$$

or

$$((z(t))^{(1-\alpha)_*}/_*(1-\alpha)_*) \Big|_{*t=x_0}^{|t=x} \le c \cdot_* (x -_* x_0),$$

or

$$(z(x))^{(1-\alpha)_*} \le c \cdot_* (1-\alpha)_* \cdot_* (x -_* x_0),$$

whereupon

$$z(x) \le (c \cdot_* (1-\alpha)_* \cdot_* (x -_* x_0))^{\frac{1}{1-\alpha}_*}.$$

Hence and (5.16), we get (5.14). This completes the proof.

Theorem 5.2.13 *Let c_0 and c_1 be multiplicative nonnegative constants, and y and a be multiplicative nonnegative continuous functions for all $x \ge 1$ satisfying*

$$y(x) \le c_0 +_* c_1 \cdot_* \int_0^x a(t) \cdot_* (y(t))^{2_*} \cdot_* d_* t, \qquad x \ge 1.$$

If

$$c_0 \cdot_* c_1 \cdot_* \int_0^x a(t) \cdot_* d_* t < e \qquad \text{for} \qquad \forall x \ge 1,$$

then

$$y(x) \le c_0 \cdot_* (e/_*(e -_* c_0 \cdot_* c_1 \cdot_* \int_{*1}^x a(t) \cdot_* d_* t)) \qquad (5.17)$$

for any $x \ge 1$.

Proof 5.2.14 *Let*

$$z(x) = c_0 +_* c_1 \cdot_* \int_0^x a(t) \cdot_* (y(t))^{2_*} \cdot_* d_* t, \quad x \ge 1.$$

Then

$$y(x) \le z(x), \quad x \ge 1. \qquad (5.18)$$

Also,

$$z(0) \ = \ c_0,$$

$$z^*(x) = c_1 \cdot_* a(x) \cdot_* (y(x))^{2*}$$

$$\leq c_1 \cdot_* a(x) \cdot_* (z(x))^{2*}, \qquad x \geq 1,$$

whereupon

$$(z^*(x)/_*(z(x))^{2*}) \leq c_1 \cdot_* a(x), \qquad x \geq 1.$$

The last inequality we multiplicative integrate on $[1,x]$, $x \geq 1$, *and we find*

$$\int_{*1}^{x}(z^*(t)/_*(z(t))^{2*}) \cdot_* d_*t \;\leq\; c_1 \cdot_* \int_{*1}^{x} a(t) \cdot_* d_*t$$

or

$$-_*(e/_*z(x)) +_* (e/_*c_0) \leq c_1 \cdot_* \int_{0}^{x} a(t) \cdot_* d_*t,$$

or

$$(e/_*c_0) -_* c_1 \cdot_* \int_{*1}^{x} a(t) \cdot_* d_*t \leq (e/_*z(x)),$$

or

$$z(x) \leq c_0 \cdot_* (e/_*(e -_* c_0 \cdot_* c_1 \cdot_* \int_{*1}^{x} a(t) \cdot_* d_*t)), \qquad x \geq 1.$$

From the last inequality and from (5.18), we get the inequality (5.17). This completes the proof.

Theorem 5.2.15 *Let y be a solution of the multiplicative initial value problem*

$$y^* = y \cdot_* f(x,y), \qquad x \in [1, \alpha],$$

$$y(1) = e$$

where f is continuous and multiplicative bounded function in the multiplicative (x,y) *plane and* $\alpha > 1$ *is a multiplicative constant. Then there exists a constant* $c > 1$ *such that*

$$|y(x)|_* \leq e^{c \cdot_* x} \quad \text{for} \quad \forall x \in [1, \alpha].$$

Proof 5.2.16 *We have that*

$$\int_{*1}^{*x} y^*(t) \cdot_* d_* t = \int_{*1}^{*x} y(t) \cdot_* f(t, y(t)) \cdot_* d_* t, \quad x \in [1, \alpha],$$

or

$$y(x) = e +_* \int_{*1}^{*x} y(t) \cdot_* f(t, y(t)) \cdot_* d_* t, \quad x \in [1, \alpha],$$

whereupon

$$|y(x)|_* = \left| e +_* \int_{*1}^{*x} y(t) \cdot_* f(t, y(t)) \cdot_* d_* t \right|_*$$

$$\leq e +_* \int_{*1}^{*x} |y(t)|_* \cdot_* |f(t, y(t))|_* \cdot_* d_* t, \quad x \in [0, \alpha].$$

$$(5.19)$$

Since f is multiplicative bounded in the multiplicative (x, y) plane, there exists a constant $c > 1$ such that

$$|f(x, y)|_* \leq c$$

in the multiplicative (x, y) plane. From the last inequality and from the inequality (5.19), we get

$$|y(x)| \leq e +_* c \cdot_* \int_{*1}^{*x} |y(t)|_* \cdot_* d_* t, \quad x \in [1, \alpha].$$

Now, we apply the multiplicative Gronwall type integral inequality (5.7) for

$$a(x) = e,$$

$$b(x) = c, \quad x \in [1, \alpha],$$

and we find

$$|y(x)|_* \leq e +_* e^{c \cdot_* \int_{*1}^{*x} \cdot_* d_* t} \cdot_* c \cdot_* \int_{*1}^{*x} e^{-* c \cdot_* \int_{*1}^{*t} \cdot_* d_* s} \cdot_* d_* t$$

$$= e +_* e^{c \cdot_* x} \cdot_* c \cdot_* \int_{*1}^{*x} e^{-* c \cdot_* t} \cdot_* d_* t$$

$$= e +_* e^{c_* x} \cdot_* (e -_* e^{-_* c_* x})$$

$$= e^{c_* x}, \qquad x \in [1, \alpha].$$

This completes the proof.

Theorem 5.2.17 *Let $\alpha > 1$, $\gamma > 1$, c_0, c_1 and c_2 be multiplicative non-negative constants and y be a multiplicative nonnegative multiplicative bounded continuous function satisfying*

$$y(x) \le c_0 \cdot_* e^{-_* \alpha \cdot_* x} +_* c_1 \cdot_* \int_{*1}^{*x} e^{-_* \alpha \cdot_* (x -_* t)} \cdot_* y(t) \cdot_* d_* t$$

(5.20)

$$+_* c_2 \cdot_* \int_{*1}^{*\infty} e^{-_* \gamma \cdot_* t} \cdot_* y(x +_* t) \cdot_* d_* t, \qquad x \ge 1.$$

If

$$\beta = (c_1 /_* (\alpha +_* (c_2 /_* \gamma)))$$

$$< e,$$

then

$$y(x) \le c_0 \cdot_* (e +_* (c_2 /_* (\alpha \cdot_* (e -_* \beta))))$$

$$\cdot_* e^{(-_* \alpha +_* c_1 \cdot_* (e +_* (c_2 /_* (\alpha \cdot_* (e -_* \beta)))))) \cdot_* x},$$

$x \ge 1.$

Proof 5.2.18 *The inequality (5.20) we can rewrite in the following form:*

$$y(x) \le c_0 \cdot_* e^{-_* \alpha \cdot_* x} +_* c_1 \cdot_* e^{-_* \alpha \cdot_* x}$$

$$\cdot_* \int_{*1}^{*x} e^{\alpha \cdot_* t} \cdot_* y(t) \cdot_* d_* t$$

$$+_* c +_* e^2 \cdot_* e^{\gamma \cdot_* x} \cdot_* \int_{*x}^{*\infty} e^{-_* \gamma \cdot_* t} \cdot_* y(t) \cdot_* d_* t, \qquad x \ge 1.$$

(5.21)

Now, we multiplicative integrate the inequality (5.21) *over* $[x,\infty)$ *and we find*

$$\int_x^\infty y(t) \cdot_* d_* t \;\leq\; c_0 \cdot_* \int_x^\infty e^{-*\alpha \cdot_* t} \cdot_* d_* t$$

$$+_* c_1 \cdot_* \int_x^\infty e^{-*\alpha \cdot_* t} \cdot_* \int_{*1}^t e^{\alpha \cdot_* s} \cdot_* y(s) \cdot_* d_* s \cdot_* d_* t$$

$$+_* c_2 \cdot_* \int_x^\infty e^{\gamma \cdot_* t} \cdot_* \int_{*t}^\infty e^{-*\gamma \cdot_* s} \cdot_* y(s) \cdot_* d_* s \cdot_* d_* t$$

$$=\; -_*(c_0/_*\alpha) \cdot_* e^{-*\alpha \cdot_* t}\Big|_{*t=x}^{t=\infty}$$

$$-_*(c_1/_*\alpha) \cdot_* e^{-*\alpha \cdot_* t} \cdot_* \int_{*1}^t e^{\alpha \cdot_* s} \cdot_* y(s) \cdot_* d_* s\Big|_{*t=x}^{t=\infty}$$

$$+_*(c_1/_*\alpha) \cdot_* \int_x^\infty e^{-*\alpha \cdot_* t} \cdot_* e^{\alpha \cdot_* t} \cdot_* y(t) \cdot_* d_* t$$

$$+_*(c_2/_*\gamma) \cdot_* e^{\gamma \cdot_* t} \cdot_* \int_{*t}^\infty e^{-*\gamma \cdot_* s} \cdot_* y(s) \cdot_* d_* s\Big|_{*t=x}^{t=\infty}$$

$$-_*(c_2/_*\gamma) \cdot_* \int_x^\infty e^{\gamma \cdot_* t} \cdot_* \left(-_* e^{-*\gamma \cdot_* t} \cdot_* y(t)\right) \cdot_* d_* t$$

$$=\; (c_0/_*\alpha) \cdot_* e^{-*\alpha \cdot_* x}$$

$$+_*(c_1/_*\alpha) \cdot_* e^{-*\alpha \cdot_* x} \int_{*1}^x e^{\alpha \cdot_* t} \cdot_* y(t) \cdot_* d_* t$$

$$+_*(c_1/_*\alpha) \cdot_* \int_x^\infty y(t) \cdot_* d_* t$$

$$-_*(c_2/_*\gamma) \cdot_* e^{\gamma \cdot_* x} \cdot_* \int_x^\infty e^{-*\gamma \cdot_* t} \cdot_* y(t) \cdot_* d_* t$$

$$+_*(c_2/_*\gamma) \cdot_* \int_x^\infty y(t) \cdot_* d_* t$$

$$\leq\; (c_0/_*\alpha) \cdot_* e^{-*\alpha \cdot_* x}$$

$$+_*(c_1/_*\alpha)\cdot_* e^{-_*\alpha\cdot_* x}\cdot_* \int_{*1}^x e^{\alpha\cdot_* t}\cdot_* y(t)\cdot_* d_* t$$

$$+_*((c_1/_*\alpha)+_*(c_2/_*\gamma))\cdot_* \int_x^\infty y(t)\cdot_* d_* t$$

$$= (c_0/_*\alpha)\cdot_* e^{-_*\alpha\cdot_* x}$$

$$+_*(c_1/_*\alpha)\cdot_* e^{-_*\alpha\cdot_* x}\cdot_* \int_{*1}^x e^{\alpha\cdot_* t}\cdot_* y(t)\cdot_* d_* t$$

$$+_*\beta\cdot_* \int_x^\infty y(t)\cdot_* d_* t, \quad x\geq 1,$$

whereupon

$$(e-_*\beta)\cdot_* \int_x^\infty y(t)\cdot_* d_* t$$

$$\leq (c_0/_*\alpha)\cdot_* e^{-_*\alpha\cdot_* x}$$

$$+_*(c_1/_*\alpha)\cdot_* e^{-_*\alpha\cdot_* x}\int_{*1}^x e^{\alpha\cdot_* t}\cdot_* y(t)\cdot_* d_* t, \quad x\geq 1,$$

or

$$\int_x^\infty y(t)\cdot_* d_* t \leq (e/_*(e-_*\beta))\cdot_* \left((c_0/_*\alpha)\cdot_* e^{-_*\alpha\cdot_* x}\right.$$

$$\left.+_*(c_1/_*\alpha)\cdot_* e^{-_*\alpha\cdot_* x}\cdot_* \int_{*1}^x e^{\alpha\cdot_* t}\cdot_* y(t)\cdot_* d_* t\right), \quad x\geq 1.$$

$$(5.22)$$

We note that for t $\geq$ x we have

$$e^{-_*\gamma\cdot_* t}\leq e^{-_*\gamma\cdot_* x}.$$

Hence and (5.21), using (5.22), we arrive at

$$y(x) \leq c_0\cdot_* e^{-_*\alpha\cdot_* x}$$

$$+_* c_1\cdot_* e^{-_*\alpha\cdot_* x}\cdot_* \int_{*1}^x e^{\alpha\cdot_* t}\cdot_* y(t)\cdot_* d_* t$$

$$+_* c_2 \cdot_* e^{\gamma_* x} \cdot_* \int_x^\infty e^{-*\gamma_* x} \cdot_* y(t) \cdot_* d_* t$$

$$= \quad c_0 \cdot_* e^{-*\alpha_* x} +_* c_1 \cdot_* e^{-*\alpha_* x} \int_{*1}^x e^{\alpha_* t} \cdot_* y(t) \cdot_* d_* t$$

$$+_* c_2 \cdot_* \int_x^\infty y(t) \cdot_* d_* t$$

$$\leq \quad c_0 \cdot_* e^{-*\alpha_* x}$$

$$+_* c_1 \cdot_* e^{-*\alpha_* x} \cdot_* \int_{*1}^x e^{\alpha_* t} \cdot_* y(t) \cdot_* d_* t$$

$$+_* (c_2/_*(e-_*\beta)) \cdot_* \Big((c_0/_*\alpha) \cdot_* e^{-*\alpha_* x}$$

$$+_* (c_1/_*\alpha) \cdot_* e^{-*\alpha_* x} \cdot_* \int_{*1}^x e^{\alpha_* t} \cdot_* y(t) \cdot_* d_* t \Big)$$

$$= \quad c_0 \cdot_* (e +_* (c_2/_*\alpha) \cdot_* (e -_* \beta)) \cdot_* e^{-*\alpha_* x}$$

$$+_* c_1 \cdot_* (e +_* (c_2/_*\alpha) \cdot_* (e -_* \beta)) \cdot_* e^{-*\alpha_* x} \int_{*1}^x e^{\alpha_* t} \cdot_* y(t) \cdot_* d_* t,$$

$x \geq 1$, *whereupon*

$$e^{\alpha_* x} \cdot_* y(x) \quad \leq \quad c_0 \cdot_* (e +_* (c_2/_*\alpha) \cdot_* (e -_* \beta)))$$

$$+_* c_1 \cdot_* (e +_* (c_2/_*(\alpha(e -_* \beta)))) \cdot_* \int_{*1}^x e^{\alpha_* t} \cdot_* y(t) \cdot_* d_* t,$$

$$(5.23)$$

$x \geq 1$. *Let*

$$A \quad = \quad c_0 \cdot_* (e +_* (c_2/_*(\alpha(e -_* \beta)))) ,$$

$$B \quad = \quad c_1 \cdot_* (e +_* (c_2/_*(\alpha \cdot_* (e -_* \beta)))) .$$

Then (5.23) takes the form

$$e^{\alpha_* x} \cdot_* y(x) \leq A +_* B \cdot_* \int_{*1}^x e^{\alpha_* t} \cdot_* y(t) \cdot_* d_* t.$$

Now, we apply the multiplicative Gronwall type integral inequality (5.7) for $a(x) = A$ and $b(x) = B$. and we get

$$e^{\alpha \cdot_* x} \cdot_* y(x) \ \leq \ A +_* e^{\int_{*1}^x B \cdot_* d_* t} \cdot_* \int_{*1}^x A \cdot_* B \cdot_* e^{-_* \int_{*1}^t B \cdot_* d_* s} \cdot_* d_* t$$

$$= \ A +_* A \cdot_* B \cdot_* e^{B \cdot_* x} \cdot_* \int_{*1}^x e^{-_* B \cdot_* t} \cdot_* d_* t$$

$$= \ A +_* e^{B \cdot + _* x} \cdot_* A \cdot_* \left(-_* e^{-_* B \cdot_* t} \Big|_{t=1}^{t=x} \right)$$

$$= \ A +_* e^{B \cdot_* x} \cdot_* A \cdot_* \left(e -_* e^{-_* B \cdot_* x} \right)$$

$$= \ A \cdot_* e^{B \cdot_* x}$$

$$= \ c_0 \cdot_* \left(e +_* \left(c_2 /_* \left(\alpha \cdot_* \left(1 -_* \beta \right) \right) \right) \right)$$

$$\cdot_* e^{c_1 \cdot_* \left(e +_* \left(c_2 /_* \left(\alpha \cdot_* \left(e -_* \beta \right) \right) \right) \right) \cdot_* x}, \qquad x \geq 1.$$

Hence,

$$y(x) \ \leq \ c_0 \cdot_* \left(e +_* \left(c_2 /_* \left(\alpha \cdot_* \left(e -_* \beta \right) \right) \right) \right)$$

$$\cdot_* e^{\left(-_* \alpha +_* c_1 \cdot_* \left(e +_* \left(c_2 /_* \left(\alpha \cdot_* \left(e -_* \beta \right) \right) \right) \right) \right) \cdot_* x}, \qquad x \geq 1.$$

This completes the proof.

Exercise 5.2.19 *Let y and b be multiplicative nonnegative continuous functions for $x \geq \alpha$, and let*

$$y(x) \leq a \cdot_* e^{-_* \gamma (x -_* \alpha)} +_* \int_{*\alpha}^x e^{-_* \gamma (x -_* t)} \cdot_* b(t) \cdot_* y(t) \cdot_* d_* t, \qquad x \geq \alpha,$$

where $a \geq 1$ and $\gamma > 0$ are multiplicative constants. Prove that

$$y(x) \leq a \cdot_* e^{-_* \gamma \cdot_* (x -_* \alpha) +_* \int_{*\alpha}^x b(t) \cdot_* d_* t}, \qquad x \geq \alpha.$$

Hint 5.2.20 *Set $z(x) = e^{\gamma \cdot_* x} \cdot_* y(x)$.*

Theorem 5.2.21 *Let a and b be continuous functions on* $[\alpha, \infty)$ *and y be a multiplicative differentiable function on* $[\alpha, \infty)$, *and*

$$y^*(x) \le a(x) \cdot_* y(x) +_* b(x), \qquad x \ge \alpha,$$

$$y(\alpha) \le y_0.$$

Then

$$y(x) \le y_0 \cdot_* e^{\int_{*\alpha}^{x} a(t) \cdot_* d_* t} +_* \int_{\alpha}^{x} b(t) \cdot_* e^{\int_{*t}^{x} a(s) \cdot_* d_* s} \cdot_* d_* t,$$

$x \ge \alpha.$

Proof 5.2.22 *We have, for* $x \ge \alpha$,

$$y^*(x) -_* a(x) \cdot_* y(x) \le b(x),$$

whereupon

$$(y^*(x) -_* a(x) \cdot_* y(x)) \cdot_* e^{-* \int_{*\alpha}^{x} a(t) \cdot_* d_* t}$$

$$\le b(x) \cdot_* e^{-* \int_{*\alpha}^{x} a(t) \cdot_* d_* t}$$

and

$$\left(y(x) e^{-* \int_{*\alpha}^{x} a(t) \cdot_* d_* t} \right)^* \le b(x) \cdot_* e^{-* \int_{*\alpha}^{x} a(t) \cdot_* d_* t},$$

or

$$\int_{*\alpha}^{x} \left(y(t) \cdot_* e^{-* \int_{*\alpha}^{t} a(s) \cdot_* d_* s} \right)^* \cdot_* d_* t$$

$$\le \int_{*\alpha}^{x} b(t) \cdot_* e^{-* \int_{*\alpha}^{t} a(s) \cdot_* d_* s} \cdot_* d_* t,$$

or

$$y(x) \cdot_* e^{-* \int_{*\alpha}^{x} a(t) \cdot_* d_* t} -_* y(\alpha)$$

$$\le \int_{*\alpha}^{x} b(t) \cdot_* e^{-* \int_{*\alpha}^{t} a(s) \cdot_* d_* s} \cdot_* d_* t,$$

or

$$y(x) \cdot_* e^{-* \int_{*\alpha}^{x} a(t) \cdot_* d_* t}$$

$$\leq\ y(\alpha)+_*\int_{*\alpha}^x b(t)\cdot_* e^{-*\int_{*\alpha}^t a(s)\cdot_* d_* s}\cdot_* d_* t$$

$$\leq\ y_0 +_*\int_{*\alpha}^x b(t)\cdot_* e^{-*\int_{*\alpha}^t a(s)\cdot_* d_* s}\cdot_* d_* t,$$

or

$$y(x)\ \leq\ e^{\int_{*\alpha}^x a(t)\cdot_* d_* t}$$

$$\cdot_*\left(y_0 +_*\int_{*\alpha}^x b(t)\cdot_* e^{-*\int_{*\alpha}^t a(s)\cdot_* d_* s}\cdot_* d_* t\right)$$

$$=\ y_0\cdot_* e^{\int_{*\alpha}^x a(t)\cdot_* d_* t}$$

$$+_*\int_{*\alpha}^x b(t)\cdot_* e^{\int_{*t}^x a(s)\cdot_* d_* s}\cdot_* d_* t,\quad x\geq\alpha.$$

This completes the proof.

Theorem 5.2.23 *Let y and a be multiplicative nonnegative continuous functions* $[\alpha,\infty)$, $k(x,t)$ *and its partial derivative* $k_x(x,t)$ *be multiplicative nonnegative continuous functions for* $\alpha\leq t\leq x$, *and*

$$y(x)\leq a(x)+_*\int_{*\alpha}^x k(x,t)\cdot_* y(t)\cdot_* d_* t,\quad x\geq\alpha.\tag{5.24}$$

Then

$$y(x)\leq a(x)+_*\int_{*\alpha}^x A(t)\cdot_* e^{\int_{*t}^x B(s)\cdot_* d_* s}\cdot_* d_* t,\quad x\geq\alpha,\tag{5.25}$$

where

$$A(x)\ =\ k(x,x)\cdot_* a(x)+_*\int_{*\alpha}^x k_x(x,t)\cdot_* a(t)\cdot_* d_* t,$$

$$B(x)\ =\ k(x,x)+_*\int_{*\alpha}^x k_x(x,t)\cdot_* d_* t,\quad x\geq\alpha.$$

Here α *is a multiplicative constant.*

Proof 5.2.24 *Let*

$$z(x) = \int_{*\alpha}^x k(x,t) \cdot_* y(t) \cdot_* d_*t, \qquad x \geq \alpha.$$

We note that the function $z(x)$ is a nondecreasing function for $x \geq \alpha$. Then, using (5.24),

$$y(x) \leq z(x) +_* a(x), \qquad x \geq \alpha, \tag{5.26}$$

and $z(\alpha) = 0$, and

$$
\begin{aligned}
z^*(x) &= k(x,x) \cdot_* y(x) +_* \int_{*\alpha}^x k_x(x,t) \cdot_* y(t) \cdot_* d_*t \\[2mm]
&\leq k(x,x) \cdot_* (a(x) +_* z(x)) +_* \int_{*\alpha}^x k_x(x,t) \cdot_* (z(t) +_* a(t)) \cdot_* d_*t \\[2mm]
&= k(x,x) \cdot_* a(x) +_* \int_{*\alpha}^x k_x(x,t) \cdot_* a(t) \cdot_* d_*t \\[2mm]
&\quad +_* k(x,x) \cdot_* z(x) +_* \int_{*\alpha}^x k_x(x,t) \cdot_* z(t) \cdot_* d_*t \\[2mm]
&\leq A(x) +_* \left(k(x,x) +_* \int_{*\alpha}^x k_x(x,t) \cdot_* d_*t \right) \cdot_z (x) \\[2mm]
&= A(x) +_* B(x) \cdot_* z(x), \qquad x \geq \alpha.
\end{aligned}
$$

Hence and the previous theorem, we get that

$$z(x) \leq \int_{*\alpha}^x A(t) \cdot_* e^{\int_{*t}^x B(s) \cdot_* d_*s} \cdot_* d_*t, \qquad x \geq \alpha.$$

From the last inequality and from (5.26), we get the inequality (5.25). This completes the proof.

Exercise 5.2.25 *Let $y(x)$, $k(x,t)$ and its partial derivative $k_x(x,t)$ be multiplicative nonnegative continuous functions for $\alpha \leq t \leq x$ and*

$$y(x) \leq a +_* \int_\alpha^x k(x,t) \cdot_* y(t) \cdot_* d_*t, \qquad x \geq \alpha,$$

where $a \geq 1$ is a multiplicative constant. Prove that

$$y(x) \leq a \cdot_* e^{\int_{*\alpha}^x k(x,t) \cdot_* d_*t}, \qquad x \geq \alpha.$$

5.3   The Picard Method of Successive Approximations and Existence Theorems

Let $y_0(x)$ be any continuous function. We often pick $y_0(x) = y_0$, which is the initial data of the initial value problem (5.1), (5.2). We define

$$y_1(x) \;=\; y_0 +_* \int_{*x_0}^{x} f(t, y_0(t)) \cdot_* d_* t,$$

$$(5.27)$$

$$y_m(x) \;=\; y_0 +_* \int_{*x_0}^{x} f(t, y_m(t)) \cdot_* d_* t, \quad m = 0, 1, 2, \ldots.$$

If the sequence $\{y_m\}_{m=1}^{\infty}$ converges uniformly to a continuous function y in the interval J, then we may pass the limit in both sides of (5.27), to obtain

$$
\begin{aligned}
y(x) \;&=\; \lim_{m \to \infty} y_{m+_*1}(x) \\[2mm]
&=\; y_0 +_* \lim_{m \to \infty} \int_{*x_0}^{x} f(t, y_m(t)) \cdot_* d_* t \\[2mm]
&=\; y_0 +_* \int_{*x_0}^{x} f(t, y(t)) \cdot_* d_* t,
\end{aligned}
$$

so y is the solution to the initial value problem (5.1), (5.2).

Definition 5.3.1 *The method for finding the solution y is called the Picard method of successive approximations.*

Example 5.3.2 *Let us consider the initial value problem*

$$y^* \;=\; e^2 \cdot_* y, \quad x > 1,$$

$$y(1) \;=\; e,$$

which is equivalent to solving the integral equation

$$y(x) = e +_* e^2 \cdot_* \int_{*1}^{x} y(t) \cdot_* d_* t.$$

Let $y_0(x) = e$. *Then*

$$y_1(x) \;=\; e +_* e^2 \cdot_* \int_{*1}^{x} y_0 \cdot_* d_* t$$

$$=\; e +_* e^2 \cdot_* x,$$

and

$$y_2(x) \;=\; e +_* e^2 \cdot_* \int_{*1}^{x} y_1(t) \cdot_* d_* t$$

$$=\; e +_* e^2 \cdot_* \int_{*1}^{x} (e +_* e^2 \cdot_* t) \cdot_* d_* t$$

$$=\; e +_* e^2 \cdot_* x +_* ((e^2 \cdot_* x)^{2*}/_* e^2),$$

and

$$y_3(x) \;=\; y_0 +_* e^2 \cdot_* \int_{*1}^{x} y_2(t) \cdot_* d_* t$$

$$=\; e +_* e^2 \cdot_* \int_{*1}^{x} \left(e +_* e^2 \cdot_* t +_* ((e^2 \cdot_* t)^{2*}/_* e^2) \right) \cdot_* d_* t$$

$$=\; e +_* e^2 \cdot_* x +_* ((e^2 \cdot_* x)^{2*}/_* e^2)$$

$$+_* ((e^2 \cdot_* x)^{3*}/_* e^{3!*}) +_* \cdots,$$

and so on. We assume that

$$y_m(x) \;=\; e +_* (e^2 \cdot_* x) +_* ((e^2 \cdot_* x)^{2*}/_* e^2) +_* ((e^2 \cdot_* x)^{3*}/_* e^{3!*})$$

$$+_* \cdots +_* ((e^2 \cdot_* x)^{m*}/_* e^{m!*})$$

for some $m \in \mathbb{N}$. *Then*

$$y_{m+1}(x) \;=\; e +_* e^2 \cdot_* \int_{*1}^{x} y_m(t) \cdot_* d_* t$$

$$=\; e +_* e^2 \cdot_* \int_{*1}^{x} \left(e +_* e^2 \cdot_* t +_* ((e^2 \cdot_* t)^{2*}/_* e^2) \right.$$

$$+_*((e^2 \cdot_* t)^{3*}/_* e^{3!*}) +_* \cdots +_* ((e^2 \cdot_* t)^{m*}/_* e^{m!*})\Big) \cdot_* d_* t$$

$$= \quad e +_* e^2 \cdot_* x +_* ((e^2 \cdot_* x)^{2*}/_* e^{2!*}) +_* ((e^2 \cdot_* x)^{3*}/_* e^{3!*})$$

$$+_* \cdots +_* ((e^2 \cdot_* x)^{(m+1)*}/_* e^{(m+1)!*}).$$

Consequently

$$y_m(x) = \sum_{*k=0}^{m} ((e^2 \cdot_* x)^{k*}/_* e^{k!*}) \qquad \text{for} \qquad \forall m \in \mathbb{N}.$$

Hence,

$$y(x) \quad = \quad \lim_{m \to \infty} y_m(x)$$

$$= \quad \sum_{k=0}^{\infty} ((e^2 \cdot_* x)^{k*}/_* e^{k!*})$$

$$= \quad e^{e^2 \cdot_* x}$$

is a solution to the considered initial value problem.

Example 5.3.3 *We will find the first three iterations with the initial approximation* $y_0(x) = e^2 \cdot_* x$ *for the following initial value problem:*

$$y^* \quad = \quad x^{2*} -_* y^{2*} -_* e,$$

$$y(1) \quad = \quad 1.$$

We have

$$y_{m+1}(x) \quad = \quad \int_{*1}^{x} f(t, y_m(t)) \cdot_* d_* t$$

$$= \quad \int_{*1}^{x} \left(t^{2*} -_* (y_m(t))^{2*} -_* e \right) \cdot_* d_* t.$$

Then

$$y_1(x) \quad = \quad \int_{*1}^{x} \left(t^{2*} -_* (y_0(t))^{2*} -_* e \right) \cdot_* d_* t$$

$$= \int_{*1}^{x} \left(t^{2*} -_* e^4 \cdot_* t^{2*} -_* e \right) \cdot_* d_* t$$

$$= -_* e^3 \cdot_* \int_{*1}^{x} t^{2*} \cdot_* d_* t -_* \int_{*1}^{x} \cdot_* d_* t$$

$$= -_* x^{3*} -_* x,$$

and

$$y_2(x) = \int_{*1}^{x} \left(t^{2*} -_* (y_1(t))^{2*} -_* e \right) \cdot_* d_* t$$

$$= \int_{*1}^{x} \left(t^{2*} -_* (-_* t^{3*} -_* t)^{2*} -_* e \right) \cdot_* d_* t$$

$$= \int_{*1}^{x} \left(t^{2*} -_* t^{6*} -_* e^2 \cdot_* t^{4*} -_* t^{2*} -_* e \right) \cdot_* d_* t$$

$$= \int_{*1}^{x} \left(-_* t^{6*} -_* e^2 \cdot_* t^{4*} -_* e \right) \cdot_* d_* t$$

$$= -_* (x^{7*}/_* e^7) -_* e^{\frac{2}{5}} \cdot_* x^{5*} -_* x,$$

and

$$y_3(x) = \int_{*1}^{x} \left(t^{2*} -_* (y_2(t))^{2*} -_* e \right) \cdot_* d_* t$$

$$= \int_{*1}^{x} \left(t^{2*} -_* \left(-_* (t^{7*}/_* e^7) -_* e^{\frac{2}{5}} \cdot_* t^{5*} -_* t \right)^{2*} -_* e \right) \cdot_* d_* t$$

$$= \int_{*1}^{x} \left(t^{2*} -_* (t^{14*}/_* e^{49}) -_* e^{\frac{4}{25}} \cdot_* t^{10*} -_* t^{2*} \right.$$

$$\left. -_* e^{\frac{4}{35}} \cdot_* t^{12*} -_* e^{\frac{2}{7}} \cdot_* t^{8*} -_* e^{\frac{4}{5}} \cdot_* t^{6*} -_* e \right) \cdot_* d_* t$$

$$= -_* (x^{15*}/_* e^{735}) -_* e^{\frac{4}{455}} \cdot_* x^{13*} -_* e^{\frac{4}{275}} \cdot_* x^{11*}$$

$$-_* e^{\frac{2}{63}} \cdot_* x^{9*} -_* e^{\frac{4}{35}} \cdot_* x^{7*} -_* x.$$

Example 5.3.4 *We will find the first four iterations with the initial approximation* $y_0(x) = e^4$ *of the following initial value problem:*

$$y^* \;=\; x^{3*} -_* e^{4} \cdot_* y, \quad x > 1,$$

$$y(1) \;=\; 1.$$

We have

$$y_{m+1}(x) = \int_{*1}^{x} \left(t^{3*} -_* e^{4} \cdot_* y_m(t)\right) \cdot_* d_*t.$$

Then

$$y_1(x) \;=\; \int_{*1}^{x}(t^{3*} -_* e^{16}) \cdot_* d_*t$$

$$\;=\; (x^{4*}/_*e^{4}) -_* e^{16} \cdot_* x,$$

and

$$y_2(x) \;=\; \int_{*1}^{x}\left(t^{3*} -_* e^{4} \cdot_* \left((t^{4*}/_*e^{4}) -_* e^{16} \cdot_* t\right)\right) \cdot_* d_*t$$

$$\;=\; \int_{*1}^{x}\left(t^{3*} -_* t^{4*} +_* e^{64} \cdot_* t\right) \cdot_* d_*t$$

$$\;=\; (x^{4*}/_*e^{4}) -_* (x^{5*}/_*e^{5}) +_* e^{32} \cdot_* x^{2*},$$

and

$$y_3(x) \;=\; \int_{*1}^{x}\left(t^{3*} -_* e^{4} \cdot_* \left((t^{4*}/_*e^{4}) -_* (t^{5*}/_*e^{5}) +_* e^{32} \cdot_* t^{2*}\right)\right) \cdot_* d_*t$$

$$\;=\; \int_{*1}^{x}\left(t^{3*} -_* t^{4*} +_* e^{\frac{4}{5}} \cdot_* t^{5*} -_* e^{128} \cdot_* t^{2*}\right) \cdot_* d_*t$$

$$\;=\; (x^{4*}/_*e^{4}) -_* (x^{5*}/_*e^{5}) +_* e^{\frac{2}{15}} \cdot_* x^{6*} -_* e^{\frac{128}{3}} \cdot_* x^{3*},$$

and

$$y_4(x) \;=\; \int_{*1}^{x}\left(t^{3*} -_* 4\left((t^{4*}/_*e^{4}) -_* (t^{5*}/_*e^{5}) +_* e^{\frac{2}{15}} \cdot_* t^{6*}\right.\right.$$

$$-_* e^{\frac{128}{3}} \cdot_* t^{3*}\Big)\Big) \cdot_* d_* t$$

$$= \int_{*1}^x \Big(t^{3*} -_* t^{4*} +_* e^{\frac{4}{5}} \cdot_* t^{5*} -_* e^{\frac{8}{15}} \cdot_* t^{6*} +_* e^{\frac{512}{3}} \cdot_* t^{3*}\Big) \cdot_* d_* t$$

$$= -_* e^{\frac{8}{105}} \cdot_* x^{7*} +_* e^{\frac{2}{15}} \cdot_* x^{6*} -_* (e^5 \cdot_* x^{5*}) +_* e^{\frac{515}{12}} \cdot_* x^{4*}.$$

Exercise 5.3.5 *Find the first three approximations with the initial approximation $y_0(x) = x$ of the following initial value problems:*

1. $y^* = x -_* y^{3*}$, $y(1) = 1$,
2. $y^* = x^{2*} -_* e^2 \cdot_* y^{2*}$, $y(1) = e$,
3. $y^* = x -_* e^2 \cdot_* x^{2*} \cdot_* y +_* y^{2*}$, $y(1) = 1$.

Theorem 5.3.6 *(Local Existence Theorem.)* *Let f be a continuous function in the closed rectangle*

$$\overline{S} = \{(x,y) \in \mathbb{R}_*^2 : |x -_* x_0|_* \le e^a, |y -_* y_0|_* \le e^b\},$$

and

$$|f(x,y)| \le e^M$$

for all $(x,y) \in \overline{S}$ for some nonnegative constant M, f satisfies the multiplicative uniform Lipschitz condition (5.4) in $\overline{S}$, y_0 be continuous function in $|x -_* x_0|_* \le e^a$, and

$$|y_0(x) -_* y_0|_* \le e^b.$$

Then the sequence $\{y_m\}_{m=1}^\infty$ generated by (5.24) converges to the unique solution y of the initial value problem (5.1), (5.2). This solution is defined in the interval $J_h : |x -_* x_0| \le h_* = \min\{e^a, e^b/_* e^M\}$. Also, for $x \in J_h$ the following error estimate holds

$$|y(x) -_* y_m(x)|_* \le e^{Ne^{Lh}} \cdot_* \min\{e, e^{\frac{(Lh)^m}{m!}}\}, \qquad m = 0,1,\dots. \quad (5.28)$$

Here

$$\max_{x \in J_h} |y_1(x) -_* y_0(x)|_* \le e^N.$$

Proof 5.3.7 Step 1. *We will prove that the sequence $\{y_m\}_{m=1}^{\infty}$ defined by (5.27) exists as continuous function in J_h and $(x, y_m(x)) \in \overline{S}$ for all $x \in J_h$.*

Since y_0 is continuous on $|x -_* x_0|_* \leq e^a$, and

$$|y_0(x) -_* y_0|_* \leq e^b.$$

f is continuous in $\overline{S}$, the function $f(\cdot, y_0(\cdot))$ is continuous in J_h, and hence y_1 is continuous in J_h. Also, for $|x -_* x_0|_* \leq e^h$,

$$|y_1(x) -_* y_0|_* \;=\; \left| \int_{*x_0}^{x} f(t, y_0(t)) \cdot_* d_* t \right|_*$$

$$\leq\; \left| \int_{*x_0}^{x} |f(t, y_0(t))| \cdot_* d_* t \right|_*$$

$$\leq\; e^M \cdot_* |x -_* x_0|_*$$

$$\leq\; e^{Mh}$$

$$\leq\; e^b,$$

i.e., for $x \in J_h$ we have that $(x, y_1(x)) \in \overline{S}$. Assume that the assertion is valid for $y_{m-1}(x)$ for some $m \geq 2$. We will prove the assertion for $y_m(x)$. Since y_{m-1} is continuous in J_h we have that $f(\cdot, y_{m-1}(\cdot))$ and y_m are continuous functions in J_h. Also, for $x \in J_h$, we have

$$|y_m(x) -_* y_0|_* \;=\; \left| \int_{*x_0}^{x} f(t, y_{m-*1}(t)) \cdot_* d_* t \right|_*$$

$$\leq\; \left| \int_{*x_0}^{x} |f(t, y_{m-*1}(t))|_* \cdot_* d_* t \right|_*$$

$$\leq\; e^M \cdot_* |x -_* x_0|_*$$

$$\leq\; e^{Mh}$$

$$\leq\; e^b.$$

Step 2. *We will prove that* $\{y_m\}_{m=1}^{\infty}$ *converges uniformly in* J_h. *Without loss of generality we suppose that* $x \geq x_0$. *Then*

$$|y_1(x) -_* y_0|_* = \left| \int_{*x_0}^{x} f(t, y_0(t)) \cdot_* d_* t \right|_*$$

$$\leq \int_{*x_0}^{x} |f(t, y_0(t))|_* \cdot_* d_* t$$

$$\leq e^M \cdot_* (x -_* x_0),$$

and

$$|y_2(x) -_* y_1(x)| = \left| \int_{*x_0}^{x} (f(t, y_1(t)) -_* f(t, y_0(t))) \cdot_* d_* t \right|_*$$

$$\leq \int_{*x_0}^{x} |f(t, y_1(t)) -_* f(t, y_0(t))|_* \cdot_* d_* t$$

$$\leq e^L \cdot_* \int_{*x_0}^{x} |y_1(t) -_* y_0(t)|_* \cdot_* d_* t$$

$$\leq e^L \cdot_* \max_{x -_* x_0 \leq h} |y_1(x) -_* y_0(x)|_* \cdot_* (x -_* x_0)$$

$$\leq e^{LN} \cdot_* (x -_* x_0),$$

and

$$|y_3(x) -_* y_2(x)| = \left| \int_{*x_0}^{x} (f(t, y_2(t)) -_* f(t, y_1(t))) \cdot_* d_* t \right|_*$$

$$\leq \int_{*x_0}^{x} |f(t, y_2(t)) -_* f(t, y_1(t))|_* \cdot_* d_* t$$

$$\leq e^L \cdot_* \int_{*x_0}^{x} |y_2(t) -_* y_1(t)|_* \cdot_* d_* t$$

$$\leq e^{L^2 N} \cdot_* \int_{*x_0}^{x} (t -_* x_0) \cdot_* d_* t$$

$$= e^{L^2 N} \cdot_* (x -_* x_0)^{2*} /_* e^{2!}.$$

We suppose that

$$|y_m(x) -_* y_{m-1}(x)| \ \leq \ e^{NL^{m-1}} \cdot_* ((x -_* x_0)^{(m-1)_*} /_* e^{(m-1)!})$$

for some $m \in \mathbb{N}$. We will prove that

$$|y_{m+1}(x) -_* y_m(x)|_* \leq e^{NL^m} \cdot_* ((x -_* x_0)^{m_*} /_* e^{m!}).$$

Indeed,

$$
\begin{aligned}
|y_{m+1}(x) -_* y_m(x)|_* \ &= \ \left| \int_{*x_0}^x (f(t, y_m(t)) -_* f(t, y_{m-1}(t))) \cdot_* d_* t \right|_* \\
&\leq \ \int_{*x_0}^x |f(t, y_m(t)) -_* f(t, y_{m-1}(t))|_* \cdot_* d_* t \\
&\leq \ e^L \cdot_* \int_{*x_0}^x |y_m(t) -_* y_{m-1}(t)|_* \cdot_* d_* t \\
&\leq \ e^{LNL^{m-1}} \cdot_* \int_{*x_0}^x ((t -_* x_0)^{(m-1)_*} /_* e^{(m-1)!}) \\
&= \ e^{NL^m} \cdot_* ((x -_* x_0)^{m_*} /_* e^{m!}).
\end{aligned}
$$

Therefore, for any $m \in \mathbb{N}$, we have

$$|y_{m+1}(x) -_* y_m(x)|_* \leq e^{NL^m} \cdot_* ((x -_* x_0)^{m_*} /_* e^{m!}).$$

We consider the series

$$y_0(x) +_* \sum_{*m=1}^{\infty} (y_m(x) -_* y_{m-1}(x)). \tag{5.29}$$

For it, we have

$$
\begin{aligned}
\left| y_0(x) +_* \sum_{*m=1}^{\infty} (y_m(x) -_* y_{m-1}(x)) \right|_* \\
\leq |y_0(x)|_* +_* \sum_{*m=1}^{\infty} |y_m(x) -_* y_{m-1}(x)|_* \\
\leq |y_0(x)|_* +_* e^N \cdot_* \sum_{*m=1}^{\infty} e^{L^{m-1}} \cdot_* ((x -_* x_0)^{(m-1)_*} /_* e^{(m-1)!})
\end{aligned}
$$

$$\leq |y_0(x)|_* +_* e^N \cdot_* \sum_{*m=1}^{\infty} \left(e^{L^{m-1}h^{m-1}} \big/_* e^{(m-1)!} \right)$$

$$= |y_0(x)|_* +_* e^{Ne^{Lh}}$$

$$< \infty.$$

Therefore, the series (5.29) converges absolutely and uniformly in the interval J_h. Hence, its partial sums

$$y_1, \quad y_2, \quad \ldots, \quad y_m, \quad \ldots$$

converge to a continuous function in this interval, i.e.,

$$y(x) = \lim_{m \to \infty} y_m(x), \quad x \in J_h.$$

From here and from (5.27) this y is a solution of the initial value problem (5.1), (5.2).

Step 3. *We assume that z is also a solution of the problem (5.1), (5.2) which exists in the interval J_h and $(x, z(x)) \in \overline{S}$ for all $x \in J_h$. Then*

$$|y(x) -_* z(x)|_* = \left| \int_{*x_0}^x (f(t, y(t)) -_* f(t, z(t))) \cdot_* d_* t \right|_*$$

$$\leq \left| \int_{*x_0}^x |f(t, y(t)) -_* f(t, z(t))|_* \cdot_* d_* t \right|_*$$

$$\leq e^L \cdot_* \left| \int_{*x_0}^x |y(t) -_* z(t)|_* \cdot_* d_* t \right|_*.$$

From here and from the Gronwall type integral inequality we conclude that

$$|y(x) -_* z(x)|_* = 1, \quad x \in J_h.$$

Hence, $y(x) = z(x)$ for all $x \in J_h$.

Step 4. *Now we will prove the error bound (5.28). Let $n > m$. Then*

$$|y_n(x) -_* y_m(x)|_* = |y_n(x) -_* y_{n-1}(x) +_* y_{n-1}(x) -_* \cdots -_* y_m(x)|$$

$$\leq \sum_{*k=m}^{n-*1} |y_{k+1}(x) -_* y_k(x)|_*$$

$$\leq e^N \cdot_* \sum_{*k=m}^{n-1} (e^{L^k} \cdot_* |x -_* x_0|^{k_*})/_* e^{k!}$$

$$\leq e^N \cdot_* \sum_{*k=m}^{n-1} \cdot_* (e^{(Lh)^k}/_* e^{k!})$$

$$= e^{N(Lh)^m} \cdot_* \sum_{*k=0}^{n-m-1} \cdot_* (e^{(Lh)^k}/_* e^{(m+k)!})$$

$$\leq e^{N \frac{(Lh)^m}{m!}} \cdot_* \sum_{*k=0}^{n-m-1} e^{\frac{(Lh)^k}{k!}}$$

$$\leq e^{N \frac{(Lh)^m}{m!}} \cdot_* \sum_{*k=0}^{\infty} e^{\frac{(Lh)^k}{k!}}$$

$$= e^{N \frac{(Lh)^m}{m!} \cdot_* e^{e^{Lh}}} \quad ,$$

i.e.,

$$|y_n(x) -_* y_m(x)|_* \leq e^{N \frac{(Lh)^m}{m!} \cdot_* e^{e^{Lh}}} \quad , \quad x \in J_h. \tag{5.30}$$

Therefore

$$|y(x) -_* y_m(x)|_* = \lim_{n \to \infty} |y_n(x) -_* y_m(x)|_*$$

$$\leq e^{N \frac{(Lh)^m}{m!} e^{Lh}}. \tag{5.31}$$

Also, using the deduction of (5.30), we have, for $n > m$,

$$|y_n(x) -_* y_m(x)|_* \leq e^N \cdot_* \sum_{*k=m}^{n-1} e^{\frac{(Lh)^k}{k!}}$$

$$\leq e^N \cdot_* \sum_{*k=0}^{\infty} e^{\frac{(Lh)^k}{k!}}$$

$$= e^{N e^{Lh}},$$

whereupon

$$|y(x) -_* y_m(x)|_* = \lim_{n \to \infty} |y_n(x) -_* y_m(x)|_*$$

$$\leq e^{Ne^{Lh}}.$$

Combining the last estimate and (5.31) we get the error bound (5.28). This completes the proof.

Example 5.3.8 *We consider the multiplicative Cauchy problem*

$$y^* = x +_* y^{2*},$$

$$y(1) = 1$$

in

$$\overline{S} = \{(x,y) \in \mathbb{R}_*^2 : |x|_* \leq e, |y|_* \leq e^2\},$$

with the initial approximation $y_0(x) = x$. Here

$$f(x,y) = x +_* y^{2*},$$

$$a = e,$$

$$b = e^2.$$

Then

$$|f(x,y)|_* = |x +_* y^{2*}|_*$$

$$\leq |x|_* +_* |y|_*^{2*}$$

$$\leq e^1 +_* e^4$$

$$= e^5, \quad (x,y) \in S,$$

i.e., $M = e^5$. Therefore,

$$h = \min\left\{e, e^{\frac{2}{5}}\right\}$$

and

$$J_h : |x|_* \le e^{\frac{2}{5}}.$$

Also,

$$y_1(x) = \int_{*1}^x (t +_* (y_0(t))^{2*}) \cdot_* d_* t$$

$$= \int_{*1}^x (t +_* t^{2*}) \cdot_* d_* t$$

$$= x^{2*}/_* e^2 +_* (x^{3*}/_* e^3),$$

and

$$|y_1(x) -_* y_0(x)|_* = |x^{2*}/_* e^2 +_* (x^{3*}/_* e^3) -_* x|_*$$

$$\le (x^{2*}/_* e^2) +_* (|x|_*^{3*}/_* e^3) +_* |x|_*$$

$$\le e^{\frac{2}{25}} +_* e^{\frac{8}{375}} +_* e^{\frac{2}{5}}$$

$$= e^{\frac{188}{375}},$$

i.e., $N = \dfrac{188}{375}$. We note that

$$|f(x, y_1) -_* f(x, y_2)|_* = |x +_* y_1^{2*} -_* x -_* y_2^{2*}|_*$$

$$= |y_1^{2*} -_* y_2^{2*}|_*$$

$$= |y_1 -_* y_2|_* \cdot_* |y_1 +_* y_2|_*$$

$$\le (|y_1|_* +_* |y_2|_*) \cdot_* |y_1 -_* y_2|_*$$

$$\le e^4 \cdot_* |y_1 -_* y_2|_*,$$

i.e., $L = 4$. Consequently the sequence $\{y_m\}_{m=1}^{\infty}$ converges to the unique solution y of the given multiplicative initial value problem, which is defined in $|x|_* \leq e^{\frac{2}{5}}$. For $|x|_* \leq e^{\frac{2}{5}}$ we have

$$|y(x) -_* y_m(x)|_* \leq e^{\frac{188}{375}} e^{\frac{8}{5}} \min\{1, \frac{\left(\frac{8}{5}\right)^m}{m!}\}.$$

Example 5.3.9 *We consider the multiplicative initial value problem*

$$y^* = x \cdot_* y^{3*},$$

$$y(1) = e,$$

in

$$\overline{S} = \{(x,y) \in \mathbb{R}_*^2 : |x|_* \leq e, |y|_* \leq e^2\},$$

with the initial approximation $y_0(x) = x^{2*}$. Here

$$f(x,y) = x \cdot_* y^{3*},$$

$$a = 1,$$

$$b = 2, \quad (x,y) \in S.$$

Then for $(x,y) \in \overline{S}$ we have

$$|f(x,y)|_* = |x \cdot_* y^{3*}|_*$$

$$= |x|_* \cdot_* |y|_*^{3*}$$

$$\leq e^{27},$$

i.e., $M = 27$, and for (x,y_1), $(x,y_2) \in \overline{S}$ we have

$$|f(x,y_1) -_* f(x,y_2)|_* = |x \cdot_* y_1^{3*} -_* x \cdot_* y_2^{3*}|_*$$

$$= |x|_* \cdot_* |y_1^{3*} -_* y_2^{3*}|_*$$

$$= |x|_* \cdot_* |y_1 -_* y_2|_* \cdot (|y_1|_*^{2_*} +_* |y_1|_* \cdot_* |y_2|_* +_* |y_2|_*^{2_*})$$

$$\le e^{27} \cdot_* |y_1 -_* y_2|_*,$$

i.e., $L = 27$. Also,

$$h = \min\left\{1, \frac{2}{27}\right\}$$

$$= \frac{2}{27}.$$

Therefore $J_h : |x|_* \le e^{\frac{2}{27}}$. *For* $x \in J_h$ *we have*

$$y_1(x) = e +_* \int_{*1}^x t \cdot_* (y_0(t))^{3_*} \cdot_* d_* t$$

$$= e +_* \int_{*1}^x t^{7_*} \cdot_* d_* t$$

$$= e +_* (x^{8_*} /_* e^8),$$

and

$$|y_1(x) -_* y_0(x)|_* = |e +_* (x^{8_*} /_* e^8) -_* x^{2_*}|_*$$

$$\le e +_* (x^{8_*} /_* e^8) +_* x^{2_*}$$

$$\le e +_* e^{\frac{2^8}{8.27^8}} +_* e^{\frac{2^2}{27^2}}$$

$$= e^{\frac{2271833747752}{2259436291848}}.$$

Consequently, the sequence $\{y_m\}_{m=1}^\infty$ *converges to the unique solution y of the given multiplicative initial value problem, which is defined in* $|x|_* \le e^{\frac{2}{27}}$. *For* $|x|_* \le e^{\frac{2}{27}}$ *we have*

$$|y(x) -_* y_m(x)|_* \le e^{\frac{2271833747752}{2259436291848} e^2 \min\left\{1, \frac{2^m}{m!}\right\}}.$$

Example 5.3.10 *We consider the multiplicative initial value problem*

$$y^* = x +_* y^{2*} +_* x \cdot_* y,$$

$$y(1) = 1,$$

in

$$\bar{S} = \{(x,y) \in \mathbb{R}^2_* : |x|_* \leq e, |y|_* \leq e\},$$

with the initial approximation $y_0(x) = 1$. *Here*

$$f(x,y) = x +_* y^{2*} +_* x \cdot_* y,$$

$$a = 1,$$

$$b = 1, \quad (x,y) \in S.$$

Then for $(x,y) \in \bar{S}$ *we have*

$$|f(x,y)|_* = |x +_* y^{2*} +_* x \cdot_* y|_*$$

$$\leq |x|_* +_* y^{2*} +_* |x|_* \cdot_* |y|_*$$

$$\leq e^3,$$

i.e., M = 3, and

$$h = \min\left\{1, \frac{1}{3}\right\}$$

$$= \frac{1}{3},$$

and

$$J_h : |x|_* \leq e^{\frac{1}{3}}.$$

For $(x,y_1), (x,y_2) \in \bar{S}$ *we have*

$$|f(x,y_1) -_* f(x,y_2)|_* = |x +_* y_1^{2*} +_* x \cdot_* y_1 -_* x -_* y_2^{2*} -_* x \cdot_* y_2|_*$$

$$= \ \left| y_1^{2*} -_* y_2^{2*} +_* x \cdot_* (y_1 -_* y_2) \right|_*$$

$$= \ \left| y_1 -_* y_2 \right|_* \cdot_* \left| x +_* y_1 +_* y_2 \right|_*$$

$$\leq \ \left| y_1 -_* y_2 \right|_* \cdot_* \left(\left| y_1 \right|_* +_* \left| y_2 \right|_* +_* \left| x \right|_* \right)$$

$$\leq \ e^3 \cdot_* \left| y_1 -_* y_2 \right|_*,$$

i.e., L = 3. Also,

$$y_1(x) \ = \ \int_{*1}^{*x} t \cdot_* d_* t$$

$$= \ x^{2*} /_* e^2$$

and for $x \in J_h$ we have

$$\left| y_1(x) -_* y_0(x) \right|_* \ = \ \left| (x^{2*} /_* e^2) -_* 1 \right|_*$$

$$= \ (x^{2*} /_* e^2)$$

$$\leq \ e^{\frac{1}{2}},$$

i.e., $N = \dfrac{1}{2}$. Consequently, the sequence $\{y_m\}_{m=1}^{\infty}$ converges to the unique solution y of the given multiplicative initial value problem, which is defined in $|x|_* \leq e^{\frac{1}{3}}$, and

$$\left| y(x) -_* y_m(x) \right|_* \leq e^{\frac{1}{2} e \frac{1}{m!}}.$$

Exercise 5.3.11 *Let $y_0(x) = e$,*

$$\bar{S} = \{(x,y) \in \mathbb{R}_*^2 : |x|_* \leq e, |y|_* \leq e^2\}.$$

Consider the MIVP

$$y^* \ = \ x \cdot_* y,$$

$$y(1) \ = \ 1.$$

Find the existence interval J_h of the unique solution of this problem. Determine the error estimate in J_h.

Answer

$$J_h : |x|_* \leq e,$$

$$|y(x) -_* y_m(x)|_* \leq e^{\frac{3}{2}} e^{\frac{1}{m!}}.$$

Theorem 5.3.12 *Let* $f(x,y)$ *be continuous in the strip*

$$T = \{(x,y) \in \mathbb{R}^2_* : |x -_* x_0|_* \leq e^a, |y|_* < \infty\},$$

and f *satisfies the multiplicative uniform Lipschitz condition (5.4) in* T, y_0 *be continuous function in* $|x -_* x_0|_* \leq e^a$. *Then the sequence* $\{y_m\}_{m=1}^{\infty}$, *generated by (5.27), exists in the entire interval* $|x -_* x_0|_* \leq e^a$ *and converges to the unique solution* y *of the problem (5.1), (5.2).*

Proof 5.3.13 *As in the proof of the Local Existence Theorem, we have that* y_m *exists in* $|x -_* x_0|_* \leq e^a$ *and* $|y_m(x)|_* < \infty$. *Replacing* h *by* a *we have that*

$$\left| y_0(x) +_* \sum_{*m=1}^{\infty} (y_m(x) -_* y_{m-1}(x)) \right|_* \leq |y_0(x)|_* +_*^{Ne^{La}}$$

$$< \infty,$$

i.e., there exists a continuous function y *so that*

$$y(x) = \lim_{m \to \infty} y_m(x), \quad x \in T.$$

This y *is the unique solution of the problem (5.1), (5.2). Also, we have the following error estimate*

$$|y(x) -_* y_m(x)| \leq e^{Ne^{La} \min\left\{1, \frac{(La)^m}{m!}\right\}}, \qquad m = 0, 1, 2, \ldots,$$

$x \in T$. *This completes the proof.*

Example 5.3.14 *Let us consider the multiplicative initial value problem*

$$y^* \; = \; x +_* e^3 \cdot_* y,$$

$$y(1) \; = \; 1,$$

in

$$T = \{(x,y) \in \mathbb{R}_*^2 : |x|_* \le e, |y| < \infty\},$$

with the initial approximation $y_0(x) = x$. *Here*

$$f(x,y) \; = \; x +_* e^3 \cdot_* y,$$

$$a \; = \; -_* e.$$

For (x,y_1), $(x,y_2) \in T$ *we have*

$$|f(x,y_1) -_* f(x,y_2)|_* \; = \; |x +_* e^3 \cdot_* y_1 -_* x -_* e^3 \cdot_* y_2|_*$$

$$= \; e^3 \cdot_* |y_1 -_* y_2|_*,$$

i.e., $L = 3$. *Then the unique solution of the given problem exists in* $J_h : |x|_* \le e$ *and for* $|x|_* \le e$ *we have*

$$y_1(x) \; = \; \int_{*1}^x (t +_* e^3 \cdot_* t) \cdot_* d_* t$$

$$= \; e^2 \cdot_* x^{2*},$$

and

$$|y_1(x) -_* y_0(x)|_* \; = \; |e^2 \cdot_* x^{2*} -_* x|_*$$

$$\le \; e^2 \cdot_* x^{2*} +_* |x|_*$$

$$\le \; e^3,$$

i.e., $N = 3$. *Hence,*

$$|y(x) -_* y_m(x)| \le e^{3e^3 \min\{1, \frac{3^m}{m!}\}}.$$

Example 5.3.15 *Let us consider the multiplicative initial value problem*

$$y^* \ = \ x +_* (e/_*(e +_* y^{2*})),$$

$$y(1) \ = \ 1,$$

in

$$T = \{(x,y) \in \mathbb{R}^2_* : |x|_* \le e^2, |y|_* < \infty\},$$

with the initial approximations $y_0(x) = e$. *Here*

$$f(x,y) \ = \ x +_* (e/_*(e +_* y^{2*})),$$

$$a \ = \ 1, \quad (x,y) \in T.$$

We note that f *is multiplicative differentiable with respect to* y *in* T *and*

$$\partial_* f/_* \partial_* y(x,y) \ = \ -_*((e^2 \cdot_* y)/_*(e +_* y^{2*})^{2*}),$$

$$|\partial_* f/_* \partial_* y(x,y)|_* \ = \ |-_*((e^2 \cdot_* y)/_*(e +_* y^{2*})^{2*})|_*$$

$$\le \ e^2 \cdot_* (|y|_*/_*(e +_* y^{2*})^{2*}$$

$$\le \ e^2,$$

i.e., $L = 2$. *Also, for* $|x|_* \le e^2$, *we have*

$$y_1(x) \ = \ \int_{*1}^{*x} \left(t +_* (e/_* e +_* t^{2*}) \right) \cdot_* d_* t$$

$$= \ (x^{2*}/_* e^2) +_* \arctan_* x,$$

$$|y_1(x) -_* y_0(x)|_* \ = \ |(x^{2*}/_* e^2) +_* \arctan_* x -_* e|$$

$$\le \ (x^{2*}/_* e^2) +_* |\arctan_* x|_* +_* e$$

$$\leq \; e^2 +_* e^{\frac{\pi}{2}} +_* e$$

$$= \; e^3 +_* e^{\frac{\pi}{2}},$$

i.e., $N = 3 + \dfrac{\pi}{2}$. *Hence,*

$$|y(x) -_* y_m(x)| \leq e^{\left(3+\frac{\pi}{2}\right)e^4} \min\left\{1, \frac{4^m}{m!}\right\},$$

and the existence interval of the unique solution of the given MIVP is
$J_h : |x|_* \leq e^2.$

Example 5.3.16 *Now we consider the multiplicative initial value problem*

$$y^* \; = \; e^2 \cdot_* x -_* \arctan_* y,$$

$$y(1) \; = \; 1,$$

in

$$T = \{(x,y) : |x|_* \leq e, |y|_* < \infty\},$$

with the initial approximation $y_0(x) = 1$. *Here*

$$f(x,y) \; = \; e^2 \cdot_* x -_* \arctan_* y,$$

$$a \; = \; 1, \quad (x,y) \in T.$$

We note that f is multiplicative differentiable with respect to y in T and

$$\partial_* f /_* \partial_* y(x,y) \; = \; -_*(e /_*(e +_* y^{2*})),$$

$$|\partial_* f /_* \partial_* y(x,y)|_* \; = \; |-_*(e /_*(e +_* y^{2*}))|_*$$

$$\leq \; e,$$

i.e., $L = 1$. *Also, for* $|x|_* \leq e,$

$$y_1(x) \; = \; \int_{*1}^{*x} e^2 \cdot_* t \cdot_* d_* t$$

$$= x^{2*},$$

$$|y_1(x) -_* y_0(x)|_* = x^{2*}$$

$$\leq e,$$

i.e., $N = 1$. The existence interval of the unique solution of the given problem is $J_h : |x|_* \leq e$ and

$$|y(x) -_* y_m(x)| \leq e^{e\frac{1}{m!}}.$$

Exercise 5.3.17 *Consider the multiplicative initial value problem*

$$y^* = e^3 \cdot_* x -_* e^4 \cdot_* \mathrm{arccotan}_*(e^3 \cdot_* y),$$

$$y(1) = 1,$$

in

$$T = \{(x,y) \in \mathbb{R}^2_* : |x|_* \leq e^2, |y|_* < \infty\},$$

with the initial approximation $y_0(x) = 1$. Find the error estimate.

Answer

$$|y(x) -_* y_m(x)|_* \leq e^{(6+4\pi)e^{24} \min\left\{1, \frac{24^m}{m!}\right\}}.$$

Theorem 5.3.18 *Let f be continuous in $\mathbb{R}^2_*$ and satisfy the multiplicative Lipschitz condition in each strip*

$$T_a = \{(x,y) \in \mathbb{R}^2_* : |x|_* \leq e^a, |y|_* < \infty\}$$

with multiplicative Lipschitz constant e^{L_a}. Then the problem (5.1), (5.2) has a unique solution which exists for all x.

Proof 5.3.19 *For any $x \in \mathbb{R}_*$ there exists $a \geq 0$ such that $|x -_* x_0|_* \leq e^a$. Since the strip T is contained in the strip $T_{e^a +_* |x_0|_*}$, the function f satisfies the conditions of the previous theorem in the strip T. Therefore, the result follows for any x. This completes the proof.*

Theorem 5.3.20 *(Peano's Existence Theorem) Let f be continuous and bounded in the strip*

$$T = \{(x,y) \in \mathbb{R}_*^2 : |x -_* x_0|_* \le e^a, |y|_* < \infty\}.$$

Then the multiplicative initial value problem (5.1), (5.2) has at least one solution in $|x -_* x_0|_* \le e^a$.

Proof 5.3.21 *We will prove the result for $x \in [x_0, x_0 +_* e^a]$. We define the sequence $\{y_m(x)\}_{m=1}^{\infty}$ as follows.*

$$y_0(x) = y_0, \qquad x_0 \le x \le x_0 +_* e^{\frac{a}{m}},$$

$$y_m(x) = y_0 +_* \int_{*x_0}^{x -_* e^{\frac{a}{m}}} f(t, y_m(t)) \cdot_* d_* t, \quad x_0 +_* e^{k\frac{a}{m}} \le x \le x_0 +_* e^{(k+1)\frac{a}{m}},$$

$k = 1, 2, \ldots, m - 1$. Since f is multiplicative bounded in the strip T there exists positive constant M such that

$$|f(x,y)|_* \le e^M \qquad for \qquad (x,y) \in T.$$

Let $x_1, x_2 \in [x_0, x_0 +_* a]$. Then we have the following cases.

1. Case. $x_1, x_2 \in \left[x_0, x_0 +_* e^{\frac{a}{m}}\right]$. *Then*

$$y_m(x_1) = y_m(x_2)$$

and

$$|y_m(x_1) -_* y_m(x_2)|_* = 1$$

$$\le e^M \cdot_* |x_1 -_* x_2|_*.$$

2. Case. $x_1 \in \left[x_0, x_0 +_* e^{\frac{a}{m}}\right]$, $x_2 \in \left[x_0 +_* e^{l\frac{a}{m}}, x_0 +_* e^{(l+1)\frac{a}{m}}\right]$ *for some $l = 1, 2, \ldots, m - 1$. Then*

$$y_m(x_1) = y_0,$$

$$y_m(x_2) = y_0 +_* \int_{*x_0}^{x_2 -_* e^{\frac{a}{m}}} f(t, y_m(t)) \cdot_* d_* t,$$

whereupon

$$|y_m(x_2) -_* y_m(x_1)|_* = \left| \int_{*x_0}^{x_2 -_* e^{\frac{a}{m}}} f(t, y_m(t)) \cdot_* d_* t \right|_*$$

$$\leq \int_{*x_0}^{x_2 -_* e^{\frac{a}{m}}} |f(t, y_m(t))|_* \cdot_* d_* t$$

$$\leq e^M \cdot_* \left(x_2 -_* e^{\frac{a}{m}} -_* x_0 \right)$$

$$\leq e^M \cdot_* \left(x_0 +_* e^{(l+1)\frac{a}{m}} -_* e^{\frac{a}{m}} -_* x_0 \right)$$

$$= e^{Ml\frac{a}{m}}$$

$$\leq e^M \cdot_* (x_2 -_* x_1).$$

3. Case. $x_1, x_2 \in \left[x_0 +_* e^{l\frac{a}{m}}, x_0 +_* e^{(l+1)\frac{a}{m}} \right]$ *for some* $l = 1, 2, \ldots, m-1$. *Then*

$$y_m(x_1) = y_m(x_2),$$

$$|y_m(x_1) -_* y_m(x_2)|_* = 1$$

$$\leq e^M \cdot_* |x_2 -_* x_1|_*.$$

4. Case. $x_1 \in \left[x_0 +_* e^{l\frac{a}{m}}, x_0 +_* e^{(l+1)\frac{a}{m}} \right]$, $x_2 \in \left[x_0 +_* e^{k\frac{a}{m}}, x_0 +_* e^{(k+1)\frac{a}{m}} \right]$ *for some* $k, l = 1, 2, \ldots, m-1$, $k \neq l$. *We suppose without loss of generality that* $k > l$. *Then* $x_2 > x_1$ *and*

$$y_m(x_1) = y_0 +_* \int_{*x_0}^{x_1 -_* e^{\frac{a}{m}}} \cdot_* f(t, y_m(t)) \cdot_* d_* t,$$

$$y_m(x_2) = y_0 +_* \int_{*x_0}^{x_2 -_* e^{\frac{a}{m}}} \cdot_* f(t, y_m(t)) \cdot_* d_* t,$$

$$|y_m(x_2) -_* y_m(x_1)|_* \quad = \quad \left| \int_{x_1 -_* e^{\frac{a}{m}}}^{x_2 -_* e^{\frac{a}{m}}} f(t, y_m(t)) \cdot_* d_* t \right|_*$$

$$\leq \quad \int_{x_1 -_* e^{\frac{a}{m}}}^{x_2 -_* e^{\frac{a}{m}}} |f(t, y_m(t))| \cdot_* d_* t$$

$$\leq \quad e^M \cdot_* (x_2 -_* x_1).$$

Consequently, for every $x_1, x_2 \in [x_0, x_0 +_* a]$ *we have*

$$|y_m(x_2) -_* y_m(x_1)|_* \leq e^M \cdot_* |x_2 -_* x_1|_*.$$

Therefore, the sequence $\{y_m\}_{m=1}^{\infty}$ *is equicontinuous.*
Also, for $x \in [x_0, x_0 +_* a]$ *we have*

1. If $x \in \left[x_0, x_0 +_* \dfrac{a}{m} \right]$, *then*

$$|y_m(x)|_* = y_0.$$

2. If $x \in \left[x_0 +_* e^{l\frac{a}{m}}, x_0 +_* e^{(l+1)\frac{a}{m}} \right]$, $l = 1, 2, \ldots, m -_* 1$, *then*

$$|y_m(x)|_* \quad = \quad \left| y_0 +_* \int_{*x_0}^{x -_* e^{\frac{a}{m}}} f(t, y_m(t)) \cdot_* d_* t \right|_*$$

$$\leq \quad |y_0| +_* \int_{*x_0}^{x -_* e^{\frac{a}{m}}} |f(t, y_m(t))|_* \cdot_* d_* t$$

$$\leq \quad |y_0|_* +_* e^M \cdot_* \left| x -_* e^{\frac{a}{m}} -_* x_0 \right|_*$$

$$\leq \quad |y_0| +_* e^M \cdot_* \left(e^a -_* e^{\frac{a}{m}} \right).$$

Consequently, there exists $N > 0$ *such that*

$$|y_m(x)|_* \leq e^N \quad \text{for} \quad \forall m = 1, 2, \ldots, \quad \forall x \in [x_0, x_0 +_* e^a],$$

i.e., the sequence $\{y_m\}_{m=1}^{\infty}$ *is multiplicative bounded in* $[x_0, x_0 +_* e^a]$.
Hence and the Ascoli-Arzèla theorem, it follows that there exists a sub-sequence $\{y_{m_p}\}_{p=1}^{\infty}$ *of the sequence* $\{y_m(x)\}_{m=1}^{\infty}$ *that converges uni-formly in* $[x_0, x_0 +_* e^a]$ *to a continuous function* y.

1. *If* $x \in \left[x_0, x_0 +_* e^{\frac{a}{mp}} \right]$, *then*

$$y_{m_p}(x) \quad = \quad y_0,$$

$$\lim_{p \to \infty} y_{m_p}(x) \quad = \quad y(x_0)$$

$$= \quad y_0.$$

2. *If* $x \in \left[x_0 +_* e^{l\frac{a}{mp}}, x_0 +_* e^{(l+1)\frac{a}{mp}} \right]$, *then*

$$y_{m_p}(x) \quad = \quad y_0 +_* \int_{*x_0}^{x-_* e^{\frac{a}{mp}}} f(t, y_{m_p}(t)) \cdot_* d_* t$$

$$= \quad y_0 +_* \int_{*x_0}^{x} f(t, y_{m_p}(t)) \cdot_* d_* t \qquad (5.32)$$

$$-_* \int_{x-_* e^{\frac{a}{mp}}}^{x} f(t, y_{m_p}(t)) \cdot_* d_* t.$$

We have that

$$\lim_{p \to \infty} \int_{*x_0}^{x} f(t, y_{m_p}(t)) \cdot_* d_* t \quad = \quad \int_{*x_0}^{x} f(t, y(t)) \cdot_* d_* t,$$

$$(5.33)$$

and

$$\left| \int_{x-_* \frac{a}{mp}}^{x} f(t, y_{m_p}(t)) \cdot_* d_* t \right|_* \quad \leq \quad \int_{x-_* \frac{a}{mp}}^{x} |f(t, y_{m_p}(t))|_* \cdot_* d_* t$$

$$\leq \quad e^{M\frac{a}{mp}}$$

$$\to \quad 1$$

as $p \to \infty$. *Consequently*,

$$\lim_{p \to \infty} \int_{x-_* \frac{a}{mp}}^{x} f(t, y_{m_p}(t)) \cdot_* d_* t = 1.$$

From the last relation and (5.33), using (5.32) when $p \to \infty$, we obtain that

$$y(x) = \lim_{p \to \infty} y_{m_p}(x)$$

$$= y_0 +_* \int_{*x_0}^x f(t, y(t)) \cdot_* d_* t.$$

Hence, y satisfies the problem (5.1), (5.2). This completes the proof.

5.4 Uniqueness

Theorem 5.4.1 *(The Multiplicative Peano Uniqueness Theorem.) Let f be continuous in*

$$\overline{S}_+ = \{(x, y) \in \mathbb{R}_*^2 : x_0 \leq x \leq x_0 +_* e^a, |y -_* y_0|_* \leq e^b\}$$

and multiplicative nonincreasing in y for each fixed $x \in [x_0, x_0 +_* e^a]$. Then the problem (5.1), (5.2) has at most one solution in $[x_0, x_0 +_* e^a]$.

Proof 5.4.2 *Let y_1 and y_2 be two solutions of the problem (5.1), (5.2) that are different somewhere in the interval $[x_0, x_0 +_* e^a]$. Let x_1 be the greatest lower bound of the set consisting of those $x \in [x_0, x_0 +_* e^a]$ for which $y_2(x) > y_1(x)$, i.e., $y_1(x) = y_2(x)$ for $x \in [x_0, x_1]$ and $y_2(x) > y_1(x)$ for $x \in (x_1, x_1 +_* e^\varepsilon) \subset [x_0, x_0 +_* e^a]$. We note that the point x_1 exists, because the set consisting of those points $x \in [x_0, x_0 +_* e^a]$ for which $y_2(x) > y_1(x)$ is bounded below by x_0. For $x \in (x_1, x_1 +_* e^\varepsilon)$ we have*

$$f(x, y_1(x)) > f(x, y_2(x)),$$

whereupon

$$y_1^*(x) \geq y_2^*(x).$$

Therefore, the function $z = y_2 -_* y_1$ is nonincreasing in $(x_1, x_1 +_* e^\varepsilon)$. Since $z(x_1) = 1$, we have that

$$z(x) \leq 1, \quad x \in (x_1, x_1 +_* e^\varepsilon),$$

i.e.,

$$y_2(x) \leq y_1(x), \quad x \in (x_1, x_1 + _* e^{\varepsilon}),$$

which is a contradiction. This completes the proof.

5.5   Continuous Dependence on Initial Data

Theorem 5.5.1 *Let* $f(x,y)$ *be continuous function in the domain D containing the points* (x_0, y_0) *and* (x_1, y_1), $|f(x,y)|_* \leq e^M$ *for all* $(x,y) \in D$,

$$|f(x,y) -_* f(x,z)|_* \leq e^L \cdot_* |y -_* z|_* \quad \text{for} \quad \text{any} \quad (x,y),(x,z) \in D,$$

g *be a continuous function in D,* $|g(x,y)|_* \leq e^{M_1}$ *for all* $(x,y) \in D$. *Let also y be a solution of the problem* (5.1), (5.2) *and z be a solution of the problem*

$$z^* = f(x,z) +_* g(x,z),$$

$$z(x_1) = y_1,$$

which exists in an interval J containing the points x_0 *and* x_1. *Then*

$$|y(x) -_* z(x)|_* \quad \leq \left(|y_0 -_* y_1|_* +_* e^{M+M_1} \cdot_* |x_1 -_* x_0|_* +_* e^{\frac{1}{L}M_1} \right)$$
$$\cdot_* e^{e^L \cdot_* |x -_* x_0|_*} -_* e^{\frac{1}{L}M_1}. \tag{5.34}$$

Proof 5.5.2 *We have*

$$z(x) = y_1 +_* \int_{*x_1}^{x} (f(t,z(t)) +_* g(t,z(t))) \cdot_* d_*t$$

$$= y_1 +_* \int_{*x_1}^{x} f(t,z(t)) \cdot_* d_*t +_* \int_{x_1}^{x} g(t,z(t)) \cdot_* d_*t$$

$$= y_1 +_* \int_{*x_0}^{x} f(t,z(t)) \cdot_* d_*t +_* \int_{x_1}^{x_0} f(t,z(t)) \cdot_* d_*t$$

$$+_* \int_{*x_1}^x g(t,z(t)) \cdot_* d_*t,$$

and

$$y(x) = y_0 +_* \int_{*x_0}^x f(t,y(t)) \cdot_* d_*t.$$

Therefore,

$$
|y(x) -_* z(x)|_* = \left| y_0 -_* y_1 +_* \int_{*x_0}^x (f(t,y(t)) -_* f(t,z(t))) \cdot_* d_*t \right.
$$

$$
-_* \int_{x_1}^x f(t,z(t)) \cdot_* d_*t
$$

$$
\left. -_* \int_{x_1}^x g(t,z(t)) \cdot_* d_*t \right|_*
$$

$$
\leq |y_0 -_* y_1|_* +_* \left| \int_{*x_0}^x (f(t,y(t)) -_* f(t,z(t))) \cdot_* d_*t \right|_*
$$

$$
+_* \left| \int_{x_1}^{x_0} f(t,z(t)) \cdot_* d_*t \right|_*
$$

$$
+_* \left| \int_{x_1}^x g(t,z(t)) \cdot_* d_*t \right|_*
$$

$$
\leq |y_0 -_* y_1|_* +_* \left| \int_{*x_0}^x |f(t,y(t)) -_* f(t,z(t))| \cdot_* d_*t \right|_*
$$

$$
+_* \left| \int_{x_1}^{x_0} |f(t,z(t))| \cdot_* d_*t \right|_*
$$

$$
+_* \left| \int_{x_1}^x |g(t,z(t))| \cdot_* d_*t \right|_*
$$

$$
\leq |y_0 -_* y_1|_* +_* e^L \cdot_* \left| \int_{*x_0}^x |y(t) -_* z(t)| \cdot_* d_*t \right|_*
$$

$$
+_* e^M \cdot_* |x_0 -_* x_1|_*
$$

$$
+_* \left| \int_{x_1}^{x_0} |g(t,z(t))|_* \cdot_* d_*t \right|_*
$$

$$+_* \left| \int_{*x_0}^{x} |g(t,z(t))|_* \cdot_* d_* t \right|_*$$

$$\leq \ |y_1 -_* y_0|_* +_* e^L \cdot_* \left| \int_{*x_0}^{x} |y(t) -_* z(t)| \cdot_* d_* t \right|_*$$

$$+_* e^M \cdot_* |x_0 -_* x_1|_* +_* e^{M_1} \cdot_* |x_0 -_* x_1|_* +_* e^{M_1} \cdot_* |x -_* x_0|_*$$

$$= \ |y_1 -_* y_0|_* +_* e^{M+M_1} \cdot_* |x_0 -_* x_1|_*$$

$$+_* e^{M_1} \cdot_* |x -_* x_0|_* +_* e^L \cdot_* \left| \int_{*x_0}^{x} |y(t) -_* z(t)| \cdot_* d_* t \right|_*.$$

Now, we apply the inequality (5.12) for

$$c_0 \ = \ |y_1 -_* y_0|_* +_* e^{M +_* M_1} \cdot_* |x_0 -_* x_1|_*,$$

$$c_1 \ = \ M_1,$$

$$c_2 \ = \ L.$$

and we get the inequality (5.34). This completes the proof.

Exercise 5.5.3 *Let $f(x,y)$ be continuous function in the domain D containing the points (x_0, y_0) and (x_1, y_1), $|f(x,y)|_* \leq e^M$ for all $(x,y) \in D$,*

$$|f(x,y) -_* f(x,z)|_* \leq e^L \cdot_* |y -_* z|_* \qquad \text{for} \qquad \text{all} \qquad (x,y),(x,z) \in D.$$

Let also, ε be a constant, y be a solution to the problem (5.1), (5.2), z be a solution to the problem

$$z^* \ = \ f(x,z) +_* e^\varepsilon,$$

$$z(x_0) \ = \ y_0,$$

which exists in an interval J. Prove that

$$|y(x) -_* z(x)| \leq e^{\frac{|\varepsilon|}{L}} \cdot_* e^{e^L \cdot_* |x -_* x_0|_*} -_* e^{\frac{|\varepsilon|}{L}} \qquad \text{for} \qquad x \in J.$$

Example 5.5.4 *Let y be a solution to the problem*

$$y^* = \cos_*(x \cdot_* y) +_* e,$$

$$y(1) = e^2,$$

z be a solution to the problem

$$z^* = e,$$

$$z(e) = e^3,$$

in

$$\bar{S} = \{(x,y) \in \mathbb{R}_*^2 : |x|_* \leq e^2, |y|_* \leq e^3\}.$$

Here

$$f(x,y) = \cos_*(x \cdot_* y) +_* e,$$

$$x_0 = 1,$$

$$x_1 = 1.$$

We will find the function g. We have

$$f(x,y) +_* g(x,y) = e,$$

whereupon

$$\cos_*(x \cdot_* y) +_* e +_* g(x,y) = e,$$

or

$$g(x,y) = -_* \cos_*(x \cdot_* y).$$

Hence,

$$|f(x,y)|_* = |\cos_*(x \cdot_* y) +_* e|_*$$

$$\leq \ |\cos_*(x \cdot_* y)|_* +_* e$$

$$\leq \ e^2,$$

$$|g(x,y)|_* \ = \ |-_* \cos_*(x \cdot_* y)|_*$$

$$\leq \ e,$$

i.e.,

$$M \ = \ e^2 2,$$

$$M_1 \ = \ e.$$

Also, for $(x,y),(x,z) \in \overline{S}$, we have

$$|f(x,y) -_* f(x,z)|_* \ = \ |\cos_*(x \cdot_* y) -_* \cos_*(x \cdot_* z)|_*$$

$$= \ |x|_* \cdot_* |\sin_*(\xi)|_* \cdot_* |y -_* z|_*$$

$$\leq \ e^2 \cdot_* |y -_* z|_*,$$

where ξ is between $x \cdot_* y$ and $x \cdot_* z$. Consequently, $L = e^2$. Then, using (5.34) we obtain

$$|y(x) -_* z(x)|_* \ \leq \ \left(e +_* e^3 +_* e^{\frac{1}{2}} \right) \cdot_* e^{e^2 \cdot_* |x -_* x_0|_*}$$

$$-_* e^{\frac{1}{2}}$$

$$= \ e^{\frac{9}{2}} \cdot_* e^{e^2 \cdot_* |x -_* x_0|_*} -_* e^{\frac{1}{2}}.$$

Example 5.5.5 *We consider the multiplicative initial value problem*

$$y^* \ = \ x +_* e^x \cdot_* \sin_*(x \cdot_* y),$$

$$y(1) \;=\; 1$$

in

$$\bar{S} = \{(x,y) \in \mathbb{R}_*^2 : |x|_* \le e, |y|_* \le e\}.$$

Here

$$f(x,y) = x +_* e^x \cdot_* \sin_*(x \cdot_* y).$$

Then

$$
\begin{aligned}
|f(x,y)|_* \;&=\; |x +_* e^x \cdot_* \sin_*(x \cdot_* y)|_* \\[4pt]
&\le\; |x|_* +_* e^{|x|_*} |\sin_*(x \cdot_* y)|_* \\[4pt]
&\le\; e +_* e^e \qquad in \qquad \bar{S},
\end{aligned}
$$

i.e.,

$$M = e^{1+e}.$$

Also, for $(x,y), x(,z) \in \bar{S}$, *we have*

$$|f(x,y) -_* f(x,z)|_* +_* |x +_* e^x \cdot_* \sin_*(x \cdot_* y) -_* x -_* e^x \cdot_* \sin_*(x \cdot_* z)|_*$$

$$= |e^x \cdot_* (\sin_*(x \cdot_* y) -_* \sin_*(x \cdot_* z))|_*$$

$$\le e^{|x|_*} \cdot_* |\sin_*(x \cdot_* y) -_* \sin_*(x \cdot_* z)|_*$$

$$\le e^e \cdot_* |x \cdot_* y -_* x \cdot_* z|_*$$

$$= e^e \cdot_* |x|_* |y -_* z|_*$$

$$\le e^e \cdot_* |y -_* z|_*,$$

i.e., $L = e$. *Let* y *be a solution of the considered problem. It is defined in*

$$|x|_* \le e^{\frac{1}{1+e}}.$$

If the initial data $y_0 = 1$ is perturbed by $e^{0.01}$, i.e.,

$$y_1 \;=\; y_0 +_* e^{0.01}$$

$$=\; e^{0.01},$$

and z is a solution of the considered equation for which $z(1) = e^{0.01}$, then we have

$$x_1 \;=\; x_0,$$

$$g(x,y) \;=\; 1,$$

$$M_1 \;=\; 1,$$

and using the inequality (5.34), we get

$$|y(x) -_* z(x)|_* \le e^{0.01} e^{e|x|_*} \qquad \text{in} \qquad |x|_* \le e^{\frac{1}{1+e}}.$$

Example 5.5.6 *Now we consider the initial value problem*

$$y^* \;=\; x^{3*} +_* e^4 \cdot_* (\cos_*(x \cdot_* y))^{2*},$$

$$y(1) \;=\; 1$$

in

$$\bar{S} = \{(x,y) \in \mathbb{R}_*^2 : |x|_* \le e, |y|_* \le e\}.$$

Here

$$f(x,y) \;=\; x^{3*} +_* e^4 \cdot_* (\cos_*(x \cdot_* y))^{2*}$$

$$x_0 \;=\; y_0$$

$$=\; 1.$$

Then

$$|f(x,y)|_* \;=\; |x^{3*} +_* e^4 \cdot_* (\cos_*(x \cdot_* y))^{2*}|_*$$

$$\leq \; |x|^{3*}_* +_* e^4 \cdot_* |(\cos_*(x \cdot_* y))^{2*}|_*$$

$$\leq \; e +_* e^4$$

$$= \; e^5,$$

and

$$|f(x,y_1) -_* f(x,y_2)|_* +_* x^{3*} +_* e^4 \cdot_* (\cos_*(x \cdot_* y_1))^{2*}$$

$$-_* x^{3*} -_* e^4 \cdot_* (\cos_*(x \cdot_* y_2))^{2*}|_*$$

$$= e^4 \cdot_* |(\cos_*(x \cdot_* y_1))^{2*} -_* (\cos_*(x \cdot_* y_2))^{2*}|_*$$

$$= e^8 \cdot_* |x|_* |\cos_*(\xi)|_* \cdot_* |y_1 -_* y_2|_*$$

$$\leq e^8 \cdot_* |y_1 -_* y_2|_*,$$

where ξ is between $x \cdot_* y_1$ and $x \cdot_* y_2$, for (x,y), (x,y_1), $(x,y_2) \in \overline{S}$, i.e.,

$$M \; = \; 5,$$

$$L \; = \; 8.$$

Let y be a solution to the considered problem. Then it is defined in $|x|_* \leq e^{\frac{1}{5}}$. Now we will suppose that y_0 is perturbed by $e^{0.1}$ and $f(x,y)$ is perturbed by 1 in $\overline{S}$, i.e., $y_1 = e^{0.1}$ and $g(x,y) = e$ in $\overline{S}$. Let z be a solution of the problem

$$z^* \; = \; x^{3*} +_* e^4 \cdot_* (\cos_*(x \cdot_* z))^{2*} +_* e,$$

$$z(1) \; = \; e^{0.1}.$$

Here

$$M_1 \; = \; 1,$$

$$x_1 = x_0$$

$$= 1.$$

We note that z exists in $|x|_* < e^{\frac{1}{6}}$. *Hence, using the estimate* (5.34), *we obtain*

$$|y(x) -_* z(x)|_* \leq \left(e^{0.1} +_* e^{\frac{1}{8}}\right) \cdot_* e^{e^8 \cdot_* |x|_*} -_* e^{\frac{1}{8}}$$

$$= e^{\frac{9}{40}} \cdot_* e^{e^8} \cdot_* |x|_* -_* e^{\frac{1}{8}} \qquad \text{in} \qquad |x| \leq \frac{1}{6}.$$

Exercise 5.5.7 *For the multiplicative initial value problem*

$$y^* = e^3 \cdot_* x +_* e^2 \cdot_* \cos_* (x \cdot_* y),$$

$$y(1) = 1,$$

estimate the variation of the solution in the interval $[1, e]$ *if* y_0 *is perturbed by* $e^{0.1}$.

Answer

$$e^{0.1} \cdot_* e^{e^2 \cdot_* x}.$$

Theorem 5.5.8 *Let f be continuous function in the domain D containing the point* (x_0, y_0),

$$|f(x,y)|_* \leq e^M$$

in D, $\partial_* f /_* \partial_* y$ *exists, continuous in D and*

$$|\partial_* f(x,y) /_* \partial_* y|_* \leq e^L$$

in D. Let also $y(x, x_0, y_0)$ *be a solution to the problem* (5.1), (5.2) *which exists in an interval J containing* x_0. *Then* $y(x, x_0, y_0)$ *is multiplicative differentiable with respect to* y_0 *and* $z(x) = \partial_* /_* \partial_* y_0 y(x, x_0, y_0)$ *is the solution of the multiplicative initial value problem*

$$z^* = (\partial_* f /_* \partial_* y(x, y(x, x_0, y_0))) \cdot_* z$$

$$z(x_0) = e.$$

Proof 5.5.9 *Let $(x_0, y_1) \in D$ and $y(x, x_0, y_1)$ be a solution to the multiplicative initial value problem*

$$y^* = f(x, y),$$

$$y(x_0) = y_1,$$

which exists in an interval J_1. Let also, $J_2 = J \bigcap J_1$. We have that $J_2 \neq \emptyset$ because $x_0 \in J$ and $x_0 \in J_1$. Hence and the inequality (5.34) we have that

$$|y(x, x_0, y_0) -_* y(x, x_0, y_1)|_* \leq |y_0 -_* y_1|_* e^{e^{L_* *} |_* x -_* x_0|_*},$$

whereupon

$$|y(x, x_0, y_0) -_* y(x, x_0, y_1)|_* \to 1$$

as $|y_0 -_* y_1|_* \to 1$. Then for $x \in J_2$ we have

$$y(x, x_0, y_0) -_* y(x, x_0, y_1) -_* z(x) \cdot_* (y_0 -_* y_1)$$

$$= \int_{*x_0}^{x} \left(f(t, y(t, x_0, y_0)) -_* f(t, y(t, x_0, y_1)) \right.$$

$$\left. -_* \partial_* f /_* \partial_* y(t, y(t, x_0, y_0)) \cdot_* z(t) \cdot_* (y_0 -_* y_1) \right) \cdot_* d_* t$$

$$= \int_{*x_0}^{x} \partial_* f /_* \partial_* y(t, y(t, x_0, y_0)) \cdot_* \left(y(t, x_0, y_0) -_* y(t, x_0, y_1) \right.$$

$$\left. -_* z(t) \cdot_* (y_0 -_* y_1) \right) \cdot_* d_* t$$

$$+_* \int_{*x_0}^{x} \delta \{ y(t, x_0, y_0), y(t, x_0, y_1) \} \cdot_* d_* t,$$

where

$$\delta \{ y(t, x_0, y_0), y(t, x_0, y_1) \} \to 1$$

as

$$|y(t, x_0, y_0) -_* y(t, x_0, y_1)|_* \to 1,$$

i.e., as $|y_0 -_* y_1|_* \to 1$. *Hence,*

$$|y(x,x_0,y_0) -_* y(x,x_0,y_1) -_* z(x)(y_0 -_* y_1)|_*$$

$$\leq \left| \int_{*x_0}^x \partial_* f /_* \partial_* y(t,y(t,x_0,y_0)) \cdot_* \left(y(t,x_0,y_0) -_* y(t,x_0,y_1) \right. \right.$$

$$\left. \left. -_* z(t) \cdot_* (y_0 -_* y_1) \right) \cdot_* d_* t \right|_*$$

$$+_* \left| \int_{*x_0}^x \delta \{ y(t,x_0,y_0), y(t,x_0,y_1) \} \cdot_* d_* t \right|_*$$

$$\leq \left| \int_{*x_0}^x \left| \partial_* f /_* \partial_* y(t,y(t,x_0,y_0)) \right| \right.$$

$$\left. \cdot_* |y(t,x_0,y_0) -_* y(t,x_0,y_1) -_* z(t)(y_0 -_* y_1)|_* \cdot_* d_* t \right|_*$$

$$+_* o_* (|y_0 -_* y_1|_*)$$

$$\leq e^L \cdot_* \left| \int_{*x_0}^x |y(t,x_0,y_0) -_* y(t,x_0,y_1) -_* z(t)(y_0 -_* y_1)|_* \cdot_* d_* t \right|_*$$

$$+_* o_* (|y_0 -_* y_1|_*).$$

From here and from the Gronwall type inequality, we get that

$$|y(x,x_0,y_0) -_* y(x,x_0,y_1) -_* z(x) \cdot_* (y_0 -_* y_1)|_*$$

$$\leq o_* (|y_0 -_* y_1|_*) \cdot_* e^{e^L \cdot_* |x -_* x_0|_*}.$$

This completes the proof.

Example 5.5.10 *We will consider the following initial value problem:*

$$x \cdot_* y^* -_* e^2 \cdot_* y \; = \; e^2 \cdot_* x^{4*}, \quad x > e,$$

$$y(e) \; = \; e^2 = y_0.$$

We will find the general solution of the equation

$$y^* = (e^2/_*x) \cdot_* y +_* e^2 \cdot_* x^{3*}, \quad x > e,$$

which is a multiplicative linear first-order differential equation. Here

$$a(x) \;=\; -_*(e^2/_*x),$$

$$b(x) \;=\; e^2 \cdot_* x^{3*}, \quad x > e.$$

Then

$$\int_* a(x) \cdot_* d_*x \;=\; -_* e^2 \cdot_* \int_* (e/_*x) \cdot_* d_*x$$

$$=\; \log_*(e/_*x^{2*}),$$

and

$$\int_* b(x) \cdot_* e^{\int_* a(x) \cdot_* d_*x} \cdot_* d_*x \;=\; e^2 \cdot_* \int_* x^{3*} \cdot_* e^{\log_*(e/_*x^{2*})} \cdot_* d_*x$$

$$=\; e^2 \cdot_* \int_* x \cdot_* d_*x$$

$$=\; x^{2*}, \quad x > e.$$

Hence,

$$y(x) \;=\; e^{-_* \log_*(e/_*x^{2*})} \cdot_* (c +_* x^{2*})$$

$$=\; x^{2*} \cdot_* (c +_* x^2), \quad x > e,$$

and

$$y(e) = e^2.$$

Therefore,

$$c +_* e = e^2,$$

whereupon

$$c = e$$

and

$$y(x) = x^{2*} +_* x^{4*}, \quad x > e.$$

Now, we will find

$$z(x) = (\partial_*/_*\partial_* y_0)y(x, x_0, y_0).$$

Here

$$f(x,y) = e^2/_* x +_* e^2 \cdot_* x^{3*}.$$

Then

$$(\partial_* f/_*\partial_* y)(x,y) = e^2/_* x.$$

Therefore, z satisfies the multiplicative initial value problem

$$z^* = (e^2/_* x) \cdot_* z, \quad x > e.$$

$$z(e) = e.$$

From here,

$$d_* z/_* z = (e^2/_* x) \cdot_* d_* x$$

or

$$\int_* d_* z/_* z = \int_* e^2/_* x \cdot_* d_* x +_* c,$$

or

$$\log_* |z|_* = \log_*(x^{2*}) +_*,$$

or

$$z = c \cdot_* x^{2*}$$

and

$$z(e) = c$$

$$= e,$$

and

$$z(x) = x^{2*}.$$

Example 5.5.11 *Now we consider the multiplicative initial value problem*

$$y^* +_* e^2 \cdot_* y \;=\; y^{2*} \cdot_* e^x, \quad x > 1,$$

$$y(1) \;=\; e.$$

Firstly, we will find the general solution of the considered equation. For this aim we divide it by $e/_* y^{2*}$ and we get

$$(e/_* y^{2*}) \cdot_* y^* = -_*(e^2/_* y) +_* e^x, \quad x > 1.$$

We set

$$l = e/_* y.$$

Then

$$l' = -_*(e/_* y^{2*}) \cdot_* y^*$$

and

$$-_* l^* = -_* e^2 \cdot_* l +_* e^x, \quad x > 1,$$

or

$$l^* = e^2 \cdot_* l -_* e^x, \quad x > 1,$$

from where

$$
\begin{aligned}
l(x) \;&=\; e^{e^2 \cdot_* \int_* d_* x} \cdot_* \left(c -_* \int_* e^x \cdot_* e^{-_* e^2 \cdot_* x} \cdot_* d_* x \right) \\
&=\; e^{e^2 \cdot_* x} \cdot_* \left(c -_* \int_* e^{-_* x} \cdot_* d_* x \right) \\
&=\; e^{e^2 \cdot_* x} \cdot_* (c +_* e^{-_* x}), \quad x > 1.
\end{aligned}
$$

Therefore

$$y(x) = e/_* (e^{e^2 \cdot_* x} \cdot_* (c +_* e^{-_* x}))$$

and

$$y(1) \;=\; e/_* c +_* e$$

$$=\; e,$$

from where

$$c = 1$$

and

$$y(x, x_0, y_0) \quad = \quad e/_* e^x$$

$$= \quad e^{-_* x}.$$

Now, we will find

$$z(x) = (\partial_* /_* \partial_* y_0) y(x, x_0, y_0).$$

Here

$$f(x, y) = -_* e^2 \cdot_* y +_* y^{2*} \cdot_* e^x.$$

Then

$$(\partial_* /_* \partial_* y) f(x, y) \quad = \quad -_* e^2 +_* e^2 \cdot_* y \cdot_* e^x$$

and

$$(\partial_* /_* \partial_* y) f(x, y(x, x_0, y_0)) \quad = \quad -_* e^2 +_* (e^2 /_* e^x) \cdot_* e^x$$

$$= \quad 1.$$

From here,

$$z^* \quad = \quad 1,$$

$$z(1) \quad = \quad e.$$

Consequently,

$$z(x) = 1, \quad x \geq 1.$$

Example 5.5.12 *Now we consider the multiplicative initial value problem*

$$y^* \quad = \quad y +_* y^{2*} +_* x \cdot_* y^{3*}, \quad x > e^2,$$
$$y(e^2) \quad = \quad 1 = y_0.$$

We note that $y(x, x_0, y_0) = 1$ is its solution. Now we will find

$$z(x) = \partial_* / {}_* \partial_* y_0 y(x, x_0, y_0).$$

Here

$$f(x, y) = y +_* y^{2*} +_* x \cdot_* y^{3*}.$$

Then

$$(\partial_* / {}_* \partial_* y) f(x, y) = e +_* e^2 \cdot_* y +_* e^3 \cdot_* x \cdot_* y^{2*}$$

and

$$(\partial_* / {}_* \partial_* y) f(x, y(x, x_0, y_0)) = e.$$

Therefore, z satisfies the multiplicative initial value problem

$$z^* \;=\; z,$$

$$z(e^2) \;=\; e.$$

We have

$$dz / {}_* z = dx$$

and

$$\int_* d_* z / {}_* z = \int_* d_* x +_* x,$$

and

$$\log_* |z|_* = x +_* c,$$

and

$$z = c \cdot_* e^x,$$

and

$$z(e^2) \;=\; c \cdot_* e^2$$

$$\;=\; e,$$

and

$$c = e^{-*2},$$

and

$$z(x) = e^{x -_* e^2}.$$

Exercise 5.5.13 *Find*

$$(\partial_*/_*\partial_*y)y(x,x_0,y_0)$$

of the following multiplicative initial value problems

1.

$$y^* = x\cdot_*y +_* y^{4*} +_* y^{3*} -_* x\cdot_* y^{5*},$$

$$y(1) = 1.$$

2.

$$y^* = x\cdot_* \sin_* y +_* x^{2*}\cdot_* y +_* y^{4*},$$

$$y(1) = 1.$$

3.

$$y^* = (x +_* \sin_* x)\cdot_* y +_* y^{4*},$$

$$y(e^2) = 1.$$

Answer

1.

$$e^{x^{2*}/_* e^2}.$$

2.

$$e^{x^{2*}/_* e^2} +_* (x^{3*}/_* e^3).$$

3.

$$e^{(x^{2*}/_* e^2) +_* \cos_* x -_* e^2 -_* \cos_* 2}.$$

Theorem 5.5.14 *Let $f(x,y)$ be continuous in the domain D containing the point (x_0,y_0), $|f(x,y)|_* \le e^M$ in D, $(\partial_* f(x,y)/_*\partial_* y)$ exists in D,*

$$|(\partial_* f(x,y)/_*\partial_* y)|_* \le e^L$$

in D. Let also $y(x,x_0,y_0)$ be a solution to the problem (5.1), (5.2) which exists in an interval J containing the point x_0. Then

$y(x,x_0,y_0)$ *is multiplicative differentiable with respect to x_0 and* $z(x) = (\partial_* y/_* \partial_* x_0)(x,x_0,y_0)$ *is the solution of the initial value problem*

$$z^* = (\partial_* f/_* \partial_* y)(x,y(x,x_0,y_0)) \cdot_* z$$

$$z(x_0) = -_* f(x_0,y_0).$$

Proof 5.5.15 *Let $(x_1,y_0) \in D$, $x_1 \in J$, and $y(x,x_1,y_0)$ be a solution to the problem*

$$y^* = f(x,y),$$

$$y(x_1) = y_0,$$

which exists in an interval J_1. We set $J_2 = J \bigcap J_1$. We note that $J_2 \neq \emptyset$ because $x_1 \in J$ and $x_1 \in J_1$. Using the inequality (5.34) we have

$$|y(x,x_0,y_0) -_* y(x,x_1,y_0)|_* \leq e^M \cdot_* |x_1 -_* x_0|_* \cdot_* e^{e^L \cdot_* |x -_* x_0|_*},$$

therefore

$$|y(x,x_0,y_0) -_* y(x,x_1,y_0)|_* \to 1$$

as $|x_1 -_* x_0|_* \to 1$. For $x \in J_2$ we have

$$y(x,x_0,y_0) -_* y(x,x_1,y_0) -_* z(x) \cdot_* (x_0 -_* x_1)$$

$$= y_0 +_* \int_{*x_0}^x f(t,y(t,x_0,y_0)) \cdot_* d_* t$$

$$-_* y_0 -_* \int_{x_1}^x f(t,y(t,x_1,y_0)) \cdot_* d_* t$$

$$-_* \int_{*x_0}^x (\partial_* f/_* \partial_* y)(t,t(,x_0,y_0)) \cdot_* z(t) \cdot_* (x_0 -_* x_1) \cdot_* d_* t$$

$$+_* f(x_0,y_0) \cdot_* (x_0 -_* x_1)$$

and

$$\int_{*x_0}^x \left(f(t,y(t,x_0,y_0)) -_* f(t,y(t,x_1,y_0)) \right)$$

$$-_*(\partial_*f/_*\partial_*y)(t,y(t,x_0,y_0))\cdot_*z(t)\cdot_*(x_0-_*x_1)\Big)\cdot_*d_*t$$

$$-_*\int_{x_1}^{x_0}f(t,y(t,x_1,y_0))\cdot_*d_*t+_*f(x_0,y_0)\cdot_*(x_0-_*x_1)$$

$$=\int_{*x_0}^{x}(\partial_*f/_*\partial_*y)(t,y(t,x_0,y_0))\cdot_*\Big(y(t,x_0,y_0)$$

$$-_*y(t,x_1,y_0)-_*z(t)\cdot_*(x_0-_*x_1)\Big)\cdot_*d_*t$$

$$+_*\int_{*x_0}^{x}\delta\{y(t,x_0,y_0),y(t,x_1,y_0)\}\cdot_*d_*t$$

$$-_*\int_{x_1}^{x_0}f(t,y(t,x_1,y_0))\cdot_*d_*t+_*f(x_0,y_0)\cdot_*(x_0-_*x_1),$$

where

$$\delta\{y(t,x_0,y_0),y(t,x_1,y_0)\}\to1$$

as

$$|y(t,x_0,y_0)-_*y(t,x_1,y_0)|_*\to1,\ \text{i.e., as } |x_1-_*x_0|_*\to1.$$

Hence,

$$|y(x,x_0,y_0)-_*y(x,x_1,y_0)-_*z(x)\cdot_*(x_0-_*x_1)|_*$$

$$\leq\left|\int_{*x_0}^{x}\left|(\partial_*f/_*\partial_*y)(t,y(t,x_0,y_0))\right|_*\right.$$

$$\left.\cdot_*|y(t,x_0,y_0)-_*y(t,x_1,y_0)-_*z(t)\cdot_*(x_0-_*x_1)|_*\cdot_*d_*t\right|_*$$

$$+_*\left|\int_{*x_0}^{x}\delta\{y(t,x_0,y_0),y(t,x_1,y_0)\}\cdot_*d_*t\right|_*$$

$$+_*\left|\int_{x_1}^{x_0}|f(t,y(t,x_1,y_0))|_*\cdot_*d_*t\right|_*+_*|f(x_0,y_0)|_*\cdot_*|x_0-_*x_1|_*$$

$$\leq\ e^L \cdot_* \left| \int_{*x_0}^{*x} |y(t,x_0,y_0) -_* y(t,x_1,y_0) -_* z(t) \cdot_* (x_0 -_* x_1)|_* \cdot_* d_*t \right|_*$$

$$+_* o(|x_0 -_* x_1|_*) +_* e^{2L} \cdot_* |x_0 -_* x_1|_*$$

$$=\ e^L \cdot_* \left| \int_{*x_0}^{*x} |y(t,x_0,y_0) -_* y(t,x_1,y_0) -_* z(t) \cdot_* (x_0 -_* x_1)|_* \cdot_* d_*t \right|_*$$

$$+_* o(|x_0 -_* x_1|_*).$$

From the last inequality and from the Gronwall type inequality, we get

$$|y(x,x_0,y_0) -_* y(x,x_1,y_0) -_* z(x) \cdot_* (x_0 -_* x_1)|_*$$

$$\leq\ o(|x_0 -_* x_1|_*) \cdot_* e^{e^L \cdot_* |x_0 -_* x_1|_*}, \qquad x \in J_2,$$

whereupon

$$|y(x,x_0,y_0) -_* y(x,x_1,y_0) -_* z(x) \cdot_* (x_0 -_* x_1)|_* \to 1$$

as $|x_0 -_* x_1|_* \to 1$, $x \in J_2$. *This completes the proof.*

Example 5.5.16 *We consider the multiplicative initial value problem*

$$y^* \ =\ \sin_* x \cdot_* y +_* x \cdot_* y^{2*},$$

$$y(1) \ =\ 1.$$

Here

$$x_0 \ =\ y_1$$

$$=\ 1,$$

$$y(x,x_0,y_0) \ =\ 1$$

and

$$f(x,y) \ =\ \sin_* x \cdot_* y +_* x \cdot_* y^{2*},$$

$$(\partial_* f /_* \partial_* y)(x, y) \quad = \quad \sin_* x +_* e^2 \cdot_* x \cdot_* y,$$

$$f(x_0, y_0) \quad = \quad 1,$$

$$(\partial_* f /_* \partial_* y)(x, y(x, x_0, y_0)) \quad = \quad \sin_* x.$$

Therefore,

$$z(x) = (\partial_* y /_* \partial_* x_0)(x, x_0, y_0)$$

satisfies the initial value problem

$$z^* \quad = \quad \sin_* x \cdot_* z,$$

$$z(1) \quad = \quad 1.$$

Hence,

$$(\partial_* y /_* \partial_* x_0)(x, x_0, y_0) = 1.$$

Example 5.5.17 *We consider the multiplicative initial value problem*

$$y^* \quad = \quad y +_* 1,$$

$$y(1) \quad = \quad 1.$$

We will find its solution. We have

$$y(x) \quad = \quad e^x \cdot_* \left(c +_* \int_* e^{-*x} \cdot_* d_* x \right)$$

$$= \quad e^x \cdot_* \left(c -_* e^{-*x} \right)$$

$$= \quad c \cdot_* e^x -_* e,$$

and

$$y(0) = c -_* e = e,$$

and

$$c = e^2,$$

and

$$y(x, x_0, y_0) = e^2 \cdot_* e^x -_* e,$$

where $x_0 = 1, y_0 = e$. Also,

$$f(x, y) = y +_* e,$$

$$(\partial_* f /_* \partial_* y)(x, y) = e,$$

$$f(x_0, y_0) = e^2.$$

Consequently,

$$z(x) = (\partial_* y /_* \partial_* x_0)(x, x_0, y_0)$$

satisfies the multiplicative initial value problem

$$z^* = z,$$

$$z(1) = -_* e^2.$$

Therefore,

$$z(x) = c \cdot_* e^x,$$

$$z(1) = c$$

$$= -_* e^2,$$

i.e.,

$$(\partial_* y /_* \partial_* x_0)(x, x_0, y_0) = -_* e^2 \cdot_* e^x.$$

Example 5.5.18 *We consider the multiplicative initial value problem*

$$x \cdot_* y^{2*} \cdot_* y^* = x^{2*} +_* y^{3*},$$

$$y(e) = e.$$

We set

$$y_*^3 = g.$$

Then

$$e^3 \cdot_* y^{2*} \cdot_* y^* = g^*,$$

or

$$y^{2*} \cdot_* y^* = (g^*/_* e^3).$$

Therefore,

$$x \cdot_* (g^*/_* e^3) = x^{2*} +_* g,$$

or

$$g^* = e^3 \cdot_* x +_* (e^3/_* x) \cdot_* g.$$

Hence,

$$
\begin{aligned}
g(x) &= x^{3*} \cdot_* \left(c +_* e^3 \cdot_* \int_* (x/_* x^{3*}) \cdot_* d_* x \right) \\
&= x^{3*} \cdot_* \left(c -_* (e^3/_* x) \right) \\
&= c \cdot_* x^{3*} -_* e^3 \cdot_* x^{2*},
\end{aligned}
$$

or

$$y^{3*} = c \cdot_* x^{3*} -_* x^{2*},$$

whereupon

$$y = (c \cdot_* x^{3*} -_* x^{2*})^{\frac{1}{3}*},$$

and

$$
\begin{aligned}
y(e) &= (c -_* e)^{\frac{1}{3}*} \\
&= e,
\end{aligned}
$$

and

$$c = e^2,$$

and

$$y(x) = (e^2 \cdot_* x^{3*} -_* x^{2*})^{\frac{1}{3}*}$$

$$= \; y(x,x_0,y_0).$$

Here

$$x_0 \; = \; y_0$$

$$= \; e$$

and

$$f(x,y) \; = \; ((x^{2*} +_* y^{3*})/_*(x\cdot_* y^{2*}))$$

$$= \; (x/_* y^{2*}) +_* (y/_* x),$$

$$f(x_0,y_0) \; = \; e +_* e$$

$$= \; e^2,$$

and

$$(\partial_* f/_*\partial_* y)(x,y) \; = \; -_* e^2 \cdot_* (x/_* y^{3*}) +_* (e/_* x),$$

$$(\partial_* f/_*\partial_* y)(x,y(x,x_0,y_0)) \; = \; -_*((e^2 \cdot_* x)/_*(e^2 \cdot_* x^{3*} -_* x^{2*})) +_* (e/_* x)$$

$$= \; -_*(e^2/_*(e^2 \cdot_* x^{2*} -_* x)) +_* (e/_* x)$$

$$= \; -_*(e^2/_*(x\cdot_* (e^2 \cdot_* x -_* e))) +_* (e/_* x)$$

$$= \; -_*\left((e^2/_*(e^2 \cdot_* x -_* e)) -_* (e/_* x)\right)$$

$$+_*(e/_* x)$$

$$= \; -_*(e^2/_*(e^2 \cdot_* x -_* e)) +_* (e^2/_* x).$$

Consequently,

$$z(x) = (\partial_* y/_*\partial_* x_0)(x,x_0,y_0)$$

satisfies the multiplicative initial value problem

$$z^* = \left(-_*(e^2/_*(e^2 \cdot_* x -_* e)) +_* (e^2/_*x)\right) \cdot_* z(x),$$

$$z(e) = -_*e^2.$$

Then we have

$$d_*z/_*z = \left(-_*(e^2(e^2 \cdot_* x -_* e)) +_* (e^2/_*x)\right) dx$$

and

$$\int_* d_*z/_*z = \int_* \left(-_*(e^2/_*(e^2 \cdot_* x -_* e)) +_* (e^2/_*x)\right) \cdot_* d_*x +_* c,$$

and

$$\log_* |z|_* = -_* \log_* |e^2 \cdot_* x -_* e|_* +_* \log_* x^{2*} +_* c,$$

and

$$z = c \cdot_* (x^{2*}/_*(e^2 \cdot_* x -_* e)),$$

or

$$z \cdot_* (e^2 \cdot_* x -_* e) = c \cdot_* x^{2*},$$

and

$$-_*e^2 \cdot_* (e^2 -_* e) = c,$$

or

$$c = -_*e^2,$$

and

$$z \cdot_* (e^2 \cdot_* x -_* e) = -_*e^2 \cdot_* x^{2*},$$

and

$$e^2 \cdot_* x^{2*} +_* z \cdot_* (e^2 \cdot_* x -_* e) = 0,$$

and

$$z = -_*((e^2 \cdot_* x^{2*})/_*(e^2 \cdot_* x -_* e)).$$

Exercise 5.5.19 *Let* $y(x,x_0,y_0)$ *be the solution of the problem*

$$y^* = \cos_* x \cdot_* y +_* y^{2*} +_* y^{4*},$$

$$y\left(e^{\frac{\pi}{2}}\right) = 1.$$

Find $(\partial_* y/_* \partial_* x_0)(x,x_0,y_0).$

Answer

$$-_*e^x.$$

Theorem 5.5.20 *Let $f(x,y,\lambda)$ be a continuous function in a domain $D_1 \subset \mathbb{R}^3_*$ containing the point (x_0,y_0,λ_0), $|f(x,y,\lambda)|_* \le e^M$ in D_1, $(\partial_* f/_*\partial_* y)(x,y,\lambda)$ and $(\partial_* f/_*\partial_*\lambda)(x,y,\lambda)$ exist and continuous in D_1,*

$$|(\partial_* f/_*\partial_* y)(x,y,\lambda)|_* \le e^L$$

and

$$|(\partial_* f/_*\partial_*\lambda)(x,y,\lambda)|_* \le e^L_1$$

in D_1. Then

1. there exist positive constants h and ε such that for given λ for which $|\lambda -_* \lambda_0|_* \le e^\varepsilon$, there exists a unique solution $y(x,\lambda)$ of the initial value problem

$$y^* = f(x,y,\lambda),$$

$$y(x_0) = y_0,$$

(5.35)

in the interval $|x -_* x_0|_* \le e^h$,

2. for every λ_1 and λ_2 in the interval $|\lambda -_* \lambda_0|_* \le e^\varepsilon$, and x in $|x -_* x_0|_* \le e^h$, the following inequality holds

$$|y(x,\lambda_1) -_* y(x,\lambda_2)|_* \le ((e^{L_1} \cdot_* |\lambda_1 -_* \lambda_2|_*)/_*e^L) \cdot_* \left(e^{e^L \cdot_* |x -_* x_0|_*} -_* e\right),$$

(5.36)

3. the solution $y(x,\lambda)$ is multiplicative differentiable with respect to λ and $z(x,\lambda) = (\partial_* y/_*\partial_*\lambda)(x,\lambda)$ is the solution of the initial value problem

$$z^*(x,\lambda) = (\partial_* f/_*\partial_* y)(x,y(x,\lambda),\lambda) \cdot_* z(x,\lambda)$$

$$+_* (\partial_* f/_*\partial_*\lambda)(x,y(x,\lambda),\lambda),$$

$$z(x_0,\lambda) = 1.$$

Proof 5.5.21   *1. We note that*

$$|f(x,y_1,\lambda) -_* f(x,y_2,\lambda)|_* \le e^L \cdot_* |y_1 -_* y_2|_*,$$

and

$$|f(x,y,\lambda_1) -_* f(x,y,\lambda_2)|_* \le e^{L_1} \cdot_* |\lambda_1 -_* \lambda_2|_*$$

for every $(x,y,\lambda_1), (x,y,\lambda_2), (x,y_1,\lambda), (x,y_2,\lambda) \in D_1$. Also, every solution $y(x,\lambda)$ of the problem (5.35) satisfies the integral equation

$$y(x,\lambda) = y_0 +_* \int_{*x_0}^x f(t,y(t,\lambda)) \cdot_* d_* t$$

and conversely. Let $a > 0$, $b > 0$, $\varepsilon > 0$ be such that

$$\bar{S}_\lambda = \left\{ (x,y,\lambda) \in \mathbb{R}^3 : |x -_* x_0|_* \le e^a, \quad |y -_* y_0|_* \le e^b, \right.$$

$$\left. |\lambda -_* \lambda_0|_* \le e^\varepsilon \right\} \subset D_1.$$

Let

$$y_0(x,\lambda) = y_0,$$

$$y_m(x,\lambda) = y_0 +_* \int_{*x_0}^x f(t,y_{m-*1}(t,\lambda),\lambda) \cdot_* d_* t, \quad m = 1,2,\dots.$$

Let also,

$$J_{h,\lambda} = \left\{ |x -_* x_0|_* \le e^h = \min\left\{ a, \frac{b}{M} \right\}, \right.$$

$$\left. |\lambda -_* \lambda_0|_* \le e^\varepsilon \right\}.$$

As in the Local Existence Theorem, we have that the sequence $\{y_m(x,\lambda)\}_{m=1}^\infty$ exists as continuous functions in $J_{h,\lambda}$ and $(x,y_m(x,\lambda),\lambda) \in \bar{S}_\lambda$ for $(x,\lambda) \in J_{h,\lambda}$, and for $(x,\lambda) \in J_{h,\lambda}$ we have that

$$|y_m(x,\lambda) -_* y_0|_* \le e^b,$$

and $\{y_m(x,\lambda)\}_{m=1}^{\infty}$ *converges uniformly in* $J_{h,\lambda}$ *to a contin-uous function* $y(x,\lambda)$ *in* $J_{h,\lambda}$, *which is the unique solution of the problem (5.35).*

2. *Let* λ_1,λ_2 *be in the interval* $|\lambda -_* \lambda_0|_* \leq e^{\varepsilon}$, *x be in the interval* $|x -_* x_0|_* \leq e^h$. *Then*

$$y(x,\lambda_1) \ = \ -_*y_0 +_* \int_{*x_0}^{x} f(t,y(t,\lambda_1),\lambda_1) \cdot_* d_*t,$$

$$y(x,\lambda_2) \ = \ y_0 +_* \int_{*x_0}^{x} f(t,y(t,\lambda_2),\lambda_2) \cdot_* d_*t,$$

and

$$|y(x,\lambda_1) -_* y(x,\lambda_2)|_* \ = \ \left| \int_{*x_0}^{x} \Big(f(t,y(t,\lambda_1),\lambda_1) \right.$$

$$\left. -_*f(t,y(t,\lambda_2),\lambda_2) \Big) \cdot_* d_*t \right|_*$$

$$\leq \ \left| \int_{*x_0}^{x} \Big| f(t,y(t,\lambda_1),\lambda_1) \right.$$

$$-_*f(t,y(t,\lambda_2),\lambda_1)$$

$$+_*f(t,y(t,\lambda_2),\lambda_1)$$

$$\left. -_*f(t,y(t,\lambda_2),\lambda_2)\Big|_* \cdot_* d_*t \right|_*$$

$$\leq \ \left| \int_{*x_0}^{x} |f(t,y(t,\lambda_1),\lambda_1) \right.$$

$$\left. -_*f(t,y(t,\lambda_2),\lambda_1)|_* \cdot_* d_*t \right|_*$$

$$+_* \left| \int_{*x_0}^{x} |f(t,y(t,\lambda_2),\lambda_1) \right.$$

$$\left. -_*f(t,y(t,\lambda_2),\lambda_2)|_* \cdot_* d_*t \right|$$

$$\leq\ e^{L}\cdot_{*}\left|\int_{*x_0}^{x}|y(t,\lambda_1)-_*y(t,\lambda_2)|\cdot_*d_*t\right|_{*}$$

$$+_*e^{L_1}\cdot_*|x-_*x_0|\cdot_*|\lambda_1-_*\lambda_2|.$$

From the last inequality and from the Gronwall type inequality we get

$$|y(x,\lambda_1)-_*y(x,\lambda_2)|_* \leq ((e^{L_1}\cdot_*|\lambda_1-_*\lambda_2|_*)/_*e^{L})$$

$$\cdot_* e^{e^{L}\cdot_*|x-_*x_0|} -_* ((e^{L_1}\cdot_*|\lambda_1-_*\lambda_2|_*)/_*e^{L}).$$

3. *Let* $(x_0,y_0,\lambda), (x_0,y_0,\lambda_1) \in D_1,\ |\lambda-_*\lambda_0|_* \leq e^{\varepsilon},\ |x-_*$ $x_0|_* \leq e^{h},\ y(x,x_0,y_0,\lambda_1)$ *be the solution of the problem*

$$y^* = f(x,y,\lambda),$$

$$y(x_0) = y_0,$$

which exists in $J_{h,\lambda}$, $y(x,x_0,y_0,\lambda_1)$ *be the solution of the problem*

$$y^* = f(x,y,\lambda_1),$$

$$y(x_0) = y_0,$$

which exists in J_{h,λ_1}. *Let*

$$J_{\lambda,\lambda_1} = J_{h,\lambda}\bigcap J_{h,\lambda_1}.$$

We note that $J_{\lambda,\lambda_1} \neq \emptyset$. *From the inequality* (5.36) *we have*

$$|y(x,x_0,y_0,\lambda)-_*y(x,x_0,y_0,\lambda_1)|_* \leq ((e^{L_1}|_*|\lambda_1-_*\lambda_2|_*)/_*e^{L})$$

$$\cdot_*\left(e^{e^{L}\cdot_*|x-_*x_0|_*} -_* e\right),$$

whereupon

$$|y(x,x_0,y_0,\lambda)-_*y(x,x_0,y_0,\lambda_1)|_* \to 1$$

as $|\lambda -_* \lambda_1|_* \to 1$. *For* $x \in J_{\lambda,\lambda_1}$, *we have*

$$y(x,x_0,y_0,\lambda) -_* y(x,x_0,y_0,\lambda_1) -_* z(x,\lambda)(\lambda -_* \lambda_1)$$

$$= \int_{*x_0}^x \Big(f(t,y(t,\lambda),\lambda) -_* f(t,y(t,\lambda_1),\lambda_1)$$

$$-_* (\partial_* f /_* \partial_* y)(t,y(t,\lambda),\lambda)$$

$$-_* (\partial_* f /_* \partial_* \lambda)(t,y(t,\lambda),\lambda) \Big) \cdot_* d_* t$$

$$= \int_{*x_0}^x \Big(f(t,y(t,\lambda),\lambda) -_* f(t,y(t,\lambda_1),\lambda)$$

$$+_* f(t,y(t,\lambda_1),\lambda)$$

$$-_* f(t,y(t,\lambda_1),\lambda_1)$$

$$-_* (\partial_* f /_* \partial_* y)(t,y(t,\lambda),\lambda) \cdot_* z(t,\lambda) \cdot_* (\lambda -_* \lambda_1)$$

$$-_* (\partial_* f /_* \partial_* \lambda)(t,y(t,\lambda),\lambda) \cdot_* (\lambda -_* \lambda_1) \Big) \cdot_* d_* t$$

$$= \int_{*x_0}^x \Big(\frac{\partial_* f}{\partial_* y}(t,y(t,\lambda),\lambda)$$

$$\cdot_* \Big(y(t,\lambda) -_* y(t,\lambda_1) -_* z(t,\lambda)(\lambda -_* \lambda_1) \Big)$$

$$+_* \Big((\partial_* f /_* \partial_* \lambda)(t,y(t,\lambda_1),\lambda)$$

$$-_* (\partial_* f /_* \partial_* \lambda)(t,y(t,\lambda),\lambda) \Big) \cdot_* (\lambda -_* \lambda_1) \Big) \cdot_* d_* t$$

$$+_* \int_{*x_0}^x \delta \{y(t,x_0,y_0,\lambda),y(t,x_0,y_0,\lambda_1)\} \cdot_* d_* t,$$

where

$$\delta \{y(t,x_0,y_0,\lambda),y(t,x_0,y_0,\lambda_1)\} \to 1$$

as $|y(t,x_0,y_0,\lambda) -_* y(t,x_0,y_0,\lambda_1)|_* \to 1$, i.e., as $|\lambda -_* \lambda_1|_* \to 1$. Hence, for $x \in J_{\lambda,\lambda_1}$. we have

$$|y(x,x_0,y_0,\lambda) -_* y(x,x_0,y_0,\lambda_1) -_* z(x,\lambda)(\lambda -_* \lambda_1)|_*$$

$$\leq e^L \cdot_* \left| \int_{*x_0}^x |y(t,x_0,y_0,\lambda) -_* y(t,x_0,y_0,\lambda_1)|_* \cdot_* d_* t \right|_*$$

$$+_* o(|\lambda -_* \lambda_1|_*).$$

From the last inequality and from the Gronwall type inequality we get

$$|y(x,x_0,y_0,\lambda) -_* y(x,x_0,y_0,\lambda_1) -_* z(x,\lambda)(\lambda -_* \lambda_1)|_*$$

$$\leq o(|\lambda -_* \lambda_1|_*) \cdot_* e^{e^L \cdot_* |x -_* x_0|_*},$$

i.e.,

$$|y(x,x_0,y_0,\lambda) -_* y(x,x_0,y_0,\lambda_1) -_* z(x,\lambda)(\lambda -_* \lambda_1)|_* \to 1$$

as $|\lambda -_* \lambda_1|_* \to 1$. *This completes the proof.*

Example 5.5.22 *We consider the multiplicative initial value problem*

$$y^* = y +_* \lambda \cdot_* (x +_* y^{2*})$$

$$y(1) = e.$$

We will find

$$z(x,0) = (\partial_* y /_* \partial_* \lambda)\Big|_{\lambda=1}.$$

Here

$$f(x,y,\lambda) = y +_* \lambda \cdot_* (x +_* y^{2*})$$

$$= y +_* \lambda \cdot_* x +_* \lambda \cdot_* y^{2*},$$

and

$$(\partial_* f /_* \partial_* y)(x,y,\lambda) = e +_* e^2 \cdot_* \lambda \cdot_* y,$$

$$(\partial_* f/_* \partial_* \lambda)(x, y, 1) \;=\; e,$$

and

$$(\partial_* f/_* \partial_* \lambda)(x, y, \lambda) \;=\; x +_* y^{2_*},$$

$$(\partial_* f/_* \partial_* \lambda)(x, y, 0) \;=\; x +_* y^{2_*}.$$

Therefore

$$y^*(x, 1) \;=\; y,$$

$$y(1, 1) \;=\; e.$$

Then

$$y(x, 1) = e^x.$$

Also,

$$z^*(x, 1) \;=\; z(x, 1) +_* x +_* e^{e^{2_*} \cdot_* x},$$

$$z(1, 1) \;=\; 1.$$

Hence,

$$z(x, 1) \;=\; e^x \cdot_* \left(c +_* \int_* e^{-_* x} \cdot_* \left(x +_* e^{e^{2_*} \cdot_* x} \right) \cdot_* d_* x \right)$$

$$=\; e^x \left(c +_* \int_* x \cdot_* e^{-_* x} \cdot_* d_* x +_* e^x \right)$$

$$=\; e^x \cdot_* \left(c -_* x \cdot_* e^{-_* x} -_* e^{-_* x} +_* e^x \right),$$

and

$$z(1, 1) = 1,$$

whereupon

$$c = 1$$

and

$$z(x, 1) = -_* x -_* e +_* e^{e^{2_*} \cdot_* x}.$$

Example 5.5.23 *We consider the multiplicative initial value problem*

$$y^* = e^2 \cdot_* x +_* \lambda \cdot_* y^{2*},$$

$$y(1) = \lambda -_* e.$$

We will find

$$z(x,1) = (\partial_* y /_* \partial_* \lambda) \big|_{\lambda=1}.$$

We have that

$$y^*(x,1) = e^2 \cdot_* x,$$

$$y(1,1) = -_* e.$$

Therefore,

$$y(x,1) = x^{2*} -_* e.$$

Also,

$$(\partial_* f /_* \partial_* y)(x,y,\lambda) = e^2 \cdot_* \lambda \cdot_* y,$$

$$(\partial_* f /_* \partial_* \lambda)(x,y,\lambda) = y^{2*},$$

$$(\partial_* f /_* \partial_* y)(x,y,1) = 1,$$

$$(\partial_* f /_* \partial_* \lambda)(x,y,1) = y^{2*}$$

$$= (x^{2*} -_* e)^{2*}.$$

Therefore,

$$z^*(x,1) = (x^{2*} -_* e)^{2*},$$

$$z(1,1) = 1.$$

Hence,

$$z(x,1) = (x^{5*} /_* e^5) -_* e^{\frac{2}{3}} \cdot_* x^{3*} +_* x +_* c,$$

$$z(1,1) \;=\; c$$

$$=\; 1,$$

i.e.,

$$z(x,1) = (x^{5*}/_* e^5) -_* e^{\frac{2}{3}} \cdot_* x^{3*} +_* x.$$

Example 5.5.24 *We consider the multiplicative initial value problem*

$$y^* \;=\; \lambda \cdot_* y^{2*} +_* e,$$

$$y(1) \;=\; 1.$$

We will find

$$z(x,1) = (\partial_* y /_* \partial_* \lambda)\Big|_{\lambda=1}.$$

We have

$$y^*(x,1) \;=\; e,$$

$$y(1,1) \;=\; 1.$$

Then

$$y(x,1) = x.$$

Also,

$$f(x,y,\lambda) \;=\; \lambda \cdot_* y^{2*} +_* e,$$

$$(\partial_* f /_* \partial_* y)(x,y,\lambda) \;=\; e^2 \cdot_* \lambda \cdot_* y,$$

$$(\partial_* f /_* \partial_* \lambda)(x,y,\lambda) \;=\; y^{2*},$$

$$(\partial_* f /_* \partial_* y)(x,y,1) \;=\; 1,$$

$$(\partial_* f/_* \partial_* \lambda)(x, y, 1) \;=\; (y(x, 1))^{2*}$$

$$=\; x^{2*}.$$

From here,

$$z^*(x, 1) \;=\; x^{2*},$$

$$z(1, 1) \;=\; 1.$$

Hence,

$$z(x, 1) = (x^{3*}/_* e^3).$$

Exercise 5.5.25 *Find* $(\partial_* y/_* \partial_* \lambda)\big|_{\lambda=1}$ *for the following initial value problems:*

1.

$$y^* \;=\; \lambda^{2*} \cdot_* y^{2*} +_* \lambda \cdot_* y +_* e,$$

$$y(1) \;=\; 1.$$

2.

$$y^* \;=\; \lambda \cdot_* y +_* \lambda^{2*} \cdot_* x +_* e^2,$$

$$y(1) \;=\; 1,$$

3.

$$y^* \;=\; \lambda \cdot_* y +_* x,$$

$$y(1) \;=\; 1.$$

Answer

1. $(x^{2*}/_* e^2)$.
2. x^{2*}.
3. $(x^{3*}/_* e^6)$.

5.6 Advanced Practical Problems

Problem 5.6.1 *Prove that the multiplicative initial value problem*

$$y^{*(n)} = f(x,y)$$

$$y(x_0) = y_0, \quad y^*(x_0) = y_1, \ldots, y^{*(n-1)}(x_0) = y_{n-1},$$

where f is continuous in a domain $D \subset \mathbb{R}_*^2$ containing the point (x_0, y_0), is equivalent to the integral equation

$$y(x) = y_0 +_* y_1 \cdot_* (x -_* x_0) +_* y_2 \cdot_* ((x -_* x_0)^{2*}/_* e^2) +_* \cdots$$

$$+_* y_{n-1} \cdot_* ((x -_* x_0)^{(n-1)*}/_* e^{(n-1)!})$$

$$+_* \int_{*x_0}^{x} ((x -_* t)^{(n-1)*}/_* e^{(n-*1)!}) \cdot_* f(t, y(t)) \cdot_* d_* t.$$

Problem 5.6.2 *Let $D = \{(x,y) \in \mathbb{R}_*^2 : |x -_* e|_* \leq e^3, |y|_* \leq e^4\}$. Prove that the function*

$$f(x,y) = x^{2*} -_* y^{3*} -_* x \cdot_* y^{2*} +_* x \cdot_* y +_* e^3$$

satisfies the multiplicative Lipschitz condition in L. Find the multiplicative Lipschitz constant L.

Answer $L = e^{800}$.

Problem 5.6.3 *Let $D = \{(x,y) \in \mathbb{R}_*^2 : |x|_* \leq e^a, |y|_* \leq e^b\}$,*

$$f(x,y) = p(x) \cdot_* e^{x \cdot_* y^{2*}},$$

where p is continuous function in $|x|_* \leq e^a$. Prove that the function f satisfies the multiplicative Lipschitz condition in D. Find the multiplicative Lipschitz constant L.

Answer

$$L = e^{2abe^{ab^2} \max_{|x| \leq a} |p(x)|}.$$

Problem 5.6.4 *Let* $D = \{(x,y) \in \mathbb{R}_*^2 : |x|_* \le e^a, |y| < \infty\}$,

$$f(x,y) = \begin{cases} ((x^{2*} \cdot_* y)/_* (x^{6*} +_* y^{6*})), & (x,y) \ne (1,1), (x,y) \in D, \\ 1, & (x,y) = (1,1). \end{cases}$$

Prove that the function f *doesn't satisfy the multiplicative Lipschitz condition in D.*

Problem 5.6.5 *Let* y *be multiplicative nonnegative continuous function in* $[1,e]$ *and*

$$y(x) \le e^{-*x} +_* e^2 \cdot_* \int_{*1}^x y(t) \cdot_* d_* t, \qquad x \in [1,e].$$

Prove that

$$y(x) \le e^{\frac{1}{3}} \cdot_* e^{-*x} +_* e^{\frac{2}{3}} \cdot_* e^{e^{2} \cdot_* x}, \qquad x \in [1,e].$$

Problem 5.6.6 *Let* a, b, c *and* y *be continuous functions in* $[e^\alpha, e^\beta]$, *let* b *and* c *be multiplicative nonnegative in* $[e^\alpha, e^\beta]$ *and*

$$y(x) \le a(x) +_* \int_{*e^\alpha}^x (b(t) \cdot_* y(t) +_* c(t)) \cdot_* d_* t, \qquad x \in [e^\alpha, e^\beta].$$

Prove that

$$y(x) \;\le\; \left(\sup_{t \in [\alpha, x]} a(t) +_* \int_{e^\alpha}^x c(t) \cdot_* d_* t \right)$$

$$\cdot_* e^{\int_{e^\alpha}^x b(t) \cdot_* d_* t}, \qquad x \in [e^\alpha, e^\beta].$$

Problem 5.6.7 *Let* y *and* b *be continuous functions in* $[e^\alpha, e^\beta]$, *and let* a *and* c *be multiplicative Riemann integrable functions in* $[e^\alpha, e^\beta]$. *Suppose that* b *and* c *are multiplicative nonnegative in* $[e^\alpha, e^\beta]$.

1. If

$$y(x) \le a(x) +_* c(x) \cdot_* \int_{*e^\alpha}^x b(t) \cdot_* y(t) \cdot_* d_* t, \qquad x \in [e^\alpha, e^\beta],$$

prove that

$$y(x) \leq a(x) +_* c(x) \cdot_* \int_{*e^\alpha}^x a(t) \cdot_* b(t)$$

$$\cdot_* e^{\int_{*t}^x b(s) \cdot_* c(s) \cdot_* d_* s} \cdot_* d_* t, \qquad x \in [e^\alpha, e^\beta].$$

2. *If $\leq$ is replaced by $\geq$ in both inequalities, prove that the result in 1 remains valid.*

3. *Both 1 and 2 remain valid if $\int_{*e^\alpha}^x$ is replaced by $\int_x^{*e^\beta}$ and $\int_{*t}^x$ is replaced by $\int_x^{*t}$ throughout.*

Problem 5.6.8 *Let y, b and c be continuous functions in $[e^\alpha, e^\beta]$, let b and c be multiplicative nonnegative in $[e^\alpha, e^\beta]$, and*

$$y(x) \leq a +_* \int_{*e^\alpha}^x b(t) \cdot_* y(t) \cdot_* d_* t$$

$$+_* \int_{*e^\alpha}^{e^\beta} c(t) \cdot_* y(t) \cdot_* d_* t, \qquad x \in [e^\alpha, e^\beta],$$

where a is a multiplicative constant. Let also,

$$q = \int_{*e^\alpha}^{e^\beta} c(t) \cdot_* e^{\int_{*e^\alpha}^t b(s) \cdot_* d_* s} \cdot_* d_* t < e.$$

Prove that

$$y(x) \leq (a/_*(e -_* e^q)) \cdot_* e^{\int_{*e^\alpha}^x b(t) \cdot_* d_* t}, \qquad x \in [e^\alpha, e^\beta].$$

Problem 5.6.9 *Find the first four approximations with the initial approximation $y_0(x) = x^{2*} -_* e$ of the following initial value problems.*

1.

$$y^* = \sin_* x,$$

$$y(1) = 1.$$

2.

$$y^* = y \cdot_* \cos_* x,$$

$$y(1) = e.$$

3.

$$y^* = y^{2*} -_* x,$$

$$y(1) = 1.$$

Problem 5.6.10 *Let* $y_0(x) = x,$

$$\bar{S} = \{(x,y) \in \mathbb{R}^2_* : |x|_* \le e^2, |y|_* \le e^3\}.$$

Consider the multiplicative initial value problem

$$y^* = e^2 \cdot_* x -_* e^3 \cdot_* y,$$

$$y(1) = 1.$$

Find the existence interval J_h *of the unique solution of this problem.*
Determine the error estimate in J_h.

Answer

$$J_h : |x|_* \le e^{\frac{3}{13}},$$

and

$$|y(x) -_* y_m(x)|_* \le e^{4e^{\frac{9}{13}} \frac{\left(\frac{9}{13}\right)^m}{m!}}.$$

Problem 5.6.11 *Consider the multiplicative initial value problem*

$$y^* = e^{-_* y^{2*}},$$

$$y(1) = 1,$$

in $T = \{(x,y) \in \mathbb{R}^2_* : |x|_* \le e^3, |y|_* < \infty\}$, *with the initial approximation*
$y_0(x) = e$. *Find the error estimate.*

Answer

$$|y(x) -_* y_m(x)|_* \le e^{\left(1+_* \frac{3}{e}\right)e^6 \min\left\{1, \frac{6^m}{m!}\right\}}.$$

Problem 5.6.12 *Find the maximum interval of existence of the solution of the problem*

$$y^* +_* e^3 \cdot_* y^{\frac{4}{3}} *_* \sin_* x = 1,$$

$$y\left(e^{\frac{\pi}{2}}\right) = e^{\frac{1}{8}}.$$

Answer $\mathbb{R}$.

Problem 5.6.13 *Prove that*

$$y(x) = \cos_* x -_* \sin_* x$$

is a solution to the differential inequality

$$y^* \le y^{4*} +_* e^5, \qquad x \in [1, e^{2\pi}].$$

Problem 5.6.14 *Let y be the solution of the equation*

$$y(x) = \int_{*1}^{x} y(t) \cdot_* d_* t +_* x +_* e.$$

Prove that y is defined in $[0, \infty)$ *and*

$$e \le y(x) \le e^3 \cdot_* e^x, \qquad x \in [1, \infty).$$

Problem 5.6.15 *For the initial value problem*

$$y^* = x^{2*} +_* e^4 \cdot_* \arctan_* y,$$

$$y(1) = e,$$

estimate the variation of the solution y(x) in $[1, e]$ *if y_0 is perturbed by* $e^{0.001}$.

Problem 5.6.16 *Find* $(\partial_* / {}_* \partial_* y_0) y(x, x_0, y_0)$ *of the following initial value problems.*

    1.

$$y^* \;=\; \cos_* y -_* e +_* x \cdot_* y,$$

$$y(1) \;=\; 1.$$

    2.

$$y^* \;=\; x \cdot_* \sin_* y +_* y^{5_*},$$

$$y(e^2) \;=\; 1.$$

    3.

$$y^* \;=\; y +_* y^{2_*} +_* y^{3_*} +_* y^{4_*},$$

$$y(e^3) \;=\; 1.$$

Answer

    1. $e^{(x^{2_*} /_* e^2)}$.

    2. $e^{(x^{2_*} /_* e^2)} -_* e^2$.

    3. $e^{x -_* e^3} \cdot_* x$.

Problem 5.6.17 *Let* $y(x, x_0, y_0)$ *be the solution of the initial value problem*

$$y^* \;=\; e^2 \cdot_* y +_* x^{2_*},$$

$$y(1) \;=\; 1.$$

Find $\dfrac{\partial_* y}{\partial_* x_0}(x, x_0, y_0)$.

Answer

$$-_*e^2 \cdot_* e^x.$$

Problem 5.6.18 *Find* $(\partial_* y /_* \partial_* \lambda)\big|_{\lambda=1}$ *for the following initial value problems.*

1.

$$y^* = \lambda \cdot_* y +_* x^{2*} +_* x,$$

$$y(1) = 1.$$

2.

$$y^* = \lambda^{2*} \cdot_* y^{2*} +_* \lambda \cdot_* y +_* x,$$

$$y(1) = 1.$$

3.

$$y^* = \lambda^{3*} \cdot_* y +_* e,$$

$$y(1) = 1.$$

Answer

1.
$$(x^{4*}/_* e^{12}) +_* (x^{3*}/_* e^6).$$

2.
$$(x^{3*}/_* e^6).$$

3.
$$(x^{2*}/_* e^2).$$

Bibliography

[1] D. Aniszewska, Multiplicative Runge-Kutta method, Nonlinear Dynamics 50 (1–2)(2007) 265–272.

[2] A. Bashirov, E. Kurpinar, A. Özyapici, Multiplicative Calculus and Its Applications, Journal of Mathematical Analysis and Its Applications 337 (1) (2008) 36–48.

[3] F. Córdova-Lepe, The Multiplicative Derivative as a Measure of Elasticity in Economics, TEMAT-Theaeteto Atheniensi Mathematica 2(3) (2006), online.

[4] S. Georgiev. Focus on Calculus, Nova Science Publisher, 2020.

[5] B. Gompertz. On the Nature of the Function Expressive of the Law of Human Mortality, and on a New Mode of Determining the Value of Life Contingencies, Philosophical Transactions of the Royal Society of London 115 (1825) 513–585.

[6] M. Grossman, R. Katz, Non-Newtonian Calculus, Pigeon Cove, Lee Press, Massachusats, 1972.

[7] M. Grossman, Bigeometric Calculus: A System with a Scale-Free Derivative, Archimedes Foundation, Rockport, Massachusats, 1983.

[8] W. Kasprzak, B. Lysik, M. Rybaczuk, Dimensions, Invariants Models and Fractals, Ukrainian Society on Fracture Mechanics, SPOLOM, Wroclaw-Lviv, Poland, 2004.

[9] R.R. Meginniss, Non-Newtonian Calculus Applied to Probability, Utility, and Bayesian Analysis, Manuscript of the report for delivery at the 20th KDKR- KSF Seminar on Bayesian Inference in Econometrics, Purdue University, West Lafayette, Indiana, May 23, 1980.

[10] M. Riza, A. Özyapici, E. Misirli, Multiplicative Finite Difference Methods, Quarterly of Applied Mathematics, 64 (4) (2009), 745–754.

[11] M. Rybaczuk, A. Kedzia, W. Zielinski, The Concepts of Physical and Fractional Dimensions II. The Differential Calculus in Dimensional Spaces, Chaos Solutions Fractals 12 (2001), 2537–2552.

[12] D. Stanley, A Multiplicative Calculus, Primus IX (4) (1999) 310–326.

Index

For Product Safety Concerns and Information please contact our EU
representative GPSR@taylorandfrancis.com
Taylor & Francis Verlag GmbH, Kaufingerstraße 24, 80331 München, Germany

www.ingramcontent.com/pod-product-compliance
Ingram Content Group UK Ltd.
Pitfield, Milton Keynes, MK11 3LW, UK
UKHW021115180425
457613UK00005B/102